Principles of Cognitive Radio

Widely regarded as one of the most promising emerging technologies for driving the future development of wireless communications, cognitive radio has the potential to mitigate the problem of increasing radio spectrum scarcity through dynamic spectrum allocation.

Drawing on fundamental elements of information theory, network theory, propagation, optimization, and signal processing, a team of leading experts present a systematic treatment of the core physical and networking principles of cognitive radio, and explore key design considerations for the development of new cognitive radio systems.

Containing all the underlying principles you need to develop practical applications in cognitive radio, this book is an essential reference for students, researchers, and practitioners alike in the field of wireless communications and signal processing.

Ezio Biglieri is an Adjunct Professor at the Universitat Pompeu Fabra, Barcelona, Spain, University of California Los Angeles, and King Saud University, Riyadh, KSA.

Andrea J. Goldsmith is a Professor of Electrical Engineering at Stanford University, California.

Larry J. Greenstein is a Research Associate at the Wireless Information Network Laboratory, Rutgers University, New Jersey.

Narayan B. Mandayam is a Professor, and the Peter D. Cherasia Faculty Scholar, at Rutgers University, New Jersey.

H. Vincent Poor is the Dean of Engineering and Applied Science, and Michael Henry Strater University Professor of Electrical Engineering, at Princeton University, New Jersey.

"Solidly global perspective on the foundations and advanced topics of one of the most important developments in radio systems engineering in the past several decades – this book is a landmark publication, a compelling page-turner in presentation and a valuable reference resource for practicing radio engineers."

Dr. Joseph Mitola III, Mitola's STATISfaction

"This is the first book to cover the topic of cognitive radio from a fundamental viewpoint. It provides new insights on propagation, spectrum sensing, system optimization and networking that will be invaluable to theoretical researchers as well as practitioners in the field."

Venugopal V. Veeravalli, University of Illinois at Urbana-Champaign

"Much has been written on the ever-expanding subject of cognitive radio. The new book entitled "Principles of Cognitive Radio", written by well-respected authorities, fills a gap by presenting detailed discussions of different aspects of this fascinating subject. I recommend the book for use by researchers who are already working in cognitive radio or planning to move into it."

Simon Haykin, McMaster University, Canada

"Principles of Cognitive Radio" is a comprehensive treatment of many of the fundamental issues that will impact cognitive wireless communications networks of the future. The authors have done an excellent job bringing together many modern concepts that are at the forefront of wireless, while reviewing the fundamentals along the way. As a result, the authors have created an excellent reference text, bringing a wide range of problems, contributions, and viewpoints together in a single, well-written book."

Theodore S. Rappaport, Polytechnic Institute of New York University and Director, NYU WIRELESS

Principles of Cognitive Radio

EZIO BIGLIERI
Universitat Pompeu Fabra, Barcelona, Spain

ANDREA J. GOLDSMITH
Stanford University

LARRY J. GREENSTEIN
Rutgers University

NARAYAN B. MANDAYAM
Rutgers University

H. VINCENT POOR
Princeton University

Shaftesbury Road, Cambridge CB2 8EA, United Kingdom

One Liberty Plaza, 20th Floor, New York, NY 10006, USA

477 Williamstown Road, Port Melbourne, VIC 3207, Australia

314–321, 3rd Floor, Plot 3, Splendor Forum, Jasola District Centre, New Delhi – 110025, India

103 Penang Road, #05–06/07, Visioncrest Commercial, Singapore 238467

Cambridge University Press is part of Cambridge University Press & Assessment, a department of the University of Cambridge.

We share the University's mission to contribute to society through the pursuit of education, learning and research at the highest international levels of excellence.

www.cambridge.org
Information on this title: www.cambridge.org/9781107028753

© Cambridge University Press & Assessment 2013

This publication is in copyright. Subject to statutory exception and to the provisions of relevant collective licensing agreements, no reproduction of any part may take place without the written permission of Cambridge University Press & Assessment.

First published 2013

A catalogue record for this publication is available from the British Library

Library of Congress Cataloging-in-Publication data
Principles of cognitive radio / Ezio Biglieri . . . [et al.].
   p. cm.
Includes bibliographical references and index.
ISBN 978-1-107-02875-3
1. Cognitive radio networks. 2. Radio frequency allocation. 3. Software radio.
I. Biglieri, Ezio.
TK5103.4815.P75   2012
621.384–dc23

                                               2012028036

ISBN   978-1-107-02875-3   Hardback

Cambridge University Press & Assessment has no responsibility for the persistence or accuracy of URLs for external or third-party internet websites referred to in this publication and does not guarantee that any content on such websites is, or will remain, accurate or appropriate.

To our families

Contents

Contributors		*page* xiii
Preface		xv
Acknowledgments		xviii
Notation		xix

1 The concept of cognitive radio — 1
- 1.1 Motivation for cognitive radios: spectrum is underutilized — 1
- 1.2 What is cognitive radio? — 2
  - 1.2.1 Agile radios and dynamic spectrum access — 2
  - 1.2.2 User hierarchy in cognitive radio networks — 3
  - 1.2.3 Usage scenarios for cognitive radio — 3
  - 1.2.4 Cognitive radio bands — 4
- 1.3 Spectrum policy: present and future — 5
  - 1.3.1 Role of spectrum policy — 5
- 1.4 Data explosion: future spectrum implications — 7
- 1.5 Applications of cognitive radio — 8
  - 1.5.1 Dynamic spectrum access in cellular systems — 9
  - 1.5.2 Cellular data boost — 9
  - 1.5.3 Machine-to-machine communications — 10
  - 1.5.4 Distribution and backhaul — 11
  - 1.5.5 Cognitive digital home — 12
  - 1.5.6 Long range vehicle-to-vehicle network — 12
- 1.6 Cognitive radio network design — 13
  - 1.6.1 Global control plane — 13
  - 1.6.2 Spectrum servers, spectrum brokers, and network information servers — 15
  - 1.6.3 Security aspects of cognitive radio — 18
- 1.7 Hardware and system design considerations — 19
  - 1.7.1 Design tradeoffs in usage scenarios — 19
  - 1.7.2 Antenna design in cognitive radio systems — 20
  - 1.7.3 Analog-to-digital converters — 21
  - 1.7.4 Wideband channels and noncontiguous transmission — 21
- 1.8 Spectrum coexistence in cognitive radio networks — 25

			1.8.1 Spectrum pooling and bandwidth exchange	26

        1.8.1 Spectrum pooling and bandwidth exchange   26
        1.8.2 Cross-layer scheduling in cognitive radio networks 29
   1.9 Prototyping   30
   1.10 Standardization activity in cognitive radio   33
   1.11 Organization of this book   35
   References   35

2 Capacity of cognitive radio networks — 41

   2.1 Introduction   41
   2.2 Cognitive radio network paradigms   41
        2.2.1 Underlay paradigm   42
        2.2.2 Overlay paradigm   43
        2.2.3 Interweave paradigm   45
        2.2.4 Comparison of cognitive radio paradigms   46
   2.3 Fundamental performance limits of wireless networks   47
        2.3.1 Performance metrics   48
        2.3.2 Mathematical definition of capacity   55
        2.3.3 Capacity region of wireless networks   59
   2.4 Interference channels without cognition   63
        2.4.1 K-user interference channels   63
        2.4.2 Two-user interference channel capacity   64
        2.4.3 Interference channel techniques for cognitive radios 68
   2.5 Underlay cognitive radio networks   69
        2.5.1 Underlay capacity region   70
        2.5.2 Capacity results for specific scenarios   72
   2.6 Interweave cognitive radio networks   76
        2.6.1 Shannon capacity   77
        2.6.2 Random switch model for secondary channels   80
        2.6.3 Scaling laws for interweave networks   83
   2.7 Overlay cognitive radio networks   84
        2.7.1 Cognitive encoder for the two-user overlay channel 85
        2.7.2 Capacity results   89
        2.7.3 K-user overlay networks   92
   2.8 Summary   93
   2.9 Further reading   96
   References   96

3 Propagation issues for cognitive radio — 102

   3.1 Introduction   102
        3.1.1 Propagation in the cognitive radio bands   102
        3.1.2 Impact of propagation on sensing   103
        3.1.3 Impact of propagation on transmission   104
        3.1.4 Outline of the chapter   105

3.2	Generic channel response	105
3.3	Introduction to path loss	107
	3.3.1 Free-space path loss	107
	3.3.2 Path loss in CR scenarios	107
3.4	Path loss models for wireless channels	108
	3.4.1 General formulation	108
	3.4.2 Shadow fading, S	111
	3.4.3 Median path loss, PL_{med}	112
	3.4.4 Antenna gain and the gain reduction factor	114
3.5	Path loss models for tower-based scenarios	115
	3.5.1 Transmissions from TV towers	115
	3.5.2 Tower-to-tower paths at low-to-moderate heights	117
3.6	Small-scale fading and the Ricean K-factor	118
	3.6.1 Spatial variation of field strength	118
	3.6.2 Temporal fading on mobile radio links	122
	3.6.3 Temporal fading on fixed wireless links	122
3.7	Small-scale fading and the Doppler spectrum	123
	3.7.1 Doppler frequency	123
	3.7.2 The angle-of-arrival and Doppler spectra	124
	3.7.3 The autocorrelation function, $A(\Delta t)$	125
	3.7.4 The Doppler spectrum for fixed terminals	126
	3.7.5 Dispersion	127
3.8	Delay dispersion	128
	3.8.1 "Narrowband" vs. "wideband"	128
	3.8.2 Wideband channels	128
	3.8.3 Time-variant impulse response	129
	3.8.4 The power delay profile, $\mathcal{P}(\tau)$	130
	3.8.5 The frequency correlation function, $F(\Delta f)$	131
	3.8.6 A model and values for the delay spread	132
	3.8.7 Ultra-wideband (UWB) channels	133
3.9	Angle dispersion	134
	3.9.1 Directions of arrival and departure	134
	3.9.2 Models for the APS shape and angular spread	136
	3.9.3 Joint dispersions	137
3.10	Polarization	138
3.11	Special environments	139
	3.11.1 Vehicle-to-vehicle (V2V) propagation	139
	3.11.2 Wireless sensor networks (WSNs)	140
3.12	Summary of key model parameters	141
	3.12.1 Path loss models	141
	3.12.2 Ricean K-factor models	141
	3.12.3 Delay dispersion models	141
	3.12.4 Frequency dispersion models	142
	3.12.5 Comprehensive models	142

		3.12.6 Usage of models	142
	3.13	Summary	142
	3.14	Further reading	143
	References		144

4 Spectrum sensing — 150

- 4.1 Introduction — 150
- 4.2 Interference temperature for cognitive underlaying — 151
- 4.3 White-space detection for cognitive interweaving — 153
  - 4.3.1 Energy sensing — 157
  - 4.3.2 Coherent detection — 160
  - 4.3.3 Cyclostationarity-based detection — 161
  - 4.3.4 Autocorrelation-based detection — 163
- 4.4 An application: spectrum sensing with OFDM — 166
  - 4.4.1 Neyman–Pearson detection — 167
  - 4.4.2 Detection based on second-order statistics — 169
- 4.5 Effects of imperfect knowledge of noise power — 170
  - 4.5.1 Energy sensing — 171
  - 4.5.2 Pilot-tone-aided coherent sensing — 172
  - 4.5.3 Cyclostationarity-based detection — 173
- 4.6 Effects of an inaccurate model of interference — 174
  - 4.6.1 Basics of moment-bound theory — 174
  - 4.6.2 Energy sensing — 176
  - 4.6.3 Pilot-tone-aided coherent sensing — 178
- 4.7 Summary — 180
- 4.8 Further reading — 180
- References — 181

5 Spectrum exploration and exploitation — 184

- 5.1 Introduction — 184
  - 5.1.1 Chapter motivation — 184
  - 5.1.2 Preview of the chapter — 186
- 5.2 Advanced spectrum sensing techniques — 187
  - 5.2.1 Distributed detection in spectrum sensing — 187
  - 5.2.2 Sequential and quickest detection — 202
- 5.3 Optimized spectrum exploration and exploitation: sensing and access policy design — 212
  - 5.3.1 Optimization techniques — 213
  - 5.3.2 Bandit problems — 220
  - 5.3.3 Reinforcement learning — 227
  - 5.3.4 Game-theoretic approaches — 237
  - 5.3.5 Location awareness and geolocation — 242
- 5.4 Summary — 243

5.5   Further reading 245
References 246

Bibliography 259
Index 295

Contributors

1 Introduction
Chandrasekharan Raman, *Alcatel–Lucent*

2 Capacity of cognitive radio networks
Ivana Marić, *Aviat Networks, Inc.*

3 Propagation
Andreas F. Molisch, *University of Southern California,* and Mansoor Shafi, *Telecom New Zealand*

5 Spectrum exploration and exploitation
Jarmo Lundén and Visa Koivunen, *Aalto University, Helsinki, Finland*

Preface

The radio spectrum is one of the most important resources for communications. Traditionally, spectrum governance throughout the world has tended towards exclusivity of its use in large geographic areas, allocating frequency bands for specific applications and assigning licenses to specific users or service providers. This policy has generated a shortage of frequencies available for emerging wireless products and services, as most frequencies are now assigned. Moreover, exclusivity creates underutilization of the spectrum, as very rarely can all licensees make full use of the frequencies assigned to them. These facts have motivated the search for technologies able to alleviate the artificial scarcity of spectrum by adapting to changing environmental and network-usage conditions.

What is perhaps the most natural among these technologies involves *opportunistic* use of the spectrum, whereby *secondary* (unlicensed) users are able to occupy the portions of the spectrum left temporarily free by the licensed *primary* users. The stringent requirement here is that secondary users should not interfere with the primary users, which this paradigm of operation (later called *interweaving*) achieves using the simplest form of orthogonalization, one that only requires knowledge of the state of a frequency band, i.e., whether it is free or occupied. The fact that the spectrum can be shared by primary and secondary users, with the latter exploiting their cognition of the environment in which transmission is taking place, has led to the development of the concept of *Cognitive Radio* (CR), whose idea was first introduced in [1] in 1999. Further paradigms, potentially more efficient than interweaving, have subsequently emerged. In the *underlay* paradigm, the secondary transmitter overlaps in frequency with the primary user, after making sure that the interference level it causes is below a given threshold. The *overlay* paradigm assumes that the secondary user has knowledge of the primary user's transmission scheme and of the channel, and uses this knowledge by choosing a transmission scheme that causes an irrelevant amount of interference.

In its multiple aspects, CR is now widely regarded as one of the most promising technologies for future wireless communications, a technology that may potentially mitigate, through dynamic spectrum access, the radio spectrum scarcity problem. The US Federal Communications Commission has approved the use of mobile devices in unused television bands, and there is considerable support worldwide for this new concept. Wide interest in the field has also been displayed in academic research. Since CR theory is still in its early stage, there is a need for a book describing the principles on which CR is based, and laying out in a unified way the background needed for further developments and innovative applications. This background blends propagation theory, information theory, detection theory, optimization theory, networking, and signal processing. Given the importance of the discipline, and the number of unique features that characterize it, this book, which covers the fundamental aspects of CR and especially those that are the most promising for commercial implementation, should be of considerable interest to researchers and practitioners in this field.

This book is organized into five chapters, each meant to be self-contained (with the possible exception of Chapter 5, which may need knowledge of material in Chapter 4 if the reader is not conversant with the basics of detection theory). Chapter 1 introduces the concepts of CR, illustrates the present and future of spectrum management policies, and examines network design and standardization activity. Chapter 2 is devoted to the information-theoretic aspects of CR. Specifically, it develops the fundamental capacity limits, guidelines for the spectral efficiency possible in CR using each of its three paradigms, as well as practical design ideas to optimize performance. These fundamental limits are based on the amount of side information that can be gathered about the radio environment and can be used to improve spectrum utilization as well as the complexity of the CR technique employed. Chapter 3 describes the propagation channels that are typical of CR environments, and presents models for path loss, Doppler spectrum, delay spread, and other relevant features. These models are useful to both designers and analysts, who need to characterize the spectral, temporal, and spatial variations of the channel responses in CR networks. Chapter 4 describes techniques for spectrum sensing, to be used in interweaving and underlay. Interweaving involves the classification and the performance analysis of a number of decision rules intended to identify spectral regions that are empty and hence can be used by secondary users, while underlay is treated by introducing the concept of interference temperature. Finally, Chapter 5 contains the discussion of a number of advanced topics related to spectrum exploration and exploitation. These include techniques, such as distributed, sequential and quickest detection, that are important in optimizing spectrum sensing and identifying available spectral opportunities. This chapter further explores high-level methodologies, including dynamic programming, bandit problems, reinforcement learning, and game theory, in the context of their applications to sensing and access policy design for CR systems. We hasten to observe here that space limitations prevent us from covering the whole range of CR-related topics. In particular, this book does not cover many of the networking aspects of CR, protocols, and standardization efforts.

Our presentation of the subject aspires to combine the features of a textbook and a guide through the research literature. Thus, the book is aimed at graduate students and researchers, as well as at practitioners in industry. Having an emphasis on principles, it can be used as a textbook in a beginning-graduate course entirely devoted to CR, or as a complementary book in courses devoted to wireless communications. It was written assuming no special prerequisite knowledge for reading it, beyond the usual contents of basic communication, information theory, and signal processing courses given in standard electrical engineering curricula. These basics are extended here to the theoretical aspects of CR that are specific to this new technology.

While we know that no book is so poorly written that it cannot be useful in some part, we are also aware that no book is so perfect that nothing wrong can be found in it. Thus, we would be grateful to readers who inform us of any mistakes or inaccuracies that they may find herein.

References

[1] J. Mitola III, *Cognitive Radio: Model-Based Competence for Software Radio*, Licentiate Thesis, The Royal Institute of Technology. Stockholm, Sweden, Aug. 1999.

Acknowledgments

The authors wish to express their gratitude to the following organizations whose support was invaluable in the preparation of this book. Ezio Biglieri was supported by the Project CONSOLIDER-INGENIO 2010 CSD2008-00010 "COMONSENS" and by a grant from King Saud University, KSA. Andrea Goldsmith was supported by US National Science Foundation, the US Office of Naval Research, the US Air Force Office of Scientific Research, and the US Defense Advanced Research Projects Agency. Narayan B. Mandayam was supported by the US National Science Foundation and the US Office of Naval Research. H. Vincent Poor was supported by the US Air Force Office of Scientific Research, the US Army Research Office, the US National Science Foundation, the US Office of Naval Research, and the Qatar National Research Fund.

Notation

General notation and symbols

$A(\cdot)$	Autocorrelation function
$\mathbf{A}^{\mathrm{T}}$	Transpose of the matrix $\mathbf{A}$
$\mathbf{A}^{\mathrm{H}}$	Hermitian of the matrix $\mathbf{A}$
$\mathcal{A}^{\mathrm{c}}$	Complement of the set $\mathcal{A}$
B_{c}	Coherence bandwidth
$c(\cdot)$	Cost function
C	Channel capacity
C_{erg}	Ergodic channel capacity
C_{out}	Outage channel capacity
$\mathbf{C}$	Channel capacity region
$\mathbb{C}$	Set of complex numbers
$\mathbb{E}[\cdot]$	Expectation operator
$D(\cdot)$	Doppler power spectrum
f	Frequency
f_{c}	Center frequency of the signal band
$F(\cdot)$	Frequency correlation function
$F_X(\cdot)$	Cumulative distribution function of the RV X
$g_{\mathrm{d}}(\cdot), g_{\mathrm{e}}(\cdot)$	Decoding, encoding function.
G	Power gain
$G_{\mathrm{r}}, G_{\mathrm{t}}$	Receive, transmit antenna gains, in dB
γ	Signal-to-noise ratio, path loss exponent
h	Complex channel gain
$h(t)$	Channel impulse response
$H(f)$	Complex frequency response of a channel or filter
H	Mean-square value of $|h|$
$\mathbf{H}$	Channel gain matrix
$\mathcal{H}_0$	Null hypothesis
$\mathcal{H}_1$	Alternative hypothesis
η	Efficiency
$\mathbf{I}$	Identity matrix
$\mathbf{I}_k$	$k \times k$ identity matrix
$I(X;Y)$	Mutual information between X and Y

Notation

K	Ricean K-factor
λ	RF wavelength
Λ	Likelihood ratio
$\ln x$	Natural logarithm of x
$\log_x y$	Logarithm, base x, of y
M_r, M_t	Number of receive, transmit antennas
ν	Doppler frequency
n	Discrete time index (in the subscript)
n_B	Number of frequency bands
N	Noise power
N_0	Power spectral density of white noise
$p_X(\cdot)$	Probability density function of the RV X
$p(y\|x)$	Conditional probability of y given x
$\mathbb{P}[A]$	Probability of event A
P_D	Probability of correct detection
P_e	Probability of error
P_FA	Probability of false alarm
P_MD	Probability of missed detection
$P_\text{on}, P_\text{off}$	Probability that a switch is on or off
P_out	Probability of outage
$\mathcal{P}$	Power
$\mathcal{P}_\text{r}$	Receive power
$\mathcal{P}_\text{t}$	Transmit power
$\mathcal{P}(\cdot)$	Power delay profile
$\mathcal{P}_\text{peak}$	Peak power constraint
$Q(\cdot)$	Gaussian tail function
$Q(\cdot,\cdot)$	Normalized incomplete Gamma function
$Q_N(\cdot,\cdot)$	Generalized Marcum Q-function
$\mathcal{Q}(\cdot)$	Action-value function
$\mathcal{Q}^*(\cdot)$	The optimal action-value function
ρ	Correlation
R	Transmission rate
$\mathbb{R}$	Set of real numbers
$\Re[\cdot]$	Real part
$\text{rank}(\mathbf{H})$	Rank of matrix $\mathbf{H}$
s	Channel state
$\mathcal{S}_\text{c}$	Set of all channel states
$\mathcal{S}_\text{PR}$	Set of primary receivers
$\mathcal{S}_\text{ST}$	Set of secondary transmitters
$(\mathcal{S}, \mathcal{S}^\text{c})$	A network cut between nodes in $\mathcal{S}$ and nodes in $\mathcal{S}^\text{c}$
S	Shadow fading, in dB
σ	Standard deviation of shadow fading
σ_X^2	Variance of random variable X

snr	Signal-to-noise ratio
τ	Multipath delay
τ_{rms}	rms multipath delay spread
t	Continuous time index
$u(t)$	Unit-step function
u_i	Local sensor mapping (e.g., a binary decision)
$\mathcal{V}(\,\cdot\,)$	Value function
$\mathcal{V}^*(\,\cdot\,)$	The optimal value function
V	Time-sharing random variable
$w(t)$	Continuous-time white noise process
w_n	Discrete-time white noise process
W	Bandwidth
X_c	Correlation distance of shadow fading
$\frac{\partial}{\partial x}$	Partial derivative with respect to x
$\hat{x}$	Estimate of x
$\mathbf{x}$	A vector
$\|\mathbf{x}\|$	Euclidean norm of the vector $\mathbf{x}$
x^n	A vector with elements $(x_1, \ldots, x_n)$
$x(t)$	Continuous-time transmitted signal
x_n	Discrete-time transmitted signal at time n
$|\mathcal{X}|$	Cardinality of the set $\mathcal{X}$
$X \in \mathcal{X}$	X is an element of the set $\mathcal{X}$
$X \sim \mathcal{N}(\mu, \sigma^2)$	X is a real, Gaussian RV with mean μ and variance σ^2
$X \sim \mathcal{N}_c(\mu, \sigma^2)$	X is a circularly symmetric Gaussian RV with mean μ and variance σ^2
$y(t)$	Continuous-time received signal
y_n	Discrete-time received signal at time n
$z(t)$	Continuous-time noise process
z_n	Discrete-time noise process

Acronyms and abbreviations

3G	Third-generation cellular systems
AAA	Authentication, authorization, and accouting
ACF	Autocorrelation function
ADC	Analog to digital converter
ALOHA	The ALOHA random access protocol
AoA	Angle of arrival
AOD	Angle of departure
APS	Angular power spectrum
ARQ	Automatic repeat on request
ART	Above roof top

AS	Angular spread
AWGN	Additive white Gaussian noise
BE	Bandwidth exchange
BEP	Bit-error probability
BP	Belief propagation
BS	Base station
CAF	Cyclic autocorrelation function
CAGR	Compound annual growth rate
CDF	Cumulative distribution function
CDMA	Code division multiple access
CEPT	European Conference of Postal and Telecommunications Administrations
CFAR	Constant false-alarm rate
CR	Cognitive radio
CSD	Cyclic spectral density
CSI	Channel-state information
CSMA	Carrier sense multiple access
CSMA/CA	Carrier sense multiple access with collision avoidance
CUSUM	Cumulative sum
DAB	Digital Audio Broadcasting
DAI	Dynamic allocation index
DDDPS	Double-directional delay power spectrum
DDIR	Double-directional impulse response
DFT	Discrete Fourier transform
DHCP	Dynamic host configuration protocol
DMDT	Diversity–multiplexing–delay tradeoff
DMT	Diversity–multiplexing tradeoff
DNS	Domain name service
DOA	Direction of arrival
DOD	Direction of departure
DPC	Dirty paper coding
DS	Delay spread
DSA	Dynamic spectrum access
DSL	Digital subscriber line
D-SPRT	Decentralized SPRT
DVB	Digital video broadcasting
ECC	European Communications Committee
EGC	Equal-gain combining
ETSI	European Telecommunications Standards Institute
EWSPRT	Enhanced weighted SPRT
EWSZOT	Enhanced weighted sequential zero/one test
FC	Fusion center
FCC	Federal Communications Commission
FCF	Frequency correlation function

FFT	Fast Fourier transform
FM	Frequency modulation
FPGA	Field programmable gate array
FSPL	Free-space power loss
FTTH	Fiber to the home
GCP	Global control plane
GLLR	Generalized log-likelihood ratio
GLR	Generalized likelihood ratio
GPS	Global positioning system
GRF	Gain reduction factor
HMM	Hidden Markov model
ICI	Intercarrier interference
IDFT	Inverse discrete Fourier transform
IEEE	Institute of Electrical and Electronics Engineers
IFFT	Inverse fast Fourier transform
IMT-Advanced	International Mobile Telecommunications – Advanced
IWFA	Iterative waterfilling algorithm
iid	Independent and identically distributed
IP	Internet protocol
ISI	Inter-symbol interference
ISM	Industrial, scientific and medical
ITU	International Telecommunications Union
LAN	Local-area network
LFD	Least favorable distribution
LOS	Line-of-sight
LDPC	Low-density parity-check
LLR	Log-likelihood ratio
LTE	Long-term evolution
MAC	Medium access control
MAP	Maximum a posteriori
MDP	Markov decision process
MEMS	Micro-electronic mechanical systems
MIMO	Multiple-input, multiple-output
MMSE	Minimum mean-square error
MP	Marginal productivity
MPC	Multipath component
MRC	Maximal-ratio combining
MS	Mobile station
MSE	Mean-square error
MSPRT	M-ary SPRT
NC-OFDM	Noncontiguous orthogonal frequency division multiplexing
NIS	Network interference server
NLOS	Non-line-of-sight
NTIA	National Telecommunications and Information Administration

OFDM	Orthogonal frequency-division multiplexing
P2P	Peer-to-peer
PAPR	Peak-to-average power ratio
PC	Personal computer
pdf	Probability density function
PDP	Power–delay profile
PHY	Physical layer
pmf	Probability mass function
POMDP	Partially observable Markov decision process
PSD	Power spectral density
PSK	Phase-shift keying
QoS	Quality of service
RF	Radio frequency
RMS	Root-mean-square
ROC	Receiver operating characteristics
RSS	Received signal strength
RST	Repeated significance test
RUCB	Restless UCB
RV	Random variable
RX	Receiver
SAP	Sensing assignment problem
SCM	Spatial channel model
SD	Sequential detection
SDR	Software-defined radio
SIC	Successive interference cancellation
SINR	Signal-to-interference-plus-noise ratio
SNR	Signal-to-noise ratio
SoC	System on a chip
SPRT	Sequential probability ratio test
SR	Shiryaev–Roberts
SRP	Shiryaev–Roberts–Pollak
SSCT	Sequential shifted chi-square test
SSPRT	Shiryaev SPRT
TCP	Transmission control protocol
TDFS	Time-division fair sharing
TDMA	Time-division multiple access
TV	Television
TVWS	Television white space
TX	Transmitter
UCB	Upper confidence bound
UHF	Ultrahigh frequency
UWB	Ultra-wideband
VI	Variational inequality
V2V	Vehicle-to-vehicle

WiFi	Wireless fidelity, refers to devices using IEEE 802.11 family of standards
WiMAX	Worldwide interoperability for microwave access
WS	White space
WSN	Wireless sensor network
WSPRT	Weighted SPRT
WWAN	Wireless wide area network

1 The concept of cognitive radio

Narayan B. Mandayam and Chandrasekharan Raman

1.1 Motivation for cognitive radios: spectrum is underutilized

Wireless spectrum is one of the most important resources required for radio communications. Throughout the world, spectrum utilization is regulated so that essential services can be provided and also protected from harmful interference. Traditional spectrum governance across the world has tended toward static long-term exclusivity of spectrum use in large geographic areas, often based on the radio technologies employed at the time of decision making. In particular, until recently spectrum regulatory bodies such as the Federal Communications Commission (FCC) in the US or the European Telecommunications Standards Institute (ETSI) in Europe have always allocated spectrum frequency blocks for specific uses, and assigned licenses for these blocks to specific groups or companies.

While the more or less static spectrum allocation strategy has led to many successful applications like, for example, broadcasting and cellular phones, it has also led to almost all of the prime available spectrum being assigned for various applications (see [63]). It may thus seem that there is little or no spectrum available for emerging wireless products and services.

On the other hand, there have been several studies and reports over the years that show that spectrum is in fact vastly underutilized. A report presenting statistics regarding spectrum utilization showed that even during the high demand period of a political convention such as the one held in 2004 in New York City, only about 13% of the spectrum opportunities were utilized [59]. Further, measurement on radio frequency bands from 30 MHz to 3 GHz, collected during 2009 in Vienna, Virginia (a dense suburb of Washington DC) revealed a number of bands with low spectrum occupancy [82].

These findings suggest that devices using advanced radio and signal processing technology should be able to exploit underutilized spectrum. Much of the early motivation for cognitive radio technology was indeed to accomplish such opportunistic spectrum use and to also alleviate the artificial scarcity of prime spectrum. If successful, this technology could revolutionize the way spectrum is allocated worldwide. Among other benefits, it would yield added bandwidth to support the demand for higher-quality and higher-data-rate wireless products and services well into the future.

1.2 What is cognitive radio?

1.2.1 Agile radios and dynamic spectrum access

Cognitive radio is a generic term used to describe a radio that is aware of the environment around it and can adapt its transmissions according to the interference it sees. In their simplest embodiments, cognitive radios can recognize the available systems around them and adjust their frequencies, waveforms and protocols to access those systems efficiently. Conceptually, cognitive radios include multiple domains of knowledge, model-based reasoning and negotiation [44, 62]. The knowledge and reasoning can include all aspects of any radio etiquette such as RF bands, air interfaces, protocols, and spatial as well as temporal patterns that moderate the use of the radio spectrum. An important feature that differentiates cognitive radios from normal radios is their *agility* along the following lines.

Spectrum agility or frequency agility refers to the discovery strategies for available spectrum as well as opportunistic transmission in the identified spectrum. Such operation requires the design of good algorithms and protocols for appropriate selection of transmission frequencies, coordination and cooperation. Spectrum agility also refers to advanced sensing capabilites.

Technology agility refers to operation of a single radio device across various access technologies. Such seamless interoperability can be enabled by multiplatform radios that are realized as a system-on-a-chip (SoC) and can operate for example as WiFi, Bluetooth, FM, and GPS transceivers.

Protocol agility refers to constituting a dynamically reconfigurable protocol stack on radio devices so that they can proactively and reactively adapt their protocols depending on the devices they interact with.

Cognitive radios equipped with such agility would be a first step towards making radios follow an etiquette in a society of radios. Such cognitive behavior could extend to networks of radios so that they mimic human behavior in civilized society [37]. This opens up interdisciplinary approaches to the study of cognitive radio networks that touch upon cognitive and neurosciences, economics, and sociology. Such interdisciplinary methodologies can be applied in modeling the behavior and dynamics of complex networks with cognitive radios.

While these developments offer exciting possibilities for the future, there are many fundamental engineering issues that need to be addressed before such a vision can be realized. Cognitive radio networks basically extend the software defined radio (SDR) framework to the development of dynamic spectrum access (DSA) algorithms that exploit temporal and spatial variability in the spectrum via: (a) initial cooperative neighbor discovery and association; (b) spectrum quality estimation and opportunity identification; and (c) radio bearer management. These, in turn, imply a framework that senses neighborhood conditions to identify spectrum opportunities for communication by building an awareness of spectrum policy, local network policy, and the capability of local nodes (including noncooperative or legacy nodes).

The fundamental issues that need to be addressed include understanding the information theoretic limits of such networks, constructing propagation models for such networks, devising efficient algorithms for spectrum sensing, as well as fostering mechanisms for spectrum coexistence, all of which are addressed in this book. Chapter 2 develops fundamental capacity limits and associated transmission techniques for different cognitive network paradigms. Chapter 3 comprehensively describes the propagation channel models used for design and analysis of cognitive radio systems. Spectrum sensing and basics of white space detection are discussed in Chapter 4 and Chapter 5 discusses how a cognitive radio network can optimize its spectrum sensing function to efficiently exploit the availability of spectrum.

1.2.2 User hierarchy in cognitive radio networks

Since cognitive radios and networks of these involve opportunistic use of spectrum and the associated rights of users to transmit over such spectrum, it is only natural to classify users according to multiple hierarchies. One classification is based on the ownership (license) of spectrum across users. In this scenario, the users with cognitive radios desirous of opportunistic use of the spectrum are usually referred to as the *secondary users*. The incumbent (licensed) users occupying the spectrum are referred to as *primary users*. The secondary users communicate either with infrastructure or other secondary users without interfering with the active primary users. An example of this hierarchy relates to operation of cognitive-radio-enabled secondary users in the TV bands. The primary user is the television receiver in the licensed TV band. But the spectrum may be intermittently used depending on the programming broadcast schedule of the television channel. A secondary user could sense the spectrum and use the band of frequencies in the television channel if the spectrum is unused. This paradigm can be extended to a network of secondary users coexisting with the primary users. Another classification of users may arise due to the differences in technology capability of the radio devices themselves. *Capable users* are those users that may have access to side information regarding the transmission of the noncapable users. The capable users can then make use of the side information to avoid interfering with the less capable users.

1.2.3 Usage scenarios for cognitive radio

Depending on the usage scenario, cognitive radio network operation can be classified as interweave, overlay, and underlay paradigms, as will be discussed further in Chapter 2. The *interweave* paradigm of operation was the original motivation for the idea of cognitive radio. The stringent requirement is that the secondary users should not interfere with the communication between the already active primary users. This mandates the secondary users to be able to detect (sense), with very high probability, the primary user transmissions in the network. Once the cognitive radio successfully detects the primary user transmissions, it can opportunistically communicate only if it is able to do so without harming the primary transmissions. This requires spectrum agility or the ability to transmit at different frequencies. The temporary space–time–frequency void in the transmission of primary users is referred to as a *spectrum hole* or a *white space*.

In the *overlay* paradigm, the secondary user needs to know the channel between the primary transmitter and the primary and secondary receivers as well as the channel between the secondary transmitter and the primary receiver. With the channel knowledge of both the primary and secondary users, the secondary user can then choose appropriate transmission strategies so that the communication in the secondary network causes least interference to the primary network. This paradigm represents an advanced operation by a highly sophisticated radio and associated architecture and poses many challenges. In the *underlay* paradigm, the secondary transmitter keeps the interference levels below a certain threshold. The primary receiver sees a higher noise level if the primary and secondary transmission overlap in the same band. An example system for the underlay paradigm is the ultrawideband (UWB) transmissions [22], where the transmissions are spread over a wider band to achieve power spectrum densities below the noise floor. While the primary transmission is decoded with an enhanced noise floor, the secondary receiver despreads the data to decode the transmissions.

1.2.4 Cognitive radio bands

In order to design efficient networks of cognitive radios, the propagation environment of these networks must be well understood. The knowledge of propagation will also help in identification, design, implementation, and analysis of transmission strategies for cognitive radios. A quantity that needs to be kept in mind by the cognitive radio system designer is the amount of interference the cognitive radio creates at the primary (or victim) receiver. This quantity depends on the transmission power of the cognitive radio as well as the wireless channel through which the cognitive radio signals propagate. The channel characteristics depend on the frequency band of operation as well as other properties of the environment, as discussed in detail in Chapter 3.

Cognitive radios may be deployed over a wide range of the frequency spectrum. The bands below about 3.5 GHz have lower propagation loss and are sought after by all services. These bands are therefore ideal (but not necessarily exclusive) candidates for the deployment of cognitive radio networks. They have different incumbent (primary) systems, each with its own mix of service type, architecture, bandwidth, and resilience to interference. Some typical candidate bands for cognitive radio systems are:

- *UHF bands*. These bands are currently used by broadcast television, though some conversion to wireless broadband services is in progress. Terrestrial broadcasting transmitters tend to have high antennas (hundreds of meters) and large powers (kilowatt range). In this service, the transmission is one-way, the transmitting antenna may be outside the area containing the cognitive radios and the TV customers are generally fixed. In 2010, the FCC adopted rules to allow unlicensed radio transmitters to operate in the broadcast television spectrum at locations where the spectrum is not being used by the licensed services [36]. The unused TV spectrum can be used

as white spaces. This might be one of the first spectrum ranges where innovative products and services using cognitive radio systems may appear.
- *Cellular bands.* Typical cellular bands are centered near 800/900 MHz, 1.8/1.9 GHz, 2.1 GHz, 2.3 GHz, and 2.5 GHz. The International Mobile Telecommunications-Advanced (IMT-Advanced) systems (fourth-generation cellular systems as defined by the International Telecommunications Union (ITU)) may also be deployed in the 3.5 GHz band. Cellular networks have ubiquitous coverage, with cell site antennas mounted typically at rooftop or lamp-post height. This service is two way, with the cell sites generally in the same region as the cognitive radios, and the cellular customers can be mobile.
- *Fixed wireless access bands.* These bands provide two-way broadband service and are centered near 2.5 and 3.5 GHz. Fixed wireless systems are similar in layout to cellular networks, with the customers at fixed locations, like homes and businesses.

1.3 Spectrum policy: present and future

1.3.1 Role of spectrum policy

Efficient regimes for spectrum management have been a research focus since the earliest days of radio communications, but the mix of technologists, lawyers, and economists that has emerged in recent years has produced a new and lively debate. The long employed command-and-control policy [17] of the FCC has thus tended toward static long-term exclusivity of spectrum use via licenses in large geographic areas, often based on the radio technologies employed at the time of decision making. This has led to many successful applications like television broadcasting and cellular networks, which can be cited as evidence by the proponents of spectrum property rights, but has also been criticized as inefficient in the overall use of spectrum. In addition to the static nature of such licensed spectrum allocations, the inherent political and nonpolitical inefficiencies of government controllers also play a role in the poor spectrum utilization achieved [14].

In addition to the licensed bands, some spectrum is set aside in specific frequency bands that can be used without license by radios following certain rules of operation, such as a maximum power per Hertz of a shared channel access mechanism. The FCC mandates strict operational requirements to be followed in the unlicensed bands like the industrial, scientific, and medical (ISM) radio bands, where devices must conform to the FCC Part 15 regulation [66]. The purpose of these unlicensed bands is to encourage innovation without the high cost to entry associated, for example, with purchasing licensed spectrum through auctions. The success of applications in the unlicensed bands (cordless telephony and WiFi being well-known examples) has sparked a hot debate regarding the spectrum governance employed by the FCC, and how spectrum policy can be improved to alleviate artificial spectrum scarcity, promote efficiency, and also encourage innovation.

The proposals for new governance regimes fall into two broad categories: spectrum property rights and spectrum commons. In its broadest sense, spectrum property rights refers to a governance mechanism in which portions of spectrum are owned by individuals (or companies). Such portions can be traded to other parties through monetary transactions, or used exclusively, in a flexible manner, with not many technical constraints. The spectrum commons approach, on the other hand, advocates that spectrum should be considered common property, shared by all communicating parties, based on predefined but minimal rules or standards.

The spectrum property rights approach is motivated by the landmark work of R. H. Coase [26], in which it is suggested that spectrum can be treated like land, and private ownership of spectrum is viable. The proponents of spectrum ownership believe that the spectrum should be allocated to the prospective spectrum holders through market forces. The spectrum holders would then have exclusive use of the spectrum portion they possess, without the potential of harmful interference from other parties. Alternatively, they would be able to trade their spectrum in a secondary market. The use of spectrum would be flexible, in that the authorized party could use the spectrum portion for any purpose. Thus, the focus in this approach is on transferring ownership of the spectrum from the government to private parties and substituting market forces for traditional spectrum regulation, overcoming two sources of inefficiency in the status quo regime. The common view is that, since the early 1990s, the FCC has chosen a partial implementation of this approach by employing spectrum auctions as a means of licensing.

The spectrum commons approach, encouraged by the unlicensed spectrum band experiments, argues that as smart technologies evolve, communicating devices will become able to avoid interference through mutual cooperation and coexistence, and the spectrum will become less scarce. The emergence of cognitive and software defined radio concepts, multiple antenna and multicarrier techniques, UWB technologies, and mesh network topologies provide a technology panacea that proponents of this approach use to support their arguments. Communicating devices will be able to efficiently share a specified spectrum band through the enforcement of technical restrictions and multiple access protocols, without requiring exclusive access or private ownership.

The analogy often articulated is that of a highway, which the motorists treat as a common property and can efficiently share as long as they abide by the traffic rules. The highway analogy also illustrates that in spite of all the smart radio technologies, there is still a need for a controller or enforcer. Thus, even the commons regime [43] is a form of lightly controlled shared access [14]. Even though the generic descriptions of the two proposals seem clear, the lack of precise modeling creates many unanswered questions regarding the details of implementation. The exact nature of the controller or enforcer mechanisms in both models is vaguely defined. The government's role in managing controlled access in a spectrum commons regime is not clear. This lack of clarity also pervades the many issues related to transferability and duration of transmission rights, transactions costs, and the specific mechanisms involved in allocating the spectrum when needed. It is not clear, for example, how often transmission rights are anticipated to change hands in a spectrum property rights model.

Cognitive radio technology, along with appropriate architecture and infrastructure support, holds the promise of bridging the gap between the above regimes of spectrum governance. As discussed in [49], cognitive radios can enable spectrum governance regimes that support both exclusivity of property rights and the dynamic nature of shared managed access to a spectrum commons. Thus the advent of cognitive radio technology and its innovative ability to use spectrum opportunistically suggests a renewed look at and possibly guidelines for shaping spectrum governance in the future.

1.4 Data explosion: future spectrum implications

There is yet another reason for evolving spectrum policy in the context of cognitive radio technology. The convergence of voice and data in wireless communications, triggered by the convergence of wireless and the Internet, has led to an explosion in the number of bits transmitted over the air in the first decade of the twenty-first century. With the number of mobile users steadily increasing, the number of wireless devices has increased along with mobile traffic in the past few years. With more applications being developed every day and convergence of applications in one mobile device, the mobile traffic is expected to grow exponentially over the next generation of wireless devices. According to a study by Cisco [23], global mobile data traffic in 2010 (237 petabytes per month)[1] was over three times greater than the total global Internet traffic in 2000 (75 petabytes per month). Future projections as shown in Fig. 1.1 suggest that this explosive growth will continue well into the foreseeable future. According to Cisco reports, laptops and netbooks will continue to generate a disproportionate amount of traffic, but new device categories such as machine-to-machine and tablets will begin to account for a significant portion of the traffic by 2015. In fact wireless communications devices themselves have seen an explosive growth from about 1 billion devices in 2005 to about 10 billion devices in 2010 with projections of about 100 billion devices by 2020. In order to address the bandwidth demand of these mobile data services, wireless

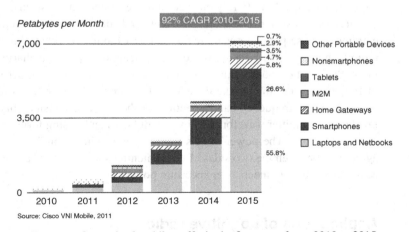

Figure 1.1 Forecast of growth of mobile traffic in the five years from, 2010 to 2015.

---

[1] A petabyte is 10^{15} bytes.

networks need to evolve so that they can handle bandwidth-hungry applications. Cognitive radio technology has the potential to enable opportunistic spectrum access in these devices and can mitigate the spectrum sharing problems posed in dense environments. Some example applications of cognitive radios are given in Section 1.5.

In future wireless systems, cognitive radio technology could serve as the foundation on which pervasive systems that solve a myriad problems can be built. Consider for example, that vehicular collisions kill forty thousand people in the US annually, and injure almost three million more. Almost one hundred thousand people die every year from medical mistakes, with a large number of these instances occurring in emergency rooms. Terror attacks and natural disasters stress our response systems, leading to unnecessary casualties and property loss and often putting responders at risk. Each of these distinct problems is complex in both technological and human terms; yet surprisingly, engineered solutions to these problems share several common attributes and challenges. Vehicles could communicate with each other and with roads and infrastructure to improve safety and manage traffic. Embedded sensors could remotely monitor patients and help medical staff manage critical information in hospital emergency rooms. Emergency workers could increase their effectiveness and personal security through wirelessly connected networks of robots and unmanned aerial vehicles. All these solutions require embedded wireless devices, allowing machines and people to interact within a local area and across the global Internet.

These solutions will be implemented in the form of pervasively deployed wireless ecosystems, which are large-scale, heterogeneous, and decentralized physical world deployments of networked cognitive radio devices including sensors, actuators, and machines, with humans in the loop. While tapping the potential of pervasive wireless ecosystems will require solutions involving general-purpose network architectures for heterogeneous mobile/sensor devices, robust computing models that include programming of the physical world, cyber-physical control, and human-centered design, fundamentally they will have to rely on spectrum coexistence in dense environments.

Wireless ecosystems require efficient and decentralized spectrum sharing in extremely dense populations of wireless devices, embedded and otherwise. Cognitive radio technology will be a cornerstone for successful spectrum sharing in these situations. Such radio nodes could sense the usage of radio spectrum across a wide range of frequencies and then make decisions about how to transmit. These decisions may be as simple as frequency band selection, or they may be considerably more complex, involving transmitter waveform design, modulation and coding formats, and even cooperation methods. The power of this transforming radio technology is even a point of agreement between the two sides of the contentious spectrum debate among regulators, economists, and engineers over spectrum policy [49].

1.5 Applications of cognitive radio

The most common application of cognitive radio is in the TV white spaces, where cognitive-radio-enabled secondary users opportunistically utilize the unused spectrum without interfering with primary incumbents of spectrum, namely TV transmitters. Cognitive radio devices can also form a secondary network in such bands where all secondary users are required to detect white spaces or spectrum holes. Cognitive radio technology finds many applications beyond TV white spaces. Over short ranges, cognitive radios find applications in wireless LANs, e.g., IEEE 802.11 family of standards, vehicle-to-vehicle communications over a range of 100 meters where data rates can be on the order of a few tens of Mbps. In the mid-range of distances spanning orders of a kilometer, cognitive radios can be used to deploy an extended range wireless LAN spanning hundreds of meters [8]. In the long range scenario, cognitive radio networks can use white spaces to meet the spectrum requirements of networks for public safety. Each of these scenarios can be classified into one of the three usage paradigms discussed earlier.

1.5.1 Dynamic spectrum access in cellular systems

With the advent of smart phones, cellular networks are undergoing rapid growth and are striving to improve spectral efficiency required to cope with the high bandwidth demands. While physical layer techniques have matured a lot in enabling spectrally efficient bit pipes, at the system level, the need for spectrally efficient transmission techniques still remains. Next-generation cellular systems are going to be more heterogeneous, with small cells embedded in large macrocells. These embedded picocells or femtocells are miniaturized versions of base stations transmitting at low power. They can form small coverage islands inside a macrocell coverage region to provide coverage and capacity to users in those coverage islands. Such coverage islands could be hot-spot areas like a shopping mall, or a sports arena. Small cells provide more cell-splitting gains, when they share the same spectrum as the entire macrocell [30]. Although the above methods require infrastructure support, schemes like DSA can exploit the unused spectrum to provide better spectral efficiency gains. Cognitive radio technology can be used to estimate the spatio-temporal characteristics of the traffic and spectrum usage in the macrocell and opportunistically use the available spectrum during periods of high data demand. Dynamic spectrum access networks can work in the interweave paradigm where the cognitive radios can detect white spaces to opportunistically transmit in that spectrum. It can also work in the overlay paradigm if overhead signaling through the backhaul can provide side information to the cognitive transmitters.

1.5.2 Cellular data boost

As a further extension of DSA in cellular networks, cognitive radios can possibly be used to opportunistically offload traffic to white spaces to provide a "data boost" to the existing cellular network. For an example scenario shown in Fig. 1.2, cognitive radios can help alleviate high loads on the cellular network. In the example scenario, an overlay mesh network of white space hot-spots can carry non-real-time or delay tolerant data like mail, content, file transfer, etc. The hot-spots can operate either in the licensed or unlicensed band. Off-loading part of the delay tolerant data traffic from the cellular network helps operators to meet the quality of service (QoS) requirements of delay

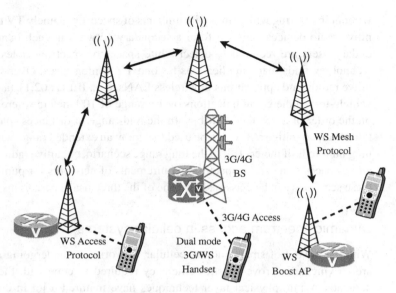

Figure 1.2 A cellular data boost network for off-loading fast-growing cellular traffic using dual-mode radio.

sensitive traffic like voice, streaming video, etc. This approach could lead to capacity boost of two to five times the existing cellular capacity, depending on the percentage coverage and service mix of the data generated in the area. The boost in the capacity comes at no additional cost, but from the detection of available white spaces and the opportunistic transmission over white spaces using the hot-spot mesh network. The end user handsets are required to support dual-mode radios equipped with the capability to transmit in the cellular band as well as in white spaces. This is an example of the interweave paradigm of cognitive radio operation.

1.5.3 Machine-to-machine communications

Machine-to-machine (M2M) communications [90] enable mechanical automation of ubiquitous sensors. The demand for wireless personal communications for exchange of voice, audio, video, emails, and photos over ad hoc networks and infrastructure-based cellular networks have empowered full automation of sensor networks. Cognitive sensor networks can be used for M2M communications in many ways. The communication could be between the sensor and the decision maker (as in meters/monitors reporting data to a decision maker), among multiple calculation agents within the decision maker (as in cloud computing) or between the decision maker and the action executer. In the future, multiple radio access networks may be integrated and managed as part of a single hierarchical network. Cognitive radios can enable efficient air interface design of M2M networks. Cognitive radios could opportunistically exploit the additional spectrum and connectivity available in different networks to improve the system capacity and device quality of service. Cognitive-radio-enabled devices can substitute

1.5 Applications of cognitive radio

the multi-radio devices to efficiently utilize the available spectrum across licensed and unlicensed bands.

1.5.4 Distribution and backhaul

Cognitive-radio-enabled white space networks can be used as a backhaul and distribution network [38, 39]. This solves not only the problem of providing wireless broadband to rural areas lacking in infrastructure but also the *last mile problem*, that has been a major concern for home-user service providers. Typically, the last mile is wired by either digital subscriber lines (DSL), coaxial cables, or fiber to the home (FTTH). White space networks can bring down the infrastructure costs of laying wires to the customer premise by serving as a backhaul network as shown in Fig. 1.3. The white space network can serve as transport network for the uplink and downlink traffic due to the content that is generated and requested by a typical user in a home. The backhauling can be done by a wireless router at each home, by routing on the white space backhaul network. Depending on the type of service, the data are processed by the appropriate service provider who operates in the core network. The user is also spared the hassle of handling and maintaining multiple interfaces for his various service needs.

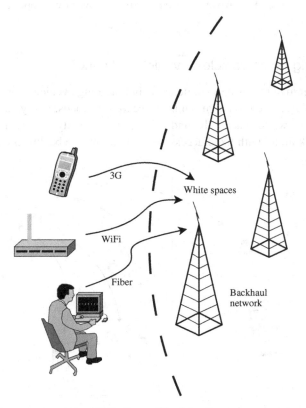

Figure 1.3 White spaces used for distribution and backhauling.

1.5.5 Cognitive digital home

The number of devices in a home has been increasing with the number of services subscribed by each user in the home. Technologies like Bluetooth, Infrared, and WiFi are wireless technologies that have eliminated the need for wired communications among the various devices. The problem, however, is that many of these technologies are not designed to operate in consonance with the other technologies and are a cause of interference for the receivers that are not part of their network. Hence, a central controller that resolves the contention and oversees the coexistence of the networks is required. A cognitive digital home can be the solution for all the interference issues due to the multiple competing technologies [55, 56]. As shown in Fig. 1.4, a cognitive digital home can operate on white spaces in unlicensed bands to carry control information between devices. A genie node acts as a central controller to coordinate the operation of these devices such that the number of contentions and collisions is reduced. This leads to efficient use of resources in the devices. The user does not have to manage the individual networks and devices and the central controller can be used to automate the management of the home network. Such central control also enhances the security in such networks where the controller can be used as a gateway to the external world. The cognitive digital home applications fit into the underlay framework of cognitive radio operation because of the low-power operation of the cognitive radio devices.

1.5.6 Long range vehicle-to-vehicle network

Cognitive radios find applications in long range vehicle-to-vehicle networks. Such networks can be used to minimize drive time on roads. Every vehicle could be mandated to have a white space radio and vehicles in a locality can form a peer-to-peer ad hoc network along with fixed access points that provide backhaul capabilities. While services

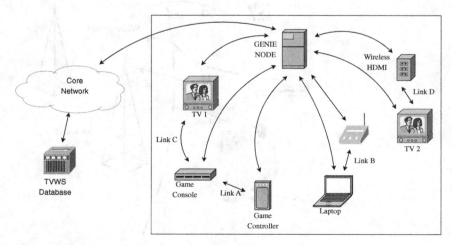

Figure 1.4 Cognitive digital home.

1.6 Cognitive radio network design

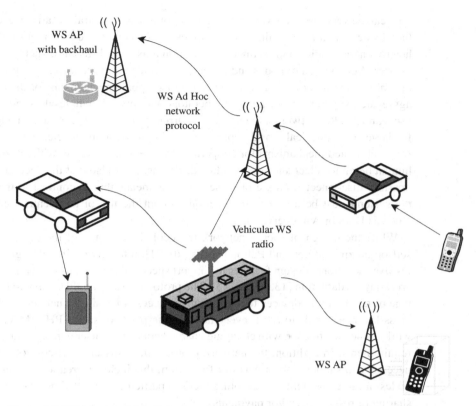

Figure 1.5 Vehicle-to-vehicle links from car radios can be used to form a high capacity emergency back up network using ad hoc mesh between cars and fixed access points.

like smart collision sensing mechanisms would require low latency and communication in a licensed spectrum, white space communication could supplement the short range communications infrastructure. The white space network can be used for traffic alerts, geographic applications services, peer-to-peer content, etc. They can also serve as a high capacity emergency back up network. Further, the flexibility provided by cognitive radio vehicular networks can also allow them to be integrated into emerging intelligent transport systems infrastructure and smart highways.

1.6 Cognitive radio network design

1.6.1 Global control plane

While the operation and advantages of an individual cognitive radio can lead to interesting applications and services, the networking aspects of multiple cognitive radios cannot be ignored. A wireless ecosystem of cognitive radios can be thought of as a heterogeneous network of networks with a multiplicity of radios and protocols similar in spirit to the global Internet. Analogous to the Internet, a wireless access network of this type can be very large in terms of the number of devices and traffic carried, even though the physical span is more limited. A wireless ecosystem will contain multiple subnets, heterogeneous radios and distinct network domains with different owners or groups of owners. Aggregation of nodes and subnetworks is also important from the point of view of scaling, and it is desirable that the protocol supports mechanisms for the creation of aggregated supergroups or network operators who provide wholesale services to end-users and smaller networks. In contrast to current Internet protocols, any viable solution for large-scale physical world deployments of sensors, mobiles, etc. must incorporate fully automated mechanisms for cooperation and resource sharing without relying on a human being to select and set up policy decisions. Note also that the decentralized and possibly disconnected nature of these networks means that in contrast to the Internet, the system must be designed to function without the use of any centralized services (such as DNS or AAA servers).

While the design of such a network protocol shares some of the characteristics of self-organizing ad hoc and P2P networks [70, 71], internetworking of cognitive radios involves additional design features to support spectrum sharing, topology discovery and cross-layer adaptation [13, 27, 32, 45]. The protocols used should be able to take advantage of nearby radio devices for opportunistic access to available networks while using cross-layer protocols to adapt every layer of the protocol stack (PHY, MAC, and network) as needed to cope with changing channel conditions, user density, and application requirements. In addition, the network protocol must disseminate control information necessary to support robust network formation; this includes creation of trust between nodes in the ecosystem and enabling them to participate in collaborative networks via sharing of resources and/or payments.

An architecture that is based on the concept of a global control plane (GCP) which is logically distinct from the data plane is the CogNet architecture [51, 78]. It provides a very general mechanism for dissemination of control information between coexisting or collaborating radio nodes. A global control architecture allows cognitive radio nodes to initialize and dynamically adapt their PHY, MAC, and network level parameters. The control plane is made up of several key components: (i) bootstrapping, (ii) discovery, (iii) cross-layer routing, and (iv) naming/addressing. See Fig. 1.6.

The radio bootstrapping function allows for detecting local links and configuring PHY/MAC parameters when cognitive radio nodes first boot up. After initialization, nodes execute a discovery protocol based on periodic reporting of local link states of neighboring nodes using a controlled one-hop broadcast mechanism. The discovery protocol also interacts with a cross-layer routing module that provides end-to-end reachability and path information across multiple hops, which are dynamically configured with cross-layer parameters including frequency, power, rate, etc. The fourth key component is the support for distributed naming and addressing by which network nodes map their permanent names to dynamically assigned network addresses which may change with network structure and mobility. The control plane generally uses a low-rate radio PHY with wider coverage than the data signal, and can thus be used to efficiently distribute control information with fewer hops than would be required during data transfer. Data plane parameters (such as frequency, power, bandwidth, rate, etc.)

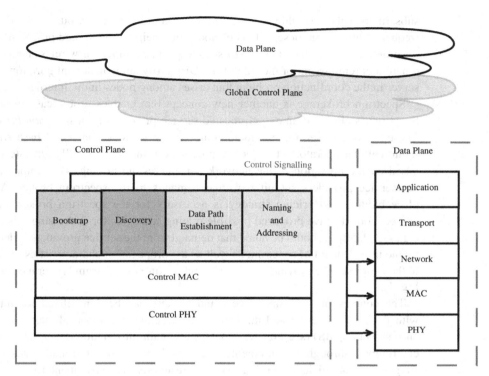

Figure 1.6 Global control plane.

can be set up at each forwarding hop to meet certain end-to-end performance criteria and/or system efficiency objectives.

1.6.2 Spectrum servers, spectrum brokers, and network information servers

Most research activities on cognitive radios are focused on the design issues of cognitive radios, related to frequency, waveform, and protocol agility. These activities, however, fail to fully reveal the ultimate limits of cognitive behavior. The cognitive radio must do more than communicate on an ad hoc basis to realize its full potential; it must develop a full awareness of a local environment that may span multiple spectrum bands and systems. This implies new discovery processes that are thorough and efficient, and even new classes of "information servers" that provide assistance in the process.

When there exist methods by which cognitive radios can independently discover local information, a variety of physical layer, system, and network layer protocols such as specified in the CogNet architecture can be applied to allow cooperation and coexistence. However, such levels of cooperation and interoperability may not be possible when multiple services and systems must coexist. In this case, efficient open access to spectrum can be aided by impartial *spectrum servers* [77] which can obtain information about the interference environment through measurements contributed by different terminals, and then offer suggestions for efficient coordination to interested service subscribers. Likely neighborhood information could include various levels of time and frequency utilization, descriptions of nodes in a neighborhood, and potentially, spatial positions as well. In fact, the role of such a spectrum server for wireless network coordination is reminiscent of the role of the DHCP (dynamic host configuration protocol) server in the coordination problem that arises among nodes in the Internet.

Spectrum brokerage is another new concept that enables timed lease of spectrum. Spectrum is allocated among co-located service providers by a *spectrum broker* [18]. These co-located service providers could be TV stations, macrocell base station, or femto-cellular operators. The service providers in turn dynamically allocate spectrum based on the operation of the network. An interesting pricing issue comes up when the service providers submit real-time demands to the spectrum broker. A market place bidding and pricing strategy is necessary for the spectrum broker. Engineers and economists have proposed possible pricing strategies for addressing this problem, e.g., see [2, 50]. It should be noted that demand from the service providers is determined by the user demand of spectrum as well as the propagation characteristics of each user to their own service providers. A simple architecture of spectrum brokerage is shown in Fig. 1.7.

The concept of spectrum servers and brokers can be generalized for integration with the future Internet and the cloud infrastructure via impartial "network interference servers" (NIS) (see Fig. 1.8) which can obtain information about the interference environment through measurements contributed by different terminals, and then offer suggestions for efficient coordination to interested service subscribers. Likely neighborhood information includes various levels of time and frequency utilization, descriptions of nodes in a neighborhood, and potentially, spatial positions as well. The most powerful engineering example of a viable open access system that exists today is the Internet. The relatively successful TCP flow coordination model among nodes in the network is reminiscent of the radio coordination problem that arises in the context of open access to spectrum. A natural question therefore arises: can we design a standardized, technologically neutral, physical, and protocol infrastructure like the backhaul and TCP of the Internet, to coordinate wireless transmissions with cognitive radios?

In anticipation of the networking environment that will support open access to spectrum, it seems reasonable to assume that all nodes will be connected via direct or multiple hops to the Internet, and that spectrum monitoring sensors will frequently be available. Thus, like the wired Internet, cognitive nodes will have access to various Internet-resident information services. One such service is the dynamic host configuration protocol (DHCP) which is used to assign local nonconflicting Internet protocol (IP) addresses. The DHCP server assigns addresses in response to requests, but also keeps track of active existing addresses through polling and listening. Extending this idea, the NIS can be established for wireless systems to facilitate efficient open access to spectrum.

The NIS can be regarded as an intelligent information-gatherer that maintains a database of wireless activity traces submitted by subscribers. Such traces might consist of wireless activity generated by nodes as well as external activity observed by nodes. The specific measurements might be closely tied to the physical layer such as signal

1.6 Cognitive radio network design

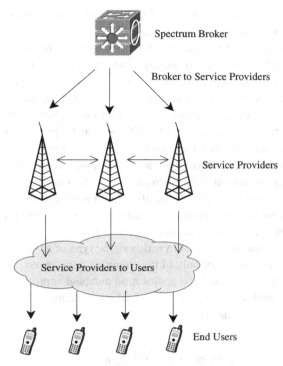

Figure 1.7 Spectrum broker.

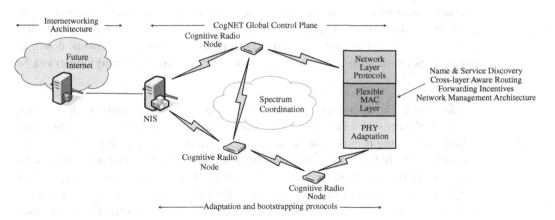

Figure 1.8 Network information servers and the cloud.

to interference levels and error rates, or could be higher level inferences such as the identities and types of nodes active in the local area, models of interference activity, and perhaps even geographic information. We also note that only partial information might be available since a given cognitive radio might not have access to all the locally used spectrum, or may be ignorant of higher level descriptions for local systems (i.e., 802.11, Bluetooth, microwave oven, etc.).

From the user perspective, a request sent to the NIS consisting of the user identity, capabilities, and any locally observed phenomena would result in advice for efficient local spectrum utilization. In simple cases, a list of preferred channels (or signal space dimensions such as codewords, hopping sequences, and the like) might suffice. In other cases, a cognitive radio might need more complex information; a list of preferred access methods (e.g., CDMA, TDMA, Bluetooth, 802.11, other) and/or information about nearby nodes which might help carry user traffic. And in yet other cases, the NIS might simply serve as a translator through which disparate cognitive radios could exchange information and negotiate. The NIS concept is therefore applicable across a wide range of levels of cognitive radio sophistication. NIS realization raises many interesting research problems relating to how a coherent picture of interference environments can be painted from measurements drawn in bits and pieces from a variety of uncoordinated sources. In addition, the type of advice the NIS can provide is related to the capabilities not only of the cognitive transceivers, but also of the underlying network. The FCC mandated geolocation database approach to regulate the coexistence of primary and secondary users is indeed a rudimentary version of the NIS-based approach to spectrum coexistence.

1.6.3 Security aspects of cognitive radio

The spectrum, protocol, and technology agility that cognitive radios are capable of also opens up new challenges for securing these and networks of these. In cognitive radio networks, the secondary user needs to maintain accurate spectrum occupancy information of all primary users in the network. As with any wireless network, a malicious user can threaten the network operation and bring the network down. In dynamic spectrum access environments, some of the following security threats can be prominent.

1 *DSA attack.* The malicious user can create signal waveforms similar to the primary users in the network and trigger a false alarm in the spectrum sensing algorithm. This kind of spoofing primary user waveforms can affect the long-term behavior of secondary users. In the primary user emulation (PUE) attack, the adversary emulates the primary transmissions in the network. The PUE attack can dupe any sensing process and reduce the channel resources available for the secondary users in the system. In order to counter this threat, a transmitter verification scheme called location-based defense has been proposed in [20]. The transmitter verification scheme consists of three steps: verifying the signal characteristics of the transmitter, measurement of received signal, and localization of the signal source. Cognitive radios can be used in innovative ways to do these tasks.
2 *Denial of service (DOS) attacks.* DoS attacks may include DSA attacks, but cover broader attack strategies. DoS attacks could also be jamming signals that can prevent

messages from being received by the victim receiver. Depending on the availability of power, the adversary may use different strategies to deny service to the secondary users. For example, jamming requires higher power while spoofing a signal to confuse the victim may be accomplished with lower power. Beyond these strategies, manual DoS attacks are possible where the victim's device interface could be misconfigured. In the worst case, the node is compromised so that it becomes a malicious participant in the cognitive radio network.

The underlying cognitive radio architecture determines to a large extent the vulnerability to different attacks. Cognitive radios sensing the spectrum in a cooperative network are less vulnerable than cognitive radios sensing the spectrum in a noncooperative network. On the other hand, a malicious cognitive radio can spoof membership in the cooperative network and intercept sensitive information in a cooperative network. This node can potentially deny service to all members in the network. The usage scenarios can also determine the vulnerability of the cognitive radio nodes. For example, underlay networks, where the secondary users spread their signals over a wideband, are inherently less vulnerable to jamming attacks [16].

1.7 Hardware and system design considerations

1.7.1 Design tradeoffs in usage scenarios

The tradeoffs associated with different cognitive radio paradigms include the hardware and system design requirements and constraints, in particular their complexity and cost. Underlay systems are required to measure the interference that a cognitive transmitter causes to a primary receiver. If a signal sent from the location of the primary receiver at a known transmit power is received at the location of the cognitive transmitter, then via reciprocity the cognitive transmitter can determine the interference it causes to that primary receiver. When there are multiple primary users, the cognitive transmitter cannot in general exploit reciprocity to determine the interference it causes to each one since it can only receive the sum of signals transmitted simultaneously from the primary receiver locations. A system-level technique to determine interference is to have multiple cognitive nodes in the vicinity of each primary node estimate the interference through cooperative sensing and detection, as described in more detail in Chapters 4 and 5. If there is no mechanism to measure interference caused to the primary users, then the cognitive radio transmission protocol must be extremely conservative, especially when there are a large number of cognitive transmitters. This typically entails putting the signal output from each cognitive transmitter far below the required interference threshold at all locations outside a very small radius around the transmitter.

Interweave systems require an extremely wideband and/or agile front end to sense the entire spectrum of the system in search of spectrum holes. In fact, current hardware capabilities place significant constraints on how fast an interweave cognitive user can scan for spectrum holes over a wide bandwidth. Cooperative sensing reduces the complexity burden of detection on any one node, but this introduces other challenges in terms of delay and overhead. Alternatively, compressive sensing can be applied to this problem to reduce complexity, albeit with some loss in performance [86].

Finally, the overlay paradigm requires the cognitive transmitter to obtain the primary users' signals, data sequences, or codebooks, along with all channel gains in the network. Moreover, complex encoding and decoding techniques must be implemented to enhance primary user performance while canceling interference on both sides to obtain the best possible data rates for both users.

1.7.2 Antenna design in cognitive radio systems

Cognitive radio networks can function efficiently and reliably only if the component radios are able to scan wide bands of spectrum and accurately detect spectrum holes. Further, they also need to be able to transmit opportunistically in the detected white spaces. This requires the cognitive radio to be equipped with wideband antennas. Using wideband antennas can only produce coarse spectrum sensing, whereas narrowband antennas are capable of accurately sensing the spectrum. Another issue with wideband antennas is that they tend to have a bigger form factor than narrowband antennas, which poses a significant problem for portability in handsets. In dynamically changing environments, frequency agility would require tunable RF filters over a wide spectral range, making their implementation complex and expensive [41, 52].

From a system design perspective, there is an increasing demand for multi-wideband antennas that can also be easily integrated with the communication system. Usually electronic reconfigurability is achieved by using switches, variable capacitors, or phase shifters in the topology of the antenna. Most frequently, lumped components such as PIN diodes, varactor diodes, or MEMS switches or varactors are used in the design of reconfigurable antennas. Typically, reconfigurable antennas are frequency reconfigurable, radiation pattern reconfigurable, or polarization reconfigurable [83].

The advantage of a frequency reconfigurable antenna is that it allows a single radio device to operate at multiple frequencies. For example, a tuning method [53] can be used to tune into multiple bands for mobile applications. The radiation patterns of such antennas remain unchanged as the frequencies are tuned from one band to another. In designing a frequency reconfigurable antenna, a single antenna will be able to tune from lower to higher bands by reusing the real estate in the antennas or tune in and out of different antenna sections. Most resonant type antennas tune the effective electrical length to achieve frequency reconfigurability.

In a radiation pattern reconfigurable antenna, the frequency band remains unchanged while the radiation pattern changes based on the system requirements. The antennas can steer their radiation beams to different directions to enhance signal reception [21, 65, 88]. Polarization reconfigurable antennas improve the signal reception performance in severe multipath fading environments using polarization as an additional degree of freedom [46]. Pattern reconfigurable antennas and polarization reconfigurable antennas can also be utilized for interference cancellation and to form rich spatial channels for multiple-input multiple-output (MIMO) systems to increase throughput and network capacity [47, 68]. It is desirable for system designers to have one of the above tuning parameters vary while keeping the other parameters constant. The implementation challenge however, comes in the device integration stage because of the interrelated parameters and the subsequent tradeoffs involved [69].

1.7.3 Analog-to-digital converters

In addition to antenna design, yet another cognitive radio design challenge is the RF front-end. As discussed in Section 1.2.4, cognitive radios are likely to operate in the UHF TV bands utilizing the unused TV channels. The UHF bands have good propagation characteristics for long range communications. For short ranges, cognitive radios find opportunities of using the 3–10 GHz band. Regardless of the operating frequency, the front-end of a cognitive radio receiver needs to be wideband, as the cognitive transmitter could potentially transmit on any of the unused bands. This necessitates the cognitive radio receiver front-end to provide for wideband signal reception, from the antenna to the analog-to-digital converter (ADC). There will be stringent requirements placed on the linearity of the RF analog circuits. The ADC will have to sample at GHz rates if it operates over GHz of bandwidth. Further, the cumulative interfering signals over a wide band could be stronger than the signal of interest, leading to SINRs as low as -50 dB. The cognitive radio signal cannot be amplified enough to achieve sufficient quantization accuracy. For example, to achieve a 20 dB signal to quantization noise ratio, the required resolution of the ADC would be on the order of 12 bits or greater if the SINR is -50 dB. This level of accuracy with GHz sampling rates results in an essentially infeasible ADC implementation if there are power and cost constraints, since the power and ADC complexity rises nearly exponentially with the number of bits. One way of increasing the SINR is to eliminate the interference by estimating it and subtracting it off [91]. Other solutions include adaptive notch filtering, banks of RF filters using MEMS technology, and spatial filtering using RF beamforming through adaptive antenna arrays [29].

1.7.4 Wideband channels and noncontiguous transmission

Increasing the transmission bandwidth is an attractive solution to meet the increasing demand for high data rate applications over wireless channels. But, as will be discussed in detail in Chapter 3, signals propagating over wireless channels undergo time-varying fading due to scattering of the transmitted signals and Doppler spread of the received signal. These effects are exacerbated in high data rate transmissions over wideband channels. In order to undo the effects of the channel on the signal, the received signal has to go through a channel equalizer. Wideband channel equalizers can be complex to implement and require high processing power to meet delay constraints of high data rate applications.

Orthogonal frequency division multiplexing (OFDM) [40, Chapter 12] is a digital multicarrier modulation technique that uses a large number of closely spaced orthogonal subcarriers to carry data over wideband channels. The data symbols modulate each subcarrier separately, and the overall modulation operation across the subcarriers result in a wideband frequency multiplexed signal. The orthogonality of the subcarriers makes the transmissions look like multiple parallel low rate transmissions and thus obviates the need for complex wideband channel equalizers. OFDM has hence emerged as a successful air-interface for the Digital Audio Broadcasting (DAB) [34], digital video broadcasting for terrestrial television (DVB-T) [54], the Third Generation Partnership Project (3GPP) Long Term Evolution (LTE) standard [81], the IEEE 802.11a/g family of wireless local area network (WLAN) [48], and the IEEE 802.16 WiMAX family of standards [5].

In OFDM, the discrete-time serial data stream x_n, $n = 0, \ldots, N-1$ (denoted by **x**) is converted into a parallel stream through a serial-to-parallel converter that splits serial data into a number of parallel channels, as shown in Fig. 1.9. The data in each parallel channel are passed through a pulse shaping filter $g(t)$ and then applied to a modulator, with n_B subcarriers $f_0, f_1, \ldots, f_{n_B-1}$. The separation between two adjacent frequencies is carefully chosen such that subcarriers are pairwise orthogonal. If Δf is the difference between adjacent subcarriers, the overall bandwidth W is $n_B \Delta f$. The n_B modulated subcarriers are combined to give an OFDM symbol $x(t)$.

In the conventional single carrier modulation approach, the modulated signal occupies the entire bandwidth W centered around the carrier frequency. A deep fade in the wireless channel extending over the duration of several bits will result in multiple data symbols being lost. In the multicarrier approach however, all n_B symbols are stretched out in time for n_B times longer duration. So, the OFDM system can recover a part of each of the n_B parallely modulated data symbols from the deep fade. Moreover, due to very low symbol durations in high data rate single carrier systems, slight delays in

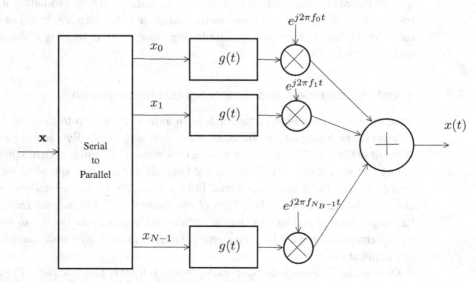

Figure 1.9 Orthogonal frequency division multiplexing scheme.

1.7 Hardware and system design considerations

arrival of multipath components at the receiver can cause severe inter-symbol interference (ISI). The advantage of stretching the symbol period is that multipath components of the same symbol arrive within a very small fraction of the new symbol duration. Hence, the effect of ISI is reduced in multicarrier systems [42]. Modulating the subcarriers using the parallel data stream is mathematically equivalent to taking the inverse discrete Fourier transform (IDFT) of the original data stream [40, Chapter 12]. The complexity of employing n_B modulators and transmit filters at the transmitter and n_B demodulators and receive filters at the receiver can be reduced by employing discrete Fourier transformers (DFT). When the number of subcarriers is chosen appropriately, the system complexity can be further reduced by using the fast Fourier transform (FFT) algorithm to compute the DFT.

In DSA networks, multicarrier techniques like OFDM provide for opportunistically filling the spectrum so that white spaces across the spectrum are effectively utilized. By effectively nulling the subcarriers, the cognitive radio users can avoid interference to the primary users, as shown in Fig. 1.10. In the link level, the transmitter and receiver are required to mutually agree on the carriers used at the transmitter and receiver. This may require link rendezvous protocols for the handshaking messages between transmitter and receiver. Such a transmission scheme is called noncontiguous OFDM (NC-OFDM) in [73, 74] and differs from conventional OFDM systems.

In contrast to single-carrier systems, OFDM systems are particularly sensitive to synchronization errors like carrier frequency offset and symbol timing errors, leading to

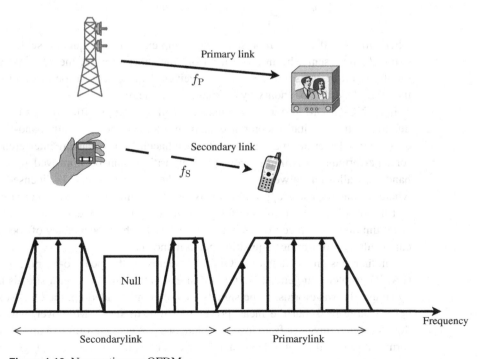

Figure 1.10 Noncontiguous OFDM.

increased intercarrier interference (ICI). Extensive work has been done in designing robust algorithms that estimate the carrier frequency and symbol timing with high accuracy [19, 80, 87] to ensure reliable communication. For NC-OFDM systems however, the imperfect estimation of timing information aggravates the synchronization problem. In some scenarios when users occupy multiple noncontiguous subbands and have large differences in the timing offsets between their transmitters and receivers, cyclic prefix-based timing acquisition algorithms can perform quite poorly. Thus, there is a need for better algorithms of reasonable complexity, or entirely different approaches to symbol timing acquisition.

In a typical OFDM transmitter with n_B subcarriers, the IDFT modulated symbols can be represented as

$$x_n = \frac{1}{n_B} \sum_{k=0}^{n_B-1} X_k e^{j2\pi kn/n_B}, \; n = 0, 1, \ldots, N-1. \tag{1.1}$$

In an NC-OFDM system, the secondary transmission is only on vacant subcarriers. The subcarriers that are occupied by the primary users are nulled, and hence $X_k = 0$ for those subcarriers.

If we assume that an NC-OFDM symbol is $n_B + p$ samples long with the first p samples, called the cyclic prefix, being the same as the last p samples. The received samples y_n in the presence of timing offset τ between transmitter and receiver is

$$y_n = \sum_{l=0}^{m} h_l x_{n-\tau-l} + I_n + w_n, \; n = 0, \ldots, N-1, \tag{1.2}$$

where h_l, $l = 0, \ldots, m$ represents the complex m-tap frequency selective channel gains, I_n represents the interference from the other users at the receiver, and w_n is the discrete-time noise process at the receiver. Timing acquisition is about estimating the OFDM symbol boundary by estimating τ at the receiver.

In an NC-OFDM system, the signals x_n and I_n occupy different sets of tones. But, due to practical limitations on pulse-shaping filters not being ideally band-limited, part of the symbol energies may spill over to the adjacent bands, causing interference. Therefore, performance of timing acquisition algorithms can be improved if wider guard bands are allowed between the signals of the different users. In licensed spectrum, where secondary users opportunistically use the primary spectrum, guard bands protect the primary users from interference caused by the secondary users. In unlicensed spectrum however, uncoordinated access can lead to high occupancy of spectrum. This can result in poor timing acquisition performance [1].

Another design challenge in OFDM systems is the peak-to-average power ratio (PAPR) problem. In general, an OFDM signal is the sum of several signals modulated by sinusoidal waveforms. If the signals are coherently added up, the OFDM signal can have a large peak but the mean power will remain small. Moreover, when the time domain signal traverses from low instantaneous power waveform to a high power waveform, large signal amplitude swings are encountered. This results in high out-of-band harmonic distortion power, unless the transmitter's power amplifier exhibits linearity across the entire signal level range. This potentially contaminates the adjacent channels with interference [42]. The PAPR of the signal in x_n, $n = 0, \ldots, N-1$ in Eq. (1.1) is given by [84]

$$\text{PAPR}(\mathbf{x}) = \frac{\max_{0 \leq n \leq N-1}\{|x_n|^2\}}{\mathbb{E}\{|x_n|^2\}}. \tag{1.3}$$

When the PAPR of the OFDM signal is large, the digital-to-analog converters and power amplifiers require high dynamic range to avoid clipping and distortion. Hence the dynamic range of digital-to-analog converters should be large enough to accommodate large peaks. One down side is that the cost of the digital-to-analog converters increases with higher dynamic range. In NC-OFDM systems, the PAPR will be larger since some of the subcarriers are nulled [75].

1.8 Spectrum coexistence in cognitive radio networks

One of the important requirements for the well-being of a cognitive radio network, for example in TV white spaces, is the healthy coexistence with the primary networks. There is a stringent requirement that secondary users should minimize the amount of interference at the primary receivers. Further, there is also the need for secondary users to be able to coexist and possibly form networks to carry information. There are many reactive methods that a cognitive radio could use to avoid interference. Typically, these involve adaptation of frequency, transmit power, and transmit waveforms. A frequency agile cognitive radio can periodically scan the entire band, identify if the channel is being used by the primary user and dynamically shift to another vacant band of the spectrum. Software defined radios can scan the spectrum either by sweeping the entire band or by frequency hopping and the RF hardware improvements have made such radios feasible. Power control is a widely used interference mitigation technique. With very minimal feedback based on the interference detected by periodic scanning, the transmit power of the secondary can be controlled to minimize the interference at the primary receiver. Orthogonalization of the signal space is another way of mitigating interference. In a time scheduling MAC of the secondary network, packets can be scheduled in time slots when primary user activity is not observed. Orthogonalizing in frequency is also a way to mitigate interference. NC-OFDM transmission discussed in the previous section is an example of orthogonalization in frequency domain. The above reactive mechanisms are limited by the amount of channel state information available at the transmitter and also due to the fact that interference is a receiver property. Reactive schemes inherently require explicit coordination protocols to obtain the network state and channel information of primary and secondary users. Without the network state information, efficient discovery, self-organization, resource management, and cooperation techniques are rendered inefficient. Exchanging link state information may overload the network and this poses a challenge for designing efficient networks using cognitive radios. This also emphasizes the need for the architecture and protocol mechanisms discussed earlier in Section 1.6.

1.8.1 Spectrum pooling and bandwidth exchange

The full power and benefits of cognitive radio technology can only be realized when cognitive radios form networks that can be used to carry information. Cooperative forwarding in wireless networks has been shown to yield rate and diversity gains, but it incurs energy costs borne by the cooperating nodes. Cognitive radio nodes have the ability to flexibly exchange the transmission bandwidth as a means of providing incentive for forwarding data, without increasing either the total bandwidth required or the total transmit power. The advent of multicarrier systems such as NC-OFDM with the ability to flexibly delegate and employ a number of subcarriers makes this approach particularly appealing compared to other incentive mechanisms that are often based on abstract notions of credit and shared understanding of worth.

Spectrum pooling is a concept where users may temporarily rent from a pool of spectral resources from primary users. The existing system of licensing spectrum need not be changed by this set-up. This system leads to a two-fold advantage: the secondary users get to access the spectral resources that they are not licensed to use, and the owners of the spectrum get to make revenue for the commodity they paid for but do not use extensively [89].

Among the many forms of cooperative communication, cooperative forwarding [12, 31] is an essential technique to enhance connectivity and throughput. As forwarding usually incurs bandwidth and energy cost, incentive mechanisms must be implemented to offset these. *Bandwidth exchange* (BE) [94], a concept similar to spectrum pooling, is an incentive mechanism that essentially enables a user to delegate a portion of its bandwidth in exchange for forwarding. Figure 1.11 illustrates the concept of BE in a network of nodes where a user incentivizes forwarding by delegating subcarriers to another user.

To motivate BE, consider Shannon's canonical channel capacity formula for an additive white Gaussian noise (AWGN) channel with a noise power spectral density of N_0,

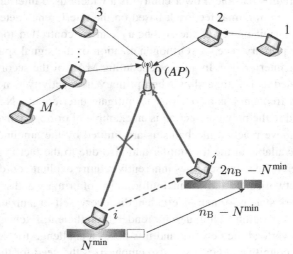

Figure 1.11 When the direct link fails, user i tries to incentivize forwarding by delegating $N - N^{\min}$ subcarriers to user j [93].

1.8 Spectrum coexistence in cognitive radio networks

$$C = W \log_2 \left(1 + \frac{\mathcal{P}}{N_0 W}\right) \text{ bits/second.} \quad (1.4)$$

In the above expression, C is only logarithmically dependent on the average transmit power $\mathcal{P}$, but nearly linearly dependent on bandwidth W, especially when W is relatively small. The largest partial derivatives with respect to these variables are given as

$$\left.\frac{\partial C}{\partial \mathcal{P}}\right|_{\mathcal{P}=0} = \frac{1}{N_0 \log 2}, \quad \left.\frac{\partial C}{\partial W}\right|_{W=0} = \infty. \quad (1.5)$$

Equation (1.5) suggests that when power or bandwidth is constrained, the marginal utility achieved by using additional bandwidth is significantly greater. However, since the total bandwidth available in the network/system needs to be preserved, one may question whether it is really beneficial to reallocate bandwidth, since after all, when one node acquires additional bandwidth, it is at the expense of some other node(s) that loses an equal amount of bandwidth. Further, as the bandwidth increases (wideband regime), the marginal increase in capacity saturates. Despite the marginal increase, a simple example can illustrate the performance gains from BE [94]. Consider the two-node network with an access point shown in Fig. 1.12. Suppose each node has a nonoverlapping bandwidth ($W_1 = W_2 = 20$ MHz) and fixed transmit power ($\mathcal{P}_1 = \mathcal{P}_2 = 20$ dBm). Also, for simplicity in illustration, suppose that the channels between the access point and the nodes as well as the channels between the nodes are determined by distance-based path loss, i.e., the rate achieved on a link is an explicit function of the bandwidth W and channel power gain G, which is parametrized by its fixed transmit power $\mathcal{P}$. Let us assume this function is given by

$$C = C(W, G) = W \log_2 \left(1 + \frac{G\mathcal{P}}{N_0 W}\right) \text{ bits/second,} \quad (1.6)$$

where the channel gain $G = \kappa d^{-3}$ with d being the distance and κ being a proportionality constant that also captures the noise power spectral density N_0. For the geometry shown in Fig. 1.12, it follows that if both nodes only use direct links for transmission, node 1 achieves a transmission rate of $R_1^{\text{dir}} = 11$ Mbps, while node 2 achieves

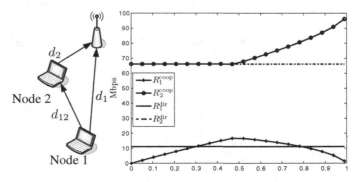

Figure 1.12 Bandwidth exchange enhances rates for both nodes simultaneously with $d_1 = 400$ m, $d_2 = 150$ m, $d_{12} = 300$ m, $\kappa = 6 \times 10^6$ MHz $\cdot$ m^3/mW.

$R_2^{\text{dir}} = 66$ Mbps. However, if node 1 uses node 2 as a forwarder and delegates a fraction α of its bandwidth to node 2, then the rates achieved through cooperation are given as

$$R_1^{\text{coop}} = \min\{C_1((1-\alpha)W_1, G_{12}), C_2(W_2 + \alpha W_1, G_{20}) - R_2^{\text{dir}}\}, \quad (1.7)$$
$$R_2^{\text{coop}} = C_2(W_2 + \alpha W_1, G_{20}) - R_1^{\text{coop}}, \quad (1.8)$$

where G_{12} and G_{20} are the channel gains from node 1 to node 2 and from node 2 to the *AP*. The functions C_1, C_2 are as defined in Eq. (1.6) and it is assumed that node 2 requires its own rate to be at least R_2^{dir} or better. Figure 1.12 shows that there is a range of values of α for which both nodes' rates are improved. While the above example illustrates BE for a two-user scenario, this incentive-based networking mechanism can be generalized to the case of M users. Specifically, it can applied in a setting for fair secondary coexistence in TV white spaces using a Nash bargaining framework [93]. In the new FCC ruling regarding TV white spaces, plenty of protections and precautions are imposed to guarantee undisturbed operation of primary users. Though spectrum sensing is no longer required, a geolocation database that registers the locations of primary users has become a mandate. Every secondary user is required to query the database through the Internet to make sure it would not produce interference to nearby primary users before it starts transmission. The query is periodic in case some primary users want to initiate operation in the vicinity. These requirements provide a reliable shield between primary users and secondary users for the purpose of their coexistence. However, the new ruling does not explicitly designate how coexistence is to be managed among the secondary users. The presence of the geolocation database can facilitate the Nash bargaining mechanism by providing global network state information to all nodes in the network.

An illustration of the performance of a BE-based networking mechanism for fair secondary coexistence in TV white spaces is shown in Fig. 1.13. It shows the average outage probability with and without BE-based cooperation [93]. As the number of users increase, the outage probability scales down exponentially, demonstrating the power of user cooperation diversity incentivized by BE.

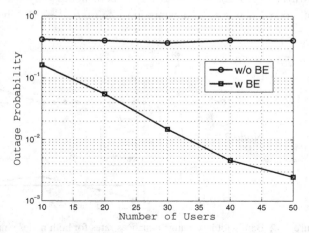

Figure 1.13 Average outage probability with and without bandwidth exchange [93].

1.8.2 Cross-layer scheduling in cognitive radio networks

The nodes of a cognitive radio network can interact in a variety of arbitrary ways. To distill these interactions, we see that each radio follows a transmission policy that results in signals that vary over time, frequency, and space. This variation may be the result of adaptation to measurements of channels or interference. The performance of a particular signaling strategy depends on each receiver's ability to resolve signals in the presence of interfering transmissions. In the realm of cognitive radio networks, two distinct sets of issues emerge. First, for a given set of transmitter and receiver technologies and a specified set of performance constraints, one must resolve the multidimensional boundaries of system performance. This is a difficult problem, even if complete system state information is available to all network nodes. Second, the collection of intelligent adaptation policies of the individual nodes constitute a large distributed system for spectrum allocation. A given set of distributed information gathering and exchange mechanisms may greatly influence the performance of the system.

These issues could be viewed separately. The boundaries of system performance can be examined under the assumption that efficient open access to spectrum can be resolved by impartial spectrum servers (as described in Section 1.6.2) that can obtain information about the interference environment through measurements contributed by different terminals, and then offer suggestions for efficient coordination to interested service subscribers. The spectrum server could play the role of a centralized scheduler that uses full knowledge of the network configuration to specify the activity patterns of the individual links. Upper-bounds on the performance of distributed adaptive routing and scheduling methods can then be provided based on the performance of the spectrum server.

Figure 1.14 shows a multihop wireless network transporting data from the source (node 1) to the destination (node 6) in a series of hops. In general, a network could have many simultaneously active sessions, each specified by an origin–destination pair. Each session may split its traffic into multiple flows through distinct routes in the network. For instance, Fig. 1.14 shows the session from node 1 to node 6 split into two flows through the network. The sessions may individually demand QoS requirements such as

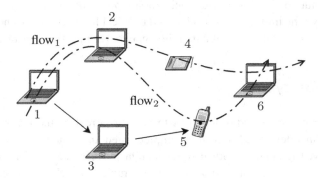

Figure 1.14 An example of a simple multihop network. The solid arrows are instances of links in the network and dotted lines show a flow through the network.

minimum average rate, maximum peak rate, limited delay, bounded jitter, etc. In order to satisfy the QoS guarantees, each layer in the protocol stack optimizes a set of utility functions with a view to achieving the global objectives (QoS agreements) across the network. There may be interdependencies across operations in each layer, but it may be in the designer's interest to bring down the complexity of the system by reducing the dependencies across the layers. For instance, the network layer specifies the appropriate routes for the individual flows in the network. The decision to choose a particular route may be based on some metric chosen according to the QoS requirements of each session. In most cases, the routing metric is a function of the service delays and the average rates supported by the links. Moreover, the routing decisions may have to be updated at regular intervals of time to account for the variation of the parameters in underlying layers.

The MAC/PHY layers solve the resource allocation problem. These layers specify the rate and power allocation, transmission strategy, and schedule for links in the network. The power allocation strategy on a link depends on the channel conditions (its knowledge or lack thereof at each transmitter node), constraints on the transmission power due to device limitations, and the overall interference of other transmitters in its neighborhood. The instantaneous rate obtained in a link depends on the underlying channel propagation parameters, modulation and coding scheme used at the transmitter, signal processing employed at the receiver, and the interference in the neighborhood of the receiver. In general, a MAC scheme specifies a set of rules for transmissions in the network. The choice of rules depends on the amount of resources allocated to them. Addressing the resource allocation problem in the MAC/PHY layer taking into account all these variables can be mathematically intractable [7].

Many individual aspects in different layers of a multihop network have been addressed by the research community by making simplifying assumptions on the rest of the network. In particular, there have been numerous advances in the physical layer technologies over the past two decades – adaptive modulation, reliable coding schemes (e.g., LDPC codes, turbo codes, etc.), hybrid ARQ schemes, OFDM, MIMO processing, cooperative diversity schemes, multiple access techniques, multiuser diversity schemes, to name a few. However, most of the cross-layer design approaches addressed in the literature make simple and specific assumptions on the underlying physical layer technology. The impact of these advanced physical layer techniques needs better understanding in a practical set-up of a multihop network supporting end-to-end flows [76].

1.9 Prototyping

For cognitive radio networks to become a reality, the theoretical concepts and algorithms developed need to be implemented and tested at scale. Towards this end, there have been many efforts in the academia and research labs for prototyping cognitive radio networks. One of the major cognitive radio networking testbeds is the cognitive-radio-enabled ORBIT radio grid testbed [67] at the Wireless Information Networks Laboratory (WINLAB) at Rutgers University. The large-scale radio grid emulator shown in

1.9 Prototyping

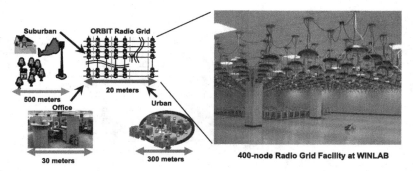

Figure 1.15 The ORBIT testbed.

Fig. 1.15 consists of an array of 20 × 20 open-access programmable nodes each with multiple 802.11a, b, g, or other (Bluetooth, Zigbee, GNU/USRP, WARP, and WINLAB WINC2R Cognitive Radio) radio cards. The radio nodes are connected to an array of backend servers over a switched Ethernet network, with separate physical interfaces for data, management, and control. Interference sources and spectrum monitoring equipment are also integrated into the radio grid. Users of the grid log into an experiment management server which executes experiments involving network topologies and protocol software specified using an ns2-like scripting language. A radio mapping algorithm that uses controllable noise sources spaced across the grid to emulate the effect of physical distance allows mapping real-world wireless network scenarios to specific nodes in the grid as shown.

Among the earliest and most comprehensive cognitive radio prototyping efforts are the activities under the DARPA Next Generation (XG) Communications program program [11]. The DARPA XG program was primarily motivated by the need to support the long-term projected spectrum demand for military, civilian, Government, and commercial wireless applications which is expected to significantly strain the limits of current static spectrum allocations. The DARPA XG program was initiated to achieve at least an order of magnitude improvement in spectrum utilization by providing access to underutilized spectrum for next-generation military and commercial networks. XG studies have focused on designing techniques to sense spectrum in real time, characterize the spectrum, react through dynamic waveforms, and adapt to changes in order to dramatically improve the efficiency of current allocations. For example, in a study conducted in 2006, the spectrum occupancy levels were examined in a field trial. XG technology was shown to provide robust networking with excellent QoS in challenging mobile scenarios. It was reported that when no spectrum was available, XG nodes avoided interfering with each other even if they had to cease operation [58]. The XG program is also involved in developing the architectural framework, protocol design, and building prototypes for dynamic spectrum access networks [10, 57].

Among cognitive radio platforms, the Universal Software Radio Peripheral (USRP) [85] has been used to perform many functionalities of the cognitive radio. The USRP is a processor board with the necessary interface that can be used to build a software radio. It can be connected to a host PC using high-speed USB or Gigabit Ethernet.

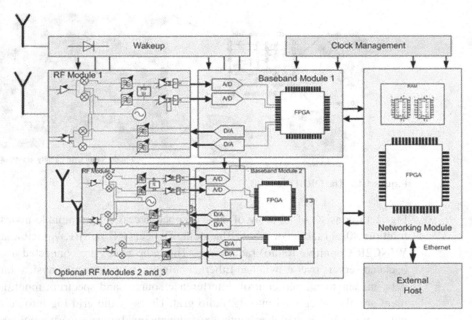

Figure 1.16 Architecture of the WiNC2R platform [60].

Using this set-up, one can build a software defined radio. Such a device is very useful for experimenting with newly developed cognitive radio algorithms. Yet another cognitive radio prototype is the WiNC2R system [60]. The architecture of the WiNC2R system is shown in Fig. 1.16. It is a flexible wireless platform that supports a range of cognitive radio network scenarios from autonomous agile radios to spectrum sharing systems. The platform design has the ability to support multiple MAC protocols, and the capability to switch protocols on-the-fly within a frame time boundary.

Rice University's Wireless Open-Access Research Platform (WARP) is a scalable and extensible programmable wireless platform [79] to prototype advanced wireless networks. It allows exchange and sharing of new physical and network layer architectures, building a true community platform. Xilinx FPGAs are used to enable programmability of both physical and network layer protocols on a single platform, which is both deployable and observable at all layers. WARP opens both hardware and software needed to research, build, and prototype next generation wireless networks. This enables a community of researchers to pool their ideas in undertaking clean-slate prototype networks. Details on the design of the WARP platform are described in [3, 4].

The KUAR (Kansas University Agile Radio) is a portable and flexible software radio architecture [61] that enables advanced research in dynamic spectrum access and cognitive radios. Other wireless testbeds in the academic community that focus on software-defined radio are the wireless networking research testbed [15] in the University of California at Riverside, the cognitive radio research testbed [28] in University of California, Los Angeles, and the Cognitive Radio Network Testbed (CORNET) in Virginia Tech [24, 64].

The KNOWS (Kognitiv Networking Over White Spaces) project at Microsoft Research is a prototyping effort of sensing in white spaces and is an example of cognitive radio prototyping in research labs. Such efforts help in showing the proof-of-concept of new ideas, later leading to commercialization. KNOWS is a hardware–software platform that includes a spectrum-aware MAC and algorithms to deal with spectrum fragmentation. The hardware implementation comprises a dual-mode scanner radio for detecting white spaces, and a reconfigurable radio for subsequent data communications [92]. As a part of the KNOWS initiative, Microsoft Research has deployed base stations on two different buildings within the Microsoft campus in Redmond, Washington. These base stations operate over the white spaces, and provide coverage to nearly all of campus. They have also deployed a mobile client on a Microsoft shuttle, which bridges packets from white spaces to WiFi within the shuttle, thereby providing Internet connectivity to existing laptops. As part of the deployment, they have also built a research prototype of a white space database [9].

Another notable initiative in the industry is the cognitive radio research at Nokia Research Center (NRC) at Helsinki, Finland [25]. Cognitive radio research in NRC is focused on methods for flexible spectrum use, devising novel ways of sensing the radio environment and location, designing distributed networks that intelligently cooperate and building low power flexible implementation of wireless in mobile devices.

1.10 Standardization activity in cognitive radio

A flurry of standardization activities for defining the air interface using white spaces started after the US Federal Communications Commission (FCC) issued an NRPM (Notice of proposed rule making) in May 2004 to provide more efficient and effective use of the TV spectrum [35]. In response to the FCC proposal, the IEEE 802.22 Working Group (WG) was formed to define a standardized air interface based on cognitive radios. The IEEE 802.22 standard is one of the earliest standardization efforts to define the MAC and PHY for opportunistic use of the television bands on a non-interfering basis. It specifies a MAC that dynamically adapts to the changes in the environment. The standard requires sensing the spectrum to detect the presence or absence of incumbent primary transmitters in the locality of the cognitive radio.

In November 2008, the FCC issued a report "Second report and order and memorandum opinion and order" to adopt rules to allow unlicensed radio transmitters to operate in TV white spaces. In spite of strong opposition from TV broadcasters, the FCC adopted the rules in order to make a significant amount of spectrum available for new and innovative products and services, including broadband data and other services for businesses and consumers. However, the FCC mandated that devices must both consult a database constructed by the FCC to determine which channels are available for use at a given location, and also monitor the spectrum locally once every minute to confirm that no legacy wireless microphones, video assist devices, or other emitters are present. If a single transmission is detected, the device may not transmit anywhere within the entire 6 MHz channel in which the transmission was received.

In September 2010, the FCC released a "Second Memorandum Opinion and Order" that finalized the rules for using unused TV bands for unlicensed wireless devices [36]. The new rules removed mandatory sensing requirements, which greatly facilitates the use of the spectrum with geolocation-based channel allocation.

Some of the other IEEE standards related to white space networks are as follows.

- The IEEE 802.11af Working Group (WG) defines changes to the IEEE 802.11 PHY and MAC for channel access and coexistence in TV White Spaces (TVWS).
- The P1900 working group of the IEEE DySPAN Standards Committee that defines standards for new methods of dynamic spectrum access including radio transmission interference and for coordination of wireless technologies including network management and information sharing among networks deploying different wireless technologies. The WG was formed to develop supporting standards dealing with new technologies and techniques being developed for cognitive radio and advanced spectrum management.
- The IEEE SCC41 standardization committee was formed with the intention of checking whether reusing the IEEE 802 PHY/MAC is optimal for white space operation and to estimate how far the performance of the system could benefit from a tailored PHY/MAC system. One of the shortcomings of the reused PHY is that OFDM has significant high power side lobes, leading to either coexistence issues or bad spectrum occupancy factor.
- The IEEE 802.19 focuses on developing standards for coexistence between wireless standards of unlicensed devices. The standards body was formed to minimize the interference between different networks belonging to various wireless standards in the unlicensed band. The Task Group IEEE 802.19.1 focuses on the coexistence in the TVWS.

The International Telecommunication Union (ITU) has the following study groups that discuss cognitive radio networks.

- Within the ITU-R Study Group 1 on Spectrum Management, dynamic spectrum issues was covered by working part 1B.
- Within the ITU-R Study Group 5 on Terrestrial Services, working part 5A has described the potential application of cognitive radio systems in the land mobile service. It also covers the technical and operational characteristics of cognitive radio systems, including potential benefits, their challenges, their deployment scenarios and their impact on the use of spectrum from a technical perspective [6].
- Within the ITU-R Study Group 5, working party 5D has started another work on "Cognitive Radio Systems Specific to IMT Systems." The scope of this work is to consider the inclusion of CRS into the IMT family of technologies.

The European Communications Committee (ECC), an autonomous committee within the European Conference of Postal and Telecommunications Administrations (CEPT) has a special Task Group working on operation of cognitive radio systems in the white spaces of the frequency band 470–790 MHz. QoSMOS (Quality of service and mobility driven cognitive radio systems) is a pan-European project with 15 organizations coming together to develop a framework for Cognitive Radio systems and to develop and prove critical technologies using a testbed [72]. Involvement in standards bodies and industrial forums is emphasized from the start, to increase the probability of adoption of QoSMOS results into standardized products. The initial focus is on opportunistic use of radio spectrum, with an early example being TV White Spaces. Another initiative is the End-to-End Efficiency (E^3) project, an ambitious German Large Scale Integrating Project (IP) aimed at integrating cognitive wireless systems in the Beyond 3G (B3G) world, evolving current heterogeneous wireless system infrastructures into an integrated, scalable, and efficiently managed B3G cognitive system framework. The key objective of the E^3 project is to design, develop, prototype, and showcase solutions to guarantee interoperability, flexibility, and scalability between existing legacy and future wireless systems, manage the overall system complexity, and ensure convergence across access technologies, business domains, regulatory domains, and geographical regions [33].

1.11 Organization of this book

The remainder of this book is organized as follows. Chapter 2 discusses fundamental capacity limits and transmission techniques for the various cognitive radio usage paradigms from an information theoretic perspective. Such fundamental performance limits provide guidelines for the design of cognitive radio networks. Chapter 3 discusses the relevant radio propagation channel models that characterize transmissions in cognitive radio networks. These channel models are critical to the analysis of interference to primary users, secondary coexistence and for establishment of interference management schemes. Sensing is a fundamental attribute of cognitive radios and Chapter 4 discusses the fundamentals of sensing as relevant to the usage paradigms for cognitive radios. Finally, Chapter 5 rounds out the material by discussing advanced spectrum sensing techniques, optimization problems related to exploration, and exploitation for cognitive radio networks and cooperative sensing.

References

[1] J. Acharya, H. Viswanathan, and S. Venkatesan, "Timing acquisition for non-contiguous OFDM based dynamic spectrum access," *IEEE International Symposium on New Frontiers in Dynamic Spectrum Access Networks, 2008*, Oct. 2008, pp. 1–10.

[2] J. Acharya and R. Yates, "Dynamic spectrum allocation for uplink users with heterogeneous utilities," *IEEE Trans. Wireless Commun.*, **8**, no. 3, 1405–1413, Mar. 2009.

[3] K. Amiri, Y. Sun, P. Murphy, C. Hunter, J. R. Cavallaro, and A. Sabharwal, "WARP, a modular testbed for configurable wireless network research at Rice," *Proc. IEEE SWRIF*, 2007.

[4] K. Amiri, Y. Sun, P. Murphy, C. Hunter, J. Cavallaro, and A. Sabharwal, "WARP, a unified wireless network testbed for education and research," in *IEEE International Conference on Microelectronic Systems Education*, June 2007, pp. 53–54.

[5] J. Andrews, A. Ghosh, and R. Muhamed, *Fundamentals of WiMAX*, Prentice-Hall, 2007.

[6] "Annex 12 to working party 5A chairman's report," Radiocommunication Study Groups, ITU, Tech. Rep., 2010.

[7] E. Arıkan, "Some complexity results about packet radio networks," *IEEE Trans. Inf. Theory*, **30**, no. 4, 681–685, July 1984.

[8] P. Bahl, R. Chandra, T. Moscibroda, R. Murty, and M. Welsh, "White space networking with WiFi like connectivity," *ACM Sigcomm*, 2009.

[9] V. Bahl, "The Promises and Challenges of the Wireless Frontier – from 600 MHz to 60 GHz," *IEEE DySPAN 2011*, Plenary lecture, May 2011.

[10] BBN Technologies, "Next generation (XG) architecture and protocol development (XAP)," Tech. Rep., Aug. 2005. www.dtic.mil/cgi-bin/GetTRDoc?Location=U2& doc=GetTRDoc.pdf&AD=ADA437096.

[11] BBN Technologies, "The XG Vision–RFC," Tech. Rep. www.ir.bbn.com/~ramanath/ pdf/rfc_vision.pdf.

[12] S. Biswas and R. Morris, "Opportunistic routing in multi-hop wireless networks," *Proc. ACM SIGCOMM*, 2004.

[13] V. Brik, E. Rozner, S. Banerjee, and P. Bahl, "DSAP: a protocol for coordinated spectrum access," *IEEE International Symposium on New Frontiers in Dynamic Spectrum Access Networks, 2005*, Nov. 2005, pp. 611–614.

[14] J. Brito, "The spectrum commons in theory and practice," Stanford Technology Law Review, Discussion Papers, 2007. stlr.stanford.edu/pdf/brito-commons.pdf.

[15] I. Broustis, J. Eriksson, S. Krishnamurthy, and M. Faloutsos, "A blueprint for a manageable and affordable wireless testbed: design, pitfalls and lessons learned," *3rd International Conference on Testbeds and Research Infrastructure for the Development of Networks and Communities, TridentCom 2007*, May 2007, pp. 1–6.

[16] T. Brown and A. Sethi, "Potential cognitive radio denial-of-service vulnerabilities and protection countermeasures: a multi-dimensional analysis and assessment," *2nd International Conference on Cognitive Radio Oriented Wireless Networks and Communications, CrownCom.*, Aug. 2007, pp. 456–464.

[17] M. Buddhikot, "Understanding dynamic spectrum access: models, taxonomy and challenges," *2nd IEEE International Symposium on New Frontiers in Dynamic Spectrum Access Networks, DySPAN*, Apr. 2007.

[18] M. Buddhikot, P. Kolodzy, S. Miller, K. Ryan, and J. Evans, "DIMSUMNet: new directions in wireless networking using coordinated dynamic spectrum access," *IEEE WoWMoM*, June 2005.

[19] B. Chen and H. Wang, "Maximum likelihood estimation of OFDM carrier frequency offset," *IEEE International Conference on Communications, 2002. ICC 2002*, **1**, May 2002, pp. 49–53.

[20] R. Chen, J.-M. Park, and J. Reed, "Defense against primary user emulation attacks in cognitive radio networks," *IEEE J. Sel. Areas Commun.*, **26**, no. 1, 25–37, Jan. 2008.

[21] S. H. Chen, J. S. Row, and K. L. Wong, "Reconfigurable square-ring patch antenna with pattern diversity," *IEEE Trans. Antennas and Propag.*, **55**, no. 2, 472–475, Feb. 2007.

[22] M. Chiani and A. Giorgetti, "Coexistence between UWB and narrow-band wireless communication systems," *Proc. IEEE*, **97**, no. 2, 231–254, Feb. 2009.

[23] Cisco, "Cisco visual networking index: Global mobile data traffic forecast update, 2010–2015," Tech. Rep., 2011. www.cisco.com/en/US/solutions/collateral/ns341/ ns525/ns537/ns705/ns827/white_paper_c11-520862.pdf.

[24] "Cognitive radio network testbed (CORNET)," cornet.wireless.vt.edu/.

[25] "Cognitive Radio Research at Nokia Research Center," www.research.nokia.com/ cognitive_radio.

[26] R. H. Coase, "The Federal Communications Commission," *Journal of Law and Economics*, **2**, 1–40, Oct. 1959.

References

[27] C. Cordeiro and K. Challapali, "C-MAC: a cognitive MAC protocol for multi-channel wireless networks," *IEEE International Symposium on New Frontiers in Dynamic Spectrum Access Networks*, Apr. 2007, pp. 147–157.

[28] D. Čabrić, "Addressing feasibility of cognitive radios," *IEEE Signal Processing Mag.*, **25**, no. 6, 85–93, Nov. 2008.

[29] D. Čabrić, S. M. Mishra, and R. W. Brodersen, "Implementation issues in spectrum sensing for cognitive radios," *Proc. 38th Asilomar Conference on Signals, Systems and Computers*, 2004.

[30] A. Damnjanović, J. Montojo, Y. Wei, *et al.* "A survey on 3GPP heterogeneous networks," *IEEE Trans. Commun.*, **18**, no. 3, 10–21, June 2011.

[31] M. Dianati, X. Ling, K. Naik, and X. Shen, "A node-cooperative ARQ scheme for wireless ad hoc networks," *IEEE Trans. Veh. Technol.*, **55**, no. 3, 1032–1044, May 2006.

[32] C. Doerr, M. Neufeld, J. Fifield, T. Weingart, D. Sicker, and D. Grunwald, "Multimac – an adaptive MAC framework for dynamic radio networking," *IEEE International Symposium on New Frontiers in Dynamic Spectrum Access Networks, 2005*, Nov. 2005, pp. 548–555.

[33] "End-to-end efficiency project," ict-e3.eu/project/overview/overview.html.

[34] European Telecommunication Standard, "Digital Audio Broadcasting (DAB) to mobile, portable and fixed receivers," ETSI, Tech. Rep., May 1997.

[35] Federal Communications Commission, "Notice of proposed rule making," FCC, ET Docket no. 04-113, May 2004. www.naic.edu/~phil/rfi/fccactions/FCC-04-113A1.pdf.

[36] Federal Communications Commission, "Second memorandum opinion and order," FCC, Tech. Rep., Sep. 2010. transition.fcc.gov/Daily_Releases/Daily_Business/2010/db0923/FCC-10-174A1.pdf.

[37] F. Fitzek and M. Katz, eds., *Cognitive Wireless Networks: Concepts, Methodologies and Visions inspiring the age of Enlightenment of Wireless Communications*, Springer, 2007.

[38] C. Gerami, *Design Methodology for Backhaul and Distribution Networks Using TV White Spaces*, M.S. Dissertation, Rutgers University, May 2011.

[39] C. Gerami, N. Mandayam, and L. Greenstein, "Backhauling in TV white spaces," *2010 IEEE Global Telecommunications Conference*, Dec. 2010, pp. 1–6.

[40] A. Goldsmith, *Wireless Communications*. Cambridge University Press, 2005.

[41] P. S. Hall, P. Gardner, J. Kelly, E. Ebrahimi, M. R. Hamid, and F. Ghanem, "Antenna challenges in cognitive radio," *Proc. ISAP 08*, Oct. 2008, pp. 141–144.

[42] L. Hanzo and T. Keller, *OFDM and MC-CDMA: a Primer*, IEEE Press, J. Wiley & Sons, 2006.

[43] G. Hardin, "Tragedy of the commons," *Science*, 1243–1248, Dec. 1968.

[44] S. Haykin, "Cognitive radio: brain-empowered wireless communications," *IEEE J. Sel. Areas Commun.*, **23**, no. 2, 201–220, Feb. 2005.

[45] B. Horine and D. Turgut, "Link rendezvous protocol for cognitive radio networks," *IEEE International Symposium on New Frontiers in Dynamic Spectrum Access Networks, 2007*, Apr. 2007, pp. 444–447.

[46] S. H. Hsu and K. Chang, "A novel reconfigurable microstrip antenna with switchable circular polarization," *IEEE Antennas and Wireless Propag. Lett.*, **6**, 160–162, 2007.

[47] G. H. Huff, J. Feng, S. Zhang, and J. T. Bernhard, "A novel radiation pattern and frequency reconfigurable single turn square spiral microstrip antenna," *IEEE Trans. Microw. Wireless Compon. Lett.*, **13**, no. 2, 57–59, Feb. 2003.

[48] IEEE 802.11, "Wireless LAN medium access control (MAC) and physical layer (PHY) specifications," Tech. Rep., June 2007.

[49] Ö. Ileri and N. Mandayam, "Dynamic spectrum access models: toward an engineering perspective in the spectrum debate," *IEEE Commun. Magazine*, **46**, no. 1, 153–160, Jan. 2008.

[50] Ö. Ileri, D. Samardzija, T. Sizer, and N. Mandayam, "Demand responsive pricing and competitive spectrum allocation via a spectrum policy server," *Proc. IEEE DySPAN*, Nov. 2005. Baltimore, MD.

[51] X. Jing and D. Raychaudhuri, "Global control plane architecture for cognitive radio networks," *IEEE International Conference on Communications, 2007. ICC '07*, June 2007, pp. 6466–6470.

[52] J. R. Kelly, E. Ebrahimi, P. S. Hall, P. Gardner, and F. Ghanem, "Combined wideband and narrowband antennas for cognitive radio applications," *IET Seminar on Cognitive Radio and Software Defined Radios: Technologies and Techniques*, Sept. 2008, pp. 1–4.

[53] M. Komulainen, M. Berg, H. Jantunen, E. T. Salonen, and C. Free, "A frequency tuning method for a planar inverted-F antenna," *IEEE Trans. Antennas and Propag.*, **56**, no. 4, 944–950, Apr. 2008.

[54] U. Ladebusch and C. Liss, "Terrestrial DVB (DVB-T): a broadcast technology for stationary portable and mobile use," *Proc. IEEE*, **94**, no. 1, 183–193, Jan. 2006.

[55] T. Li, N. Mandayam, and A. Reznik, "A framework for resource allocation in a cognitive digital home," *2010 IEEE Global Telecommunications Conference*, Dec. 2010, pp. 1–5.

[56] T. Li, N. Mandayam, and A. Reznik, "Distributed algorithms for joint channel and RAT allocation in a cognitive digital home," *International Symposium on Modeling and Optimization in Mobile, Ad Hoc and Wireless Networks (WiOpt)*, May 2011, pp. 213–219.

[57] P. Marshall, *Quantitative Analysis of Cognitive Radio and Network Performance*, Artech House, 2010.

[58] M. McHenry, E. Livsics, T. Nguyen, and N. Majumdar, "Xg dynamic spectrum access field test results," *IEEE Commun. Magazine*, **45**, no. 6, 51–57, June 2007.

[59] M. McHenry, D. McCloskey, and G. Lane-Roberts, "New York City Spectrum Occupancy Measurements Sep. 2004," Tech. Rep., Dec. 2004.

[60] Z. Miljanic, I. Seskar, K. Le, and D. Raychaudhuri, "The WINLAB network centric cognitive radio hardware platform – WiNC2R," *Mobile Networks and Applications*, **13**, no. 5, October 2008.

[61] G. Minden, J. Evans, L. Searl, *et al.*, "Cognitive radios for dynamic spectrum access – an agile radio for wireless innovation," *IEEE Commun. Magazine*, **45**, no. 5, 113–121, May 2007.

[62] J. Mitola and G. Q. Maguire, "Cognitive radio: making software radios more personal," *IEEE Pers. Commun.*, **6**, no. 4, 13–18, Aug. 1999.

[63] National Telecommunications and Information Administration (NTIA), "FCC frequency allocation chart," Tech. Rep., 2003. www.ntia.doc.gov/files/ntia/publications/2003-allochrt.pdf.

[64] T. Newman, S. Hasan, D. Depoy, T. Bose, and J. Reed, "Designing and deploying a building-wide cognitive radio network testbed," *IEEE Commun. Magazine*, **48**, no. 9, 106–112, Sept. 2010.

[65] S. Nikolaou, R. Bairavasubramanian, C. Lugo Jr., *et al.*, "Pattern and frequency reconfigurable annular slot antenna using pin diodes," *IEEE Trans. Antennas and Propag.*, **54**, no. 2, 439–448, Feb. 2006.

[66] Office of Engineering and Technology in FCC, "Understanding the FCC regulations for low-power, non-licensed transmitters," Tech. Rep., 1993. transition.fcc.gov/Bureaus/Engineering_Technology/Documents/bulletins/oet63/oet63rev.pdf.

[67] "ORBIT: open-access research testbed for next-generation wireless networks," www.orbit-lab.org/.

[68] H. Pan, J. T. Bernhard, and V. K. Nair, "Reconfigurable single-armed square spiral microstrip antenna design," *IEEE International Workshop on Antenna Technology Small Antennas and Novel Metamaterials*, **6–8**, 2006, pp. 180–183.

[69] H. K. Pan, J. Tsai, J. Martinez, S. Golden, V. K. Nair, and J. T. Bernhard, "Reconfigurable antenna implementation in multi-radio platform," *IEEE Antennas and Propagation Society International Symposium, 2008*, July 2008, pp. 1–4.
[70] C. E. Perkins, *Ad Hoc Networking*, Addison-Wesley, 2001.
[71] C. Prehofer and C. Bettstetter, "Self-organization in communication networks: principles and design paradigms," *IEEE Commun. Magazine*, 43, no. 7, 78–85, July 2005.
[72] "Quality of service and mobility driven cognitive radio systems," www.ict-qosmos.eu/home.html.
[73] R. Rajbanshi, Q. Chen, A. M. Wyglinski, G. J. Minden, and J. B. Evans, "Quantitative comparison of agile modulation techniques for cognitive radio transceivers," *IEEE Consumer Communications and Networking Conference, 2007*, Jan. 2007, pp. 1144–1148.
[74] R. Rajbanshi, A. M. Wyglinski, and G. J. Minden, "An efficient implementation of NC-OFDM transceivers for cognitive radios," *International Conference on Cognitive Radio Oriented Wireless Networks and Communications*, June 2006, pp. 1–5.
[75] R. Rajbanshi, A. M. Wyglinski, and G. J. Minden, "Peak-to-average power ratio analysis for NC-OFDM transmissions," *IEEE Vehicular Technology Conference*, Sept. 2007, pp. 1351–1355.
[76] C. Raman, *Scheduling and Relaying in Interference Limited Wireless Networks*, Ph.D. Dissertation, Rutgers University, 2010.
[77] C. Raman, R. Yates, and N. Mandayam, "Scheduling variable rate links via a spectrum server," *Proc. IEEE DySPAN*, 2005, Baltimore, MD.
[78] D. Raychaudhuri, N. Mandayam, J. Evans, B. Ewy, S. Seshan, and P. Steenkiste, "CogNet – an architectural foundation for experimental cognitive radio networks within the future internet," *Proc. MobiArch*, 2006.
[79] "Rice University WARP," warp.rice.edu/.
[80] T. Schmidl and D. Cox, "Robust frequency and timing synchronization for OFDM," *IEEE Trans. Commun.*, 45, no. 12, 1613–1621, Dec. 1997.
[81] S. Sesia, I. Toufik, and M. Baker, Eds., *LTE – The UMTS Long Term Evolution*, J. Wiley & Sons, 2009.
[82] Shared Spectrum Company, "General survey of radio frequency bands – 30 MHz to 3 GHz," Tech. Rep., Sep. 2010.
[83] H. F. A. Tarboush, S. Khan, R. Nilavalan, H. S. Al-Raweshidy, and D. Budimir, "Reconfigurable wideband patch antenna for cognitive radio," *Antennas Propagation Conference, 2009*, Nov. 2009, pp. 141–144.
[84] J. Tellado, ed., *Multicarrier Modulation with Low PAR: Applications to DSL and Wireless*, Kluwer Academic Publishers, 2000.
[85] "The Universal Software Radio Peripheral," www.ettus.com/.
[86] Z. Tian and G. Giannakis, "Compressed sensing for wideband cognitive radios," *IEEE International Conference on Acoustics, Speech and Signal Processing, ICASSP.*, 4, Apr. 2007, pp. IV-1357–IV-1360.
[87] J. van de Beek, M. Sandell, and P. Borjesson, "ML estimation of time and frequency offset in OFDM systems," *IEEE Trans. Signal Process.*, 45, no. 7, 1800–1805, July 1997.
[88] C.-J. Wang and W. T. Tsai, "A slot antenna module for switchable radiation patterns," *IEEE Antennas and Wireless Propag. Lett.*, 4, 202–204, 2005.
[89] T. Weiss and F. Jondral, "Spectrum pooling: an innovative strategy for the enhancement of spectrum efficiency," *IEEE Commun. Magazine*, 42, no. 3, S8–14, Mar. 2004.
[90] G. Wu, S. Talwar, K. Johnsson, N. Himayat, and K. Johnson, "M2M: from mobile to embedded internet," *IEEE Commun. Magazine*, 49, no. 4, 36–43, Apr. 2011.

[91] J. Yang, R. Brodersen, and D. Tse, "Addressing the dynamic range problem in cognitive radios," *IEEE International Conference on Communications, ICC.*, June 2007, pp. 5183–5188.
[92] Y. Yuan, P. Bahl, R. Chandra, *et al.*, "KNOWS: Cognitive radio networks over white spaces," *IEEE International Symposium on New Frontiers in Dynamic Spectrum Access Networks, 2007. DySPAN 2007*, Apr. 2007, pp. 416–427.
[93] D. Zhang and N. Mandayam, "Bandwidth exchange for fair secondary coexistence in TV white space," *Proceedings of International ICST Conference on Game Theory for Networks (GameNets)*, Apr. 2011.
[94] D. Zhang, R. Shinkuma, and N. Mandayam, "Bandwidth exchange: an energy conserving incentive mechanism for cooperation," *IEEE Trans. Wireless Commun.*, **9**, no. 6, 2055–2065, June 2010.

2 Capacity of cognitive radio networks

Andrea Goldsmith and Ivana Marić

2.1 Introduction

This chapter develops the fundamental capacity limits and associated transmission techniques for different cognitive radio network paradigms. These limits are based on the premise that the cognitive radios of secondary users are intelligent wireless communication devices that exploit side information about their environment to improve spectrum utilization. This side information typically consists of knowledge about the activity, channels, encoding strategies, and/or transmitted data sequences of the primary users with which the secondary users share the spectrum. Based on the nature of the available side information as well as regulatory constraints on spectrum usage, cognitive radio systems seek to underlay, overlay, or interweave the secondary users' signals with the transmissions of primary users. This chapter develops the fundamental capacity limits for all three cognitive radio paradigms. These capacity limits provide guidelines for the spectral efficiency possible in cognitive radio networks, as well as practical design ideas to optimize performance of such networks.

While the general definition of cognitive radio was provided in Chapter 1, we now interpret that definition in a mathematically precise manner that can be used in the development of cognitive radio capacity limits. Specifically, in the mathematical terminology of information theory, it is the availability and utilization of network side information that defines a cognitive radio, which we formalize as follows.

A cognitive radio is a wireless communication device that intelligently utilizes any available side information about the (a) activity, (b) channel conditions, (c) encoding strategies, or (d) transmitted data sequences of primary users with which it shares the spectrum.

In the next section we describe these different cognitive radio paradigms in more detail. The fundamental capacity limits for each of these paradigms are discussed in later sections.

2.2 Cognitive radio network paradigms

We now describe the principle features of the underlay, overlay, and interweave paradigms. The underlay paradigm allows secondary users to operate if the interference they cause to primary users is below a given threshold or meets a given bound on primary user performance degradation. In overlay systems the secondary users transmit simultaneously at the same frequencies as the primary users; however, they overhear the primary user transmissions, then use this information along with complex signal processing and coding techniques to maintain or improve the performance of primary users, while also obtaining spectral resources for their own communication. Under ideal conditions, these encoding and decoding strategies allow both the secondary and primary users to remove all or part of the interference caused by other users. In interweave systems the secondary users detect the absence of primary user signals in space, time, or frequency, and opportunistically communicate during these absences. For all three paradigms, if there are multiple secondary users then these users must share bandwidth amongst themselves as well as with the primary users, subject to their given cognitive paradigm. This gives rise to the medium access control (MAC) problem among secondary users similar to that which arises among users in conventional wireless networks. Given this similarity, MAC protocols that have been proposed for secondary users within a particular paradigm are often derived from conventional MAC protocols. In addition, multiple secondary users may transmit to a single secondary receiver, as in the uplink of a cellular or satellite system, and one secondary user may transmit to multiple secondary receivers, as in the corresponding downlink. We now describe each of the three cognitive radio paradigms in more detail, including the associated regulatory policy as well as underlying assumptions about what network side information is available, how it is used, and the practicality of obtaining this information.

2.2.1 Underlay paradigm

The underlay paradigm, illustrated in Figs. 2.1 and 2.2, mandates that concurrent primary and secondary transmissions may occur only if the interference generated by the secondary transmitters at the primary receivers is below some acceptable

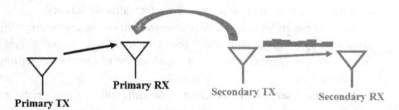

Figure 2.1 The underlay paradigm: wideband signaling (e.g., spread spectrum).

Figure 2.2 The underlay paradigm: transmit antenna array.

threshold. The interference threshold may be defined with respect to a *spectral mask*, defined as the power spectral density (PSD) of the interference over all frequencies within the underlay frequency band. Alternatively, the interference threshold may be defined with respect to an average power constraint, equal to the integral of the interference PSD. A spectral mask is used when interference constraints vary across the band, e.g., when different primary systems occupy different parts of the underlay frequency band, and these systems have different sensitivities to interference. Rather than determining the exact interference it causes, a secondary user can spread its signal over a very wide bandwidth such that the interference PSD is below the noise floor at any primary user location. These spread signals are then despread at each of their intended secondary receivers. This spreading technique is the basis of both spread spectrum and ultra-wideband (UWB) communications [87]. Alternatively, the secondary transmitter can be very conservative in its output power to ensure that its signal remains below the prescribed interference threshold. In this case, since the interference constraints in underlay systems are typically quite restrictive, this limits the secondary users to short-range low-rate communications. Both spreading and severe restriction of transmit power do not require that the secondary user interference at primary receivers is determined. Rather, these underlay approaches adopt a conservative design whereby the collective interference of all secondary transmissions is small everywhere. This collective interference, sometimes called the *interference temperature* [12], is discussed in more detail in Section 4.2. Determining the exact interference a secondary transmitter causes to a primary receiver is one of the biggest challenges in underlay systems. The secondary user can determine this interference at a given primary receiver by overhearing a transmission from that primary user if the link between them is reciprocal. For MIMO systems, a secondary user only interferes with a primary user in their overlapping spatial dimensions. If the secondary user occupies only the null space of the MIMO primary receiver, no interference is caused, and hence this falls within the interweave paradigm discussed below, whereby the primary and secondary users occupy orthogonal spatial dimensions. The underlay paradigm is most common in the licensed spectrum, where the primary users are the licensees, but it can also be used in unlicensed bands to provide different classes of service to different users.

2.2.2 Overlay paradigm

The premise for overlay systems, illustrated in Fig. 2.3, is that the secondary transmitter has knowledge of the primary user's transmitted data sequence (also called its *message*) and how this sequence is encoded (also called its *codebook*). The codebook information could be obtained, for example, if the primary users follow a uniform standard for communication based on a publicized codebook. Alternatively, the primary users could broadcast their codebooks periodically. A primary user's data sequence might be obtained by decoding it at the secondary user's receiver or in other ways, as explained further in Section 2.7.

Knowledge of a primary user's data sequence and/or codebook can be exploited by the secondary transmitter in a variety of ways to improve the performance of both

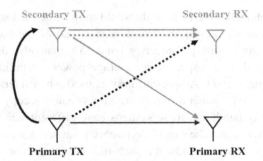

Figure 2.3 The overlay paradigm.

the secondary and the primary systems. On the one hand, this information can be used to cancel the interference due to the primary signals at the secondary receiver. Specifically, encoding techniques like dirty paper coding (DPC) [14] can be used to precode the secondary user's signal such that the known primary user interference at the secondary receiver is effectively removed. On the other hand, given knowledge of the primary user's data sequence and/or codebook, the secondary user can *cooperate* with the primary system to improve its communication. Specifically, a secondary user can assign part of its power for its own communication and the remainder of its power to assist (relay) the primary transmissions. The power assigned for cooperation boosts the primary user's signal-to-interference-plus-noise power ratio (SINR), while the power assigned for the secondary user's communication causes interference to the primary user and hence decreases its SINR. Thus, the secondary user's power allocation determines the primary user's SINR and its associated performance. Furthermore, if the primary receiver can be modified to decode both its data sequence and all or part of the secondary user's data sequence, then the interference caused by the secondary transmitter to the primary receiver can be partially or completely removed. This guarantees that the primary user's rate either remains unchanged or can be increased, while the secondary user obtains capacity based on the power it allocates for its own transmissions. When there are multiple secondary and primary users then the power allocation of the secondary users for cooperation becomes more complicated, as do the encoding and decoding techniques. In addition, a MAC protocol for each user class is required.

There are many practical hurdles that must be overcome for overlay systems to be successful. These include the technical challenges of overhearing primary user transmissions and decoding them, as well as developing encoding and decoding strategies of reasonable complexity. Moreover, sharing of primary user data sequences with secondary users, even when encrypted, will raise significant security and privacy concerns for many primary systems. These challenges may preclude overlay implementations in some types of system. However, some of these challenges do not apply in certain settings. For example, privacy and security concerns do not apply when the primary user data are public, e.g., in a cellular overlay within the TV broadcast spectrum [72]. Note that the overlay paradigm can be applied to either licensed or unlicensed band communications. In licensed bands, secondary users might be allowed to share the band with the licensed users when they do not degrade their performance but rather improve it through cooperation. In unlicensed bands secondary and primary users might have equal priority in the band, with secondary users considered *more capable*. Hence, the secondary system could provide more efficient use of spectrum by exploiting knowledge of the primary users' data sequences and encoding strategies to reduce interference to all users.

2.2.3 Interweave paradigm

The interweave paradigm is based on the idea of *opportunistic communication*, and was the original motivation for cognitive radio [62]. The idea came about after studies conducted by the FCC [27], universities [6], and industry [80] showed that a major part of the spectrum is not fully utilized most of the time. In other words, there exist temporary space–time–frequency voids, referred to as *spectrum holes* or *white spaces*, that are not in constant use in both the licensed and unlicensed bands, as shown in Fig. 2.4. The spatial spectrum holes may be in a single spatial dimension or, for MIMO devices, in the subset of spatial dimensions not occupied by the primary users (i.e., in the null space of the primary users' receivers) [101]. Spectrum holes can be exploited by secondary users to operate in orthogonal dimensions of space, time, or frequency relative to the primary user signals. Thus, the utilization of spectrum is improved by opportunistic reuse over the spectrum holes. The interweave technique requires detection of primary (licensed or unlicensed) users in one or more of the space–time–frequency dimensions. This detection is quite challenging since primary user activity changes over time and also depends on geographical location. Chapters 4 and 5 discuss spectrum hole detection by a single receiver and by multiple receivers, respectively, in more detail. Interweave techniques can also serve as a MAC protocol in networks where all users have equal priority. In this setting existing transmissions are treated as primary signals, while new users access the spectrum using interweave techniques such that they do not interfere with active transmissions.

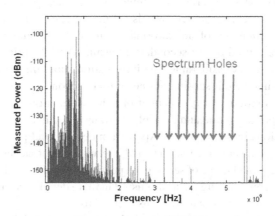

Figure 2.4 Spectrum occupancy measurements up to 6 GHz in an urban area at mid-day (Berkeley Wireless Research Center [6]).

For interweave networks with multiple secondary users, a MAC protocol is needed to share the available spectrum holes amongst them. Given the similarity of this problem with medium access control in conventional networks, the protocols that have been proposed for this setting are often derived from conventional MAC protocols such as ALOHA and CSMA [13]. Simple time-sharing mechanisms may also be used, and this can greatly simplify capacity analysis. Advanced MAC protocols for multiuser interweave networks utilize additional spatial degrees of freedom from multiple antennas, optimization based on more advanced mathematical models such as partially-observed Markov chains, or game theory and pricing mechanisms [43, 88, 101, 102]. The challenge to medium access in the interweave setting above and beyond what has been addressed in conventional MAC protocols is that the channel to be shared is unknown, since it depends on the activity of the primary users. This primary user activity will depend on the MAC protocol of the primary system, which is designed completely independently of the secondary system. Given the many variants of MAC protocols for conventional systems, developing an effective MAC protocol for secondary users remains one of the biggest challenges in interweave system design.

To summarize, an interweave cognitive radio is an intelligent wireless communication system that periodically monitors the radio spectrum, detects primary user occupancy over time, space, and frequency, and opportunistically communicates over spectrum holes with minimal interference to the primary users. Additional motivation and discussion of the signal processing challenges faced in interweave cognitive radio is presented in [41], as well as in Chapters 4 and 5.

2.2.4 Comparison of cognitive radio paradigms

Table 2.1 summarizes the differences among the underlay, overlay, and interweave cognitive radio approaches. While underlay and overlay techniques permit concurrent primary and secondary user transmissions, avoiding simultaneous transmissions with primary users in overlapping dimensions of time, space, or frequency is the main goal in the interweave technique. We also point out that the cognitive radio approaches require different *amounts* of side information: underlay systems require knowledge of the interference caused by the secondary transmitters to the primary receivers, interweave systems require considerable side information about the primary user activity (which can be obtained from sensing at one or more secondary nodes in the system) and overlay systems require a large amount of side information (knowledge of the primary user's encoding technique and possibly its transmitted data sequence, along with the channel conditions in the network). Apart from device level power limits, the secondary user's transmit power in the underlay and interweave approaches is decided by the interference constraint and range of sensing, respectively. Finally, hardware requirements vary across the different paradigms, as discussed in more detail in Section 1.7. While underlay, overlay, and interweave are three distinct approaches to cognitive radio, hybrid schemes can also be constructed that combine the advantages of different approaches. For example, the overlay and interweave approaches are combined in [96].

Underlay	Overlay	Interweave
Network side information: Secondary transmitters know interference caused to primary receivers.	**Network side information:** Secondary nodes know channel gains, encoding techniques, and possibly the transmitted data sequences of the primary users.	**Network side information:** Secondary users identify spectrum holes in space, time, and/or frequency from which the primary users are absent.
Simultaneous transmission: Secondary users can transmit simultaneously with the primary users as long as interference caused is below an acceptable limit.	**Simultaneous transmission:** Secondary users can transmit simultaneously with the primary users; the interference to the primary users can be offset by using part of the secondary users' power to relay the primary users' data sequences.	**Simultaneous transmission:** Secondary users transmit simultaneously with a primary user only when there is missed detection of the primary user activity.
Transmit power limits: Secondary user's transmit power is limited by a constraint on the interference caused to the primary users.	**Transmit power limits:** Secondary users can transmit at any power, the interference to primary users can be offset by relaying the primary users' data sequences.	**Transmit power limits:** Secondary user's transmit power is limited by the range of primary user activity it can detect (alone or via cooperative sensing).
Hardware: Secondary users must measure the interference they cause to primary users' receivers by either sounding and exploiting channel reciprocity or via cooperative sensing.	**Hardware:** Secondary users must also listen to primary user transmissions. Encoding and decoding complexity is also significantly higher than in other paradigms.	**Hardware:** Receiver must be frequency agile or have a wideband front end for spectrum hole detection.

Table 2.1 Comparison of underlay, overlay, and interweave cognitive radio techniques.

2.3 Fundamental performance limits of wireless networks

A wireless network consists of a collection of wireless devices communicating over a common wireless channel. The simplest wireless network consists of a single-user (point-to-point) channel. In general, a wireless network contains multiple source nodes, each communicating its information to a set of destination nodes. A wireless network can have a supporting infrastructure (e.g., as in cellular networks), or an ad hoc structure, where nodes self-configure into a network and control is decentralized among the nodes. The typical topologies of multiuser channels (in isolation or within one cell of a cellular system) are multiple access (many transmitters to one receiver) and broadcast (one transmitter to many receivers) channels. These channels correspond, respectively, to the uplink and downlink of a satellite system or one base station in a cellular system. In these networks, communication occurs between a group of nodes transmitting to or receiving from a single node. In an ad hoc wireless network, each node can serve as a source, destination, and/or relay forwarding data for other users.

In cognitive radio applications, primary and secondary users accessing the same spectrum form a wireless network. Primary and secondary users have different transmit/receive constraints due to interference limitations at the primary receivers, as well as possibly different transmit/receive capabilities. In cognitive radio networks the primary users can be cellular or ad hoc, whereas the secondary users are generally ad hoc and fall into the paradigms of underlay, interweave, or overlay. Hence, these two types of cognitive radio network users form a two-tier wireless network. Performance limits of wireless networks are thus of direct relevance to the performance limits of cognitive radio networks. In particular, the fundamental capacity limits of ad hoc networks dictate not only how much information can be transmitted by secondary users under a given set of network and interference conditions, but also limitations on the information exchange possible between sensing nodes to collaboratively assess spectral occupancy. In the following section we describe the broad range of performance metrics relevant to wireless networks, including their capacity. We then formally define mutual information and capacity for single-user channels as well as for general wireless networks.

2.3.1 Performance metrics

The fundamental performance limits of a wireless network define its best possible performance relative to one or more specific metrics. Many different metrics can be used to measure performance, such as capacity, throughput, outage, energy consumption, as well as combinations of these and other metrics. Since wireless networks exhibit significant dynamics (user movement, data traffic, channel variations, etc.), these dynamics must be taken into account in the definition of the network performance metrics.

The most common fundamental performance limit for time-invariant communication systems is Shannon capacity [76] – the maximum rate that can be achieved over a channel with asymptotically small probability of error. Shannon's simple yet elegant mathematics coupled with his revolutionary ideas for coding over noisy channels and bounding their fundamental data rate limits via mutual information has inspired generations of theorists and practitioners, and provided significant insights into communication system design. For single-user channels the Shannon capacity is a number, the maximum data rate of the channel, as will be defined mathematically in terms of the channel's maximum mutual information in the next section. For a multiuser (broadcast or multiple access) channel Shannon capacity is a K-dimensional region defining the maximum rates possible for all K users simultaneously. Shannon capacity of wireless single-user and multiuser channels is known in many cases, including static and time-varying single-user, broadcast and multiple access channels with noise, fading, multipath, and/or multiple antennas [4, 18, 33].

Time-varying channels are typically modeled based on the notion of a *channel state*. The channel state s lies within a given set $\mathcal{S}$ of all possible channel states, which may be discrete or continuous. For stationary and ergodic time-varying channels, at any given time the channel is assumed to be in state s with probability $p(s)$, and we denote the channel capacity in state s as $C(s)$. This model is also refered to as a *composite channel* [21]. The Shannon capacity or capacity region of a time-varying stationary and ergodic channel with channel state known at the receiver(s) is therefore called the *ergodic* capacity, since it corresponds to the data rate or rate region in a particular channel state (e.g., a particular fading value) averaged over the probability distribution of the channel states (e.g., the fading distribution):

$$C_{\text{erg}} = \int_{s \in \mathcal{S}} C(s) p(s) ds. \qquad (2.1)$$

When the channel state is known at the transmitter(s), the transmission strategy and system resources are typically adapted to the time-varying channel so that $C(s)$ denotes the capacity or capacity region of the channel in state s with transmit parameters such as power and rate optimally adapted over all channel states.

An alternative performance metric for stationary and ergodic time-varying channels is *outage capacity*, whereby transmission to one or more users is suspended in some channel states, deemed *outage states*, and a fixed transmission rate is used in the nonoutage states. The average data rate associated with outage capacity is then the maximum fixed rate that can be achieved in nonoutage states with asymptotically small error probability, multiplied by the probability of nonoutage (since the data rate is zero in outage states). The maximum fixed rate for nonoutage states is typically determined based on the worst-case channel state that is not an outage state. Specifically, for a given set of outage states $\mathcal{S}_{\text{out}}$ with $P(s \in \mathcal{S}_{\text{out}}) = P_{\text{out}}$, the nonoutage states are $\mathcal{S}_{\text{out}}^c = \mathcal{S} - \mathcal{S}_{\text{out}}$, and the outage capacity is given by

$$C_{\text{out}} = (1 - P_{\text{out}}) \left[\min_{s \in \mathcal{S}_{\text{out}}^c} C(s) \right]. \qquad (2.2)$$

The outage capacity metric is based on the underlying assumption that the transmitter knows the channel state and hence can suspend transmission during outage states. Thus, the transmission strategy is binary: no transmission in outage states and fixed-rate transmission in nonoutage states. Outage capacity cannot exceed Shannon capacity, since the latter adapts transmission parameters such as power and rate to each channel state. However, outage capacity is a useful metric for applications, such as voice and video, that require fixed-rate transmission. By allowing for suspension of transmission in some states, outage capacity achieves a higher average rate than if a single rate must be maintained in all channel states, including very poor ones.

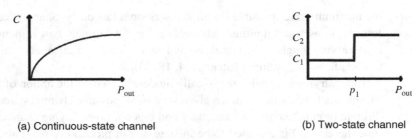

(a) Continuous-state channel   (b) Two-state channel

Figure 2.5 Capacity versus outage probability for a single-user channel.

When the channel state is unknown at the transmitter, the performance metric used is *capacity versus outage probability*. In this case the transmitter cannot adapt to channel conditions; it therefore selects a given fixed rate R (for single-user systems) or set of fixed rates **R** (for multiuser systems) to transmit to the user(s). If the channel supports these rates, i.e., the rates are within the capacity or capacity region of the channel under its realized channel state, then the data are received without error; if not, errors occur which are deemed a data outage. For single-user channels the capacity versus outage probability metric takes the form of a function characterizing the capacity C associated with each outage probility P_{out}. The capacity versus outage probability function, illustrated in Fig. 2.5a for a continuous-state single-user channel, thus corresponds to the transmitter's data rate versus the probability that this rate cannot be supported by a given channel. The plot of C versus P_{out} is nondecreasing with P_{out}, since at high outage probability more of the *bad* channel states need not support rate C, and hence a higher capacity can be achieved in the nonoutage states. Consider now a finite-state channel, where the set of channel states $\mathcal{S}$ is finite, and assume the states are ordered so that the capacity C_i in state i satisfies $C_i \leq C_j$ for $i \leq j$. Then C versus P_{out} has a staircase shape with discrete increases for each n such that $P_{\text{out}} = \sum_{i=1}^{n} p_i$, where p_i is the probability of the ith channel state. For example, in a two-state channel with capacity C_i for state i and state probability p_i, $i = 1, 2$, if $C_1 < C_2$ then capacity versus outage is C_2 for $P_{\text{out}} \geq p_1$ and C_1 for $P_{\text{out}} < p_1$, as shown in Fig. 2.5b. More details on ergodic capacity, outage capacity, and capacity versus outage can be found in [4, 32, 33, 86]. Note that when the channel is nonergodic, such that the channel state is chosen at random from the set $\mathcal{S}$ and remains constant for all time, the channel is referred to as a *compound channel*. In this case the capacity generally corresponds to achievable rates associated with the worst-case channel state [90].

Capacity results are much more limited for general wireless networks with multiple sources and multiple destinations, even for simple static models. For a K-node network where each node is both a source and a destination, the capacity is a $K \times (K-1)$-dimensional region defining the maximum rates achievable between all node pairs. Such regions are typically characterized by two-dimensional slices, which define the maximum rates between two source–destination pairs in the network. More general capacity regions whereby one source sends data to multiple destinations, also called *multicasting*, can also be analyzed but we do not consider multicast in our network models. In practice

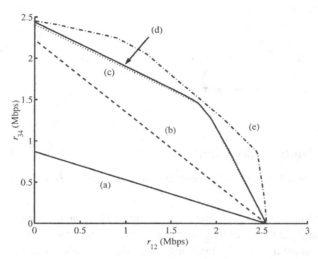

Figure 2.6 Capacity region slice for node pairs (1,2) and (3,4) of a five-node wireless network (a) Single-hop routing, no spatial reuse. (b) Multihop routing, no spatial reuse. (c) Multihop routing with spatial reuse. (d) power control added to (c). (e) Successive interference cancellation (SIC) added to (c).

wireless networks often include multihop routing via relaying, whereby intermediate nodes relay data toward their final destination. Such relaying can increase the achievable data rates for the network as well as other performance metrics, often significantly [85]. Other advanced capabilities in the system design, such as power control, multiple frequency bands to enable *frequency reuse* (the reuse of the same frequency at spatially-separate locations), and interference cancellation can further increase network performance. This is illustrated in Fig. 2.6 (from [85]), where a two-dimensional capacity region slice for a five node network is illustrated for different design assumptions about the network. We see from this figure that spatial reuse of frequencies, multihop routing, and interference cancellation all significantly increase the achievable rates within this slice.

The Shannon capacities for many of the most basic wireless networks, including the three-node relay channel and the four-node interference channel, illustrated in Fig. 2.7, have remained open problems for decades. This makes it unlikely that the capacity region can be obtained exactly for these and other similar networks, especially when the number of users is larger than in these canonical examples. Instead, capacity regions are often characterized by their upper and lower bounds rather than the exact region (where these bounds meet). Lower bounds are easier to obtain than upper bounds, as any communication scheme yields an achievable rate region that lower bounds the capacity region. Upper bounds are more difficult to obtain as they must contain all achievable rate regions. Fano's inequality is the most common tool used to obtain capacity upper bounds [34]. There has also been significant progress on deriving capacity scaling laws, which characterize how the maximum sum of user rates scales in an asymptotically large network [99]. However, these laws provide just one point, the sum-rate point,

Figure 2.7 Simple ad hoc networks for which capacity is unknown.

on the $K \times (K-1)$-dimensional network capacity region. In particular, a network's scaling law defines how the ratio of the sum-rate divided by the number of users behaves in an asymptotically large network. The sum-rate point, i.e., the point on the capacity region corresponding to the maximum sum of user rates simultaneously achievable, can also be of interest for finite-size networks, especially symmetric networks where this point defines the maximum symmetric rate per user. Similarly, *interference alignment* can achieve the sum-rate point in interference networks, but does not achieve the full capacity region [7, 57].

Cognitive radio networks are wireless networks where secondary users overhear the transmissions of primary users in the network and use that information in their encoding and decoding. From a Shannon capacity perspective, the two-user cognitive radio channel is a generalization of the two-user interference channel of Fig. 2.7 in that information about the primary user (source–destination pair 2) is assumed known by the secondary user (source–destination pair 1). In particular, for the underlay paradigm source 1 knows the amount of interference it causes to destination 2; for the interweave paradigm source 1 knows the activity of source 2 across time, space, and frequency dimensions (possibly through coordination with destination 1) and refrains from transmitting in those dimensions when the primary user is active; for the overlay paradigm source 1 is assumed to know the data sequence and encoding scheme of source 2 along with network channel gains, and uses that information in its encoding. Capacity results for the K-user interference channel are given in Section 2.4, and the capacity of the different cognitive radio paradigms are given in Sections 2.5–2.7.

Figure 2.8 illustrates a slice of the wireless network performance region (the slice is for one source–destination pair in a K-node network) where capacity is not the only performance metric of interest. Indeed, delay (average, maximum, tail probability, or the entire delay distribution) is an important metric for many applications. In addition, dynamic wireless channels may exhibit improved rates if some outage or error is allowed (Shannon capacity regions assume zero outage). To illustrate tradeoffs for a set of network performance metrics, the region in Fig. 2.8 shows a hypothetical tradeoff between data rate (capacity), average delay, and outage (measured as the percentage of outage time $100 \times P_{\text{out}}$) for a given source–destination pair in a K-node network. Since this region includes three performance metrics, the performance region for the entire

2.3 Fundamental performance limits

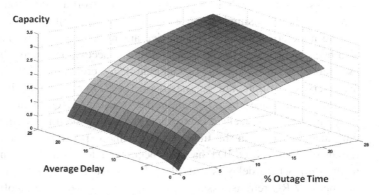

Figure 2.8 Performance region where capacity is not the only metric.

network will be of dimension $K \times (K-1) \times 3$; Fig. 2.8 shows the three-dimensional tradeoff among data rate, average delay, and outage for the selected source–destination pair in this network. The figure illustrates that as the system constraints on average delay and outage are relaxed, capacity increases. Note that transmit power is not explicit in this performance region but rather is a parameter of the underlying model. Other model parameters might include available bandwidth, number of antennas at each node, and complexity limitations. The capacity metric generally increases as delay and/or outage increase, as indicated in the figure, since this entails a relaxation of system constraints. Shannon capacity generally assumes infinite delay and zero outage, hence these dimensions in Fig. 2.8 collapse. Outage capacity and capacity versus outage have been well-studied for point-to-point and multiuser channels, but there are few fundamental outage results for general wireless networks, where outage can be declared for any subset of node pairs within the network.

For systems with multiple *degrees of freedom*, i.e., multiple dimensions over which to transmit data, the tradeoff between different performance metrics can be characterized more formally. In such systems some degrees of freedom can be used for diversity whereby the same information is sent over multiple dimensions for robustness to errors and outage. Other degrees of freedom can be used for multiplexing, whereby independent data are multiplexed over independent channels enabled by the multiple degrees of freedom. The multiple dimensions associated with degrees of freedom are typically obtained via space, time, and frequency. Time and frequency degrees of freedom are obtained by dividing the total signaling dimension into orthogonal time and frequency slots. The spatial dimension is obtained via multiple antennas at the transmitter and receiver (MIMO) systems. For single-user MIMO systems, Zheng and Tse [103] developed a fundamental diversity–multiplexing tradeoff (DMT) in the limit of asymptotically large signal-to-noise power ratio (SNR). The multiplexing gain r in this setting is defined as the number of degrees of freedom utilized for data transmission: more formally, the constant that preceeds the log function in the bandwidth-normalized capacity expression (called the capacity *pre-log*). Diversity gain d is defined as the negative of the slope of the probability of error curve as a function of SNR at a fixed transmission rate. The diversity–multiplexing tradeoff at asymptotically high SNR was shown to obey the simple expression $d(r) = (M_r - r)(M_t - r)$, where M_t and M_r are the number of transmit and receive antennas, respectively. The DMT region has also been investigated for broadcast, multiple access, and relay channels. The single-user region was also extended to include delay, creating a performance region called the diversity–multiplexing–delay tradeoff (DMDT) region [25]. In this work the delay tradeoff is introduced by automatic-repeat-request (ARQ), which provides robustness by identifying data received in error and requesting a retransmission of such data. This introduces diversity in the time domain at the expense of delay in the request for a retransmission. The DMDT has also been extended to multihop networks with ARQ in [97, 98], where delay is caused by both queueing and ARQ retransmissions. The number of ARQ retransmissions invokes a diversity–delay tradeoff, and these retransmissions must be optimally allocated between all hops in the network as well as in the end-to-end link to achieve the optimal DMDT tradeoff. The DMDT of multihop networks under hierarchical cooperation, whereby the network is stratified into tiers and cooperation takes place within a tier, has also been characterized in [67].

While capacity, delay, and outage are key performance metrics for most wireless networks, they are not the most critical metrics for every system. For example, nodes powered by nonrechargeable batteries, as is typical in sensor networks, have energy consumption as a critical metric. Shannon-theoretic analysis was used in [89] to obtain fundamental results for capacity per unit energy (cost) of point-to-point, multiple access, and interference channels. Since this landmark paper there have been many follow-on works examining capacity per unit cost under different channel conditions, different input alphabet constraints, and different single and multiuser channel models. The most relevant for wireless networks are [3, 71] (and the references therein). The first of these works develops the bits-per-joule capacity of wireless networks, a scaling law that defines the maximum total number of bits that the network can deliver per joule of transmit energy deployed into the network. This scaling law is found to be $(K/\log K)^{0.5(\gamma-1)}$ for γ the common path loss exponent of all channels and K the number of nodes in the network. The assumptions used to obtain this energy scaling law are similar to those used to develop capacity scaling laws. The second paper takes a unique approach relative to most work on minimum energy per bit; it considers total energy consumption – transmit energy plus circuit energy – as opposed to just transmit energy. In particular, [3] derives the tradeoff between total energy consumption and end-to-end data rate in wireless multihop networks, assuming interference treated as noise and orthogonal scheduling of user transmissions. The inclusion of circuit energy, which can include the energy associated with analog front-end electronics as well as signal processing hardware, can change the nature of the energy–rate tradeoff dramatically when transmit power does not dominate total energy consumption (e.g., at relatively short transmission distances). For example, sophisticated codes and multiple antenna techniques can save transmit power but increase circuit power. Similarly, in multihop routing, using intermediate nodes to forward data saves total transmit power but increases circuit power due to intermediate node processing. Thus, optimizing energy consumption in networks depends heavily on transmission distances (since transmit power dominates circuit power at large distances but not at small ones), as well as the precise models for circuit energy consumption associated with the different hardware blocks of a transceiver. Characterizing the tradeoffs between energy consumption and other network performance metrics has generally been hampered by a lack of fundamental energy consumption models for hardware. Hence, a fundamental characterization of such tradeoffs remains largely an open problem. Robustness is also important for many systems, yet it is not clear how to capture robustness in a mathematical metric. Information-theoretic tools are not always well-suited to characterizing fundamental performance limits in networks that have bounded delay, complexity, and power. In [34] a new theoretical framework is proposed to determine fundamental performance limits of wireless networks based on an interdisciplinary approach that incorporates Shannon theory along with network theory, combinatorics, optimization, stochastic control, and game theory.

We now proceed to formally define mutual information and capacity for single-user and multiuser channels and networks. These definitions will also be used in the more complex capacity analysis for cognitive networks.

2.3.2 Mathematical definition of capacity

Shannon's *Mathematical Theory of Communication* [76, 77, 79] defined the capacity of a single-user channel, denoted by C, in terms of the mutual information between the input and output of the channel. Moreover, Shannon showed that this capacity equals the maximum data rate at which bits can be transmitted with arbitrarily small error probability, assuming there are no constraints on the encoder or decoder complexity and delay. In Shannon's transmission scheme data is embedded in a channel code and the data rate corresponds to the transmission rate of the embedded information bits. Specifically, for any $R < C$ and any $P_e > 0$ Shannon proves that there exists a channel code with data rate R and error probability P_e. Moreover, codes operating at data rates $R > C$ cannot achieve an arbitrarily small error probability and, in fact, such codes have an error probability that is bounded away from zero.

The most basic discrete-time channel model for which mutual information is defined consists of a random input $X \in \mathcal{X}$ (also called a *symbol*), a random output $Y \in \mathcal{Y}$, and a probabilistic relationship between X and Y which is generally characterized by the conditional distribution of Y given X, or $p(y|x)$. For continuous random variables $p(y|x)$ is a probability distribution function (pdf) and for discrete random variables it is a probability mass function (pmf). In this notation, random variables are denoted by capital letters (e.g., X) while their realizations and probability distributions are denoted by small letters (e.g., x and $p(x)$, respectively). If the channel has memory, such that the output y_n at a given time n depends on the current as well as past inputs $x^n = (x_1, \ldots, x_n)$, then the input, output, and probability distribution are defined in terms of vectors X^n, Y^n, and $p(y^n|x^n)$. The channel is said to be *memoryless* if the channel output at any time n is independent of past inputs, i.e., if $p(y_n|x^n) = p(y_n|x_n)$. The *mutual information* of a discrete-time memoryless single-user channel, assuming continuous input and output random variables, is defined as

$$I(X;Y) \triangleq \int_{\mathcal{X},\mathcal{Y}} p(x,y) \log\left(\frac{p(x,y)}{p(x)p(y)}\right) dx dy, \qquad (2.3)$$

where the integral is taken over the set of possible values $\mathcal{X}, \mathcal{Y}$ for the random variables X and Y, respectively, which are also called the input and output *alphabets*, and $p(x)$, $p(y)$, and $p(x,y)$ denote the pdfs of the random variables. When the input and output alphabets $\mathcal{X}$ and $\mathcal{Y}$ are finite, the integral becomes a summation over their joint pmf:

$$I(X;Y) = \sum_{\mathcal{X},\mathcal{Y}} p(x,y) \log\left(\frac{p(x,y)}{p(x)p(y)}\right). \qquad (2.4)$$

The log function is typically with respect to base 2, in which case the units of mutual information are bits per channel use, since input X and output Y correspond to a single use of the channel.

Shannon proved that the capacity of a large class of single-user time-invariant channels is equal to the mutual information of the channel maximized over all possible input distributions:

$$C = \max_{p(x)} I(X;Y) = \max_{p(x)} \int_{\mathcal{X},\mathcal{Y}} p(x,y) \log\left(\frac{p(x,y)}{p(x)p(y)}\right) dx dy. \qquad (2.5)$$

For a real discrete time-invariant additive white Gaussian noise (AWGN) channel with bandwidth W, received signal power $\mathcal{P}$, and receiver noise PSD N_0, the channel is described by the input–output relationship

$$Y = X + Z, \qquad (2.6)$$

where X is the channel input, Y is the channel output, and Z is the noise introduced by the channel. The capacity-achieving input distribution $p(x)$ for this channel is Gaussian, and results in the channel capacity

$$C = \frac{1}{2} \log_2\left(1 + \frac{\mathcal{P}}{N_0 W}\right) \text{ bits/channel use.} \qquad (2.7)$$

The units of bits per channel use are alternatively referred to as bits per dimension. The dimension of a scalar channel with bandwidth W over transmission time T is $2WT$. Channels with additional dimensions, e.g., MIMO channels with M spatial dimensions, have dimension $2WTM$. Defining capacity in terms of bits per channel use or dimension is typical for multidimensional channels such as MIMO channels, and we will follow this convention. In system designs the data rates are typically given in terms of bits per second. We can convert bits per channel use (corresponding to $2WT$ dimensions) to bits per second (corresponding to $2W$ dimensions over T seconds) by multiplying (2.7) by $2W$, which yields the capacity expression

$$C = W \log_2\left(1 + \frac{\mathcal{P}}{N_0 W}\right) \text{ bits/second.} \qquad (2.8)$$

When the channel has attenuation h, so that $Y = hX + Z$, its capacity becomes

$$C = W \log_2\left(1 + \frac{|h|^2 \mathcal{P}}{N_0 W}\right) \text{ bits/second.} \qquad (2.9)$$

2.3 Fundamental performance limits

Capacity expressions will generally be in units of bits per second unless otherwise stated. For complex AWGN channels, the real and imaginary signal components comprise orthogonal signal dimensions, so capacity is double that of (2.8).

Cognitive radios generally operate in channels far more complex than the AWGN channel. As discussed in more detail in Chapter 3, these channels exhibit flat or frequency-selective fading, and multiple antenna channels exhibit angular dispersion and fading correlation across antennas. These propagation characteristics lead to a more complex characterization of channel capacity. In particular, frequency-selective fading channels give rise to a set of parallel channels across the frequency domain, as described in more detail in Section 3.8. Let us first consider a time-invariant channel with frequency response $H(f)$ known at both the transmitter and receiver. First suppose that $H(f)$ is block-fading in frequency, so that $H(f) = h_j$ is constant over subchannel j of bandwidth W. The frequency-selective fading channel thus consists of a set of independent AWGN channels in parallel, where the jth subchannel has SNR $|h_j|^2 \mathcal{P}_j/(N_0 W)$. Here $\mathcal{P}_j$ is the power allocated to the jth subchannel in this parallel set, subject to the power constraint $\sum_j \mathcal{P}_j \leq \mathcal{P}$. The capacity of this parallel set of channels is the sum of rates associated with each channel with power optimally allocated over all channels [18, 28]

$$C = \sum_{\max \mathcal{P}_j : \sum_j \mathcal{P}_j \leq \mathcal{P}} W \log_2 \left(1 + \frac{|h_j|^2 \mathcal{P}_j}{N_0 W} \right). \qquad (2.10)$$

The optimal power allocation is found by solving the Lagrangian, which leads to the optimal power allocation

$$\frac{\mathcal{P}_j}{\mathcal{P}} = \begin{cases} \frac{1}{\gamma_0} - \frac{1}{\gamma_j}, & \gamma_j \geq \gamma_0 \\ 0, & \gamma_j < \gamma_0 \end{cases} \qquad (2.11)$$

for some cutoff value γ_0, where $\gamma_j = |h_j|^2 \mathcal{P}/(N_0 W)$ is the SNR associated with the jth subchannel assuming it is allocated the entire power budget. This optimal power allocation is referred to as *waterfilling* over frequency, whereby water (power) is poured into a bowl of variable depth $1/\gamma_j$ up to the water line $1/\gamma_0$. Hence, more power is allocated to subchannels with higher gains above the cutoff value γ_0, which is dictated by the power constraint. The capacity with this optimal power allocation then becomes

$$C = \sum_{j : \gamma_j \geq \gamma_0} W \log_2(\gamma_j/\gamma_0). \qquad (2.12)$$

This capacity is achieved by sending different codes with different rates and transmit powers over each subchannel, similar to adaptive techniques used in OFDM. When $H(f)$ is continuous the capacity under power constraint $\mathcal{P}$ is similar to the case of the block-fading channel with the sum over subchannel capacities replaced by an integral of incremental capacity per frequency over the frequency domain; details can be found in [28, Section 8.5][42].

Let us now consider MIMO channels, for which the channel input is a random vector $\mathbf{X} = (X_1, \ldots, X_{M_t})$ sent from the M_t transmit antennas, the channel output is the vector $\mathbf{Y} = (Y_1, \ldots, Y_{M_r})$ obtained at the M_r receive antennas, and the channel is characterized by an $M_t \times M_r$ matrix $\mathbf{H}$ of complex gains between each transmit and each receive antenna. The multiple dimensions associated with MIMO channel inputs and outputs give rise to the multiple spatial degrees of freedom over which independent data streams can be transmitted. Assuming the channel is known at both the transmitter and receiver, capacity is achieved by optimizing the transmit power and rate allocation across these spatial degrees of freedom. Specifically, when the channel $\mathbf{H}$ is constant and known perfectly at the transmitter and receiver, the capacity (maximum mutual information) in units of bits per channel use is

$$C = \max_{\mathbf{Q}\,:\,\text{tr}(\mathbf{Q})=\mathcal{P}} \log_2 \det\left(\mathbf{I}_N + \mathbf{H}\mathbf{Q}\mathbf{H}^H\right) \qquad (2.13)$$

where the optimization is over the input covariance matrix $\mathbf{Q}$, which is $M_t \times M_t$ and must be positive semi-definite by definition. Using the singular value decomposition (SVD) of $\mathbf{H}$, the MIMO channel can be converted into rank($\mathbf{H}$) spatially parallel, non-interfering single-input/single-output channels [32, 84]. The jth spatial channel corresponding to singular value σ_j has SNR $\gamma_j = |\sigma_j|^2 \mathcal{P}_j / (N_0 W)$, where power $\mathcal{P}_j$ is optimally allocated across these spatial channels, similar to the case of frequency selective fading, which results in a water-filling power allocation over the spatial domain. The capacity formula is the same as in the frequency-selective fading case, given by (2.12): the sum of capacities across the parallel spatial channels with this optimal power allocation based on SNR per spatial dimension γ_j.

For time-varying channels Shannon capacity is given by the ergodic capacity formula (2.1). The channel inputs are based on optimal allocation of transmit power over time or, equivalently, over each channel state subject to an average power constraint:

$$\int_{s \in \mathcal{S}} \mathcal{P}(s) p(s) ds \leq \overline{\mathcal{P}}. \qquad (2.14)$$

For example, consider a flat-fading channel, where the instantaneous SNR γ varies with time according to a distribution $p(\gamma)$. A discussion of fading distributions $p(\gamma)$ under different conditions can be found in Section 3.6. If the transmit power $\mathcal{P}(\gamma)$ is adapted relative to γ, subject to an average power constraint $\overline{\mathcal{P}}$, then the flat-fading channel capacity is given by

$$C = \max_{\mathcal{P}(\gamma):\int \mathcal{P}(\gamma)p(\gamma)d\gamma = \overline{\mathcal{P}}} \int_0^\infty W \log_2 \left(1 + \frac{\mathcal{P}(\gamma)\gamma}{\overline{\mathcal{P}}}\right) p(\gamma) d\gamma. \qquad (2.15)$$

The optimal power allocation $\mathcal{P}(\gamma)$ is found by solving the Lagrangian, similar to the case of frequency selective fading. This yields optimal power allocation as a *water-filling over time*:

$$\frac{\mathcal{P}(\gamma)}{\overline{\mathcal{P}}} = \begin{cases} \frac{1}{\gamma_0} - \frac{1}{\gamma}, & \gamma \geq \gamma_0 \\ 0, & \gamma < \gamma_0 \end{cases} \qquad (2.16)$$

for some "cutoff" value γ_0 which is found via the average power constraint. If $\gamma(t)$ is below this cutoff at time t then no data are transmitted at that time. With this optimal power allocation, the capacity of the time-varying flat-fading channel becomes

$$C = \int_{\gamma_0}^{\infty} W \log_2 \left(\frac{\gamma}{\gamma_0} \right) p(\gamma) d\gamma, \qquad (2.17)$$

where the rate corresponding to instantaneous SNR γ is $W \log_2(\gamma/\gamma_0)$. Since γ_0 is constant, this means that as the instantaneous SNR increases, the data rate sent over the channel for that instantaneous SNR also increases.

There is a strong similarity between time-varying flat-fading channels, MIMO channels, and time-invariant frequency-selective fading channels in that these channels can be represented as a set of independent parallel channels in time, space, or frequency, respectively. This property can be exploited in cognitive radio paradigms, in particular the interweave paradigm. Specifically, if the cognitive radio can sense that a given dimension in time, space, or frequency is not being utilized by the primary user, it can occupy that dimension with no harm to the primary system. The capacity analysis for interweave systems is based on this concept, whereby the interweave cognitive radio channel is modeled as a channel varying over time, frequency, or space. When a given dimension is occupied by the primary user, under perfect sensing the interweave channel in that dimension is unavailable to the cognitive radio, i.e., it is assumed to have an SNR of zero. The capacity analysis above for parallel channels in time, space, or frequency can then be applied directly to determine the capacity of the interweave cognitive radio, as described in more detail in Section 2.6.

The capacities of multiuser channels and wireless networks are also built upon notions of mutual information. However, the encoding and decoding strategies and their associated mutual information become more complicated with multiple users, and hence the capacity regions are typically defined by a set of mutual information bounds that implicitly define the capacity region boundary. Before launching into capacity results for the different cognitive radio paradigms, we will first review capacity results for the interference channel. Since, as discussed in Section 2.3.1, cognitive radio networks are generalizations of the interference channel in Fig. 2.7, the capacity region and optimal encoding and decoding strategies of the interference channel will provide fundamental building blocks for obtaining capacity and design insights for cognitive networks.

2.3.3 Capacity region of wireless networks

We consider a wireless network consisting of K source–destination pairs communicating over a common wireless channel, as shown in Fig. 2.9. We assume a discrete-time network model with discrete channel inputs and outputs. At each time instant, a source s_k chooses a channel input X_k from a finite set $\mathcal{X}_k$ of possible inputs. Each destination node d_j observes a channel output Y_j from output set $\mathcal{Y}_j$. The channel is described by the conditional distribution $p(y_1, \ldots, y_K | x_1, \ldots, x_K)$, which characterizes the probability of the given set of outputs $(y_1, \ldots, y_K)$ at the destinations, for the given set of channel inputs $(x_1, \ldots, x_K)$. A source s_k wishes to communicate a data sequence or

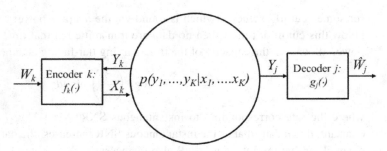

Figure 2.9 Wireless network model.

message $W_k \in \mathcal{W}_k = \{1, \ldots, 2^{nR_k}\}$ to destination d_k, at rate R_k. To do so, the source encoder maps the data sequence into a codeword X_k^n consisting of n symbols from the input alphabet, and sends it in n time instants over the channel. All data sequences are mutually independent. Upon receiving the sequence Y_j^n of length n, decoder j maps it to its estimate of the transmitted data sequence, denoted by $\hat{W}_j$. The data sequence sets $\mathcal{W}_1, \mathcal{W}_2, \ldots, \mathcal{W}_K$ along with the encoder and decoder mappings of all users define an $(R_1, R_2, \ldots, R_K, n)$ code for this channel. The encoding function at source s_k for data sequence W_k at time i is given by

$$X_{k,i} = f_{k,i}(W_k, Y_k^{i-1}), \quad k = 1, \ldots, K. \tag{2.18}$$

Note that the encoding function $f_{k,i}(\cdot)$ allows the source to use its receiver's previous observations of the channel (typically obtained via feedback) to encode W_k. This allows sources to obtain information about data sequences sent by other users and potentially forward them through the network. The decoding function at destination j at time i is given by

$$\hat{W}_j = g_j(Y_j^n), \quad j = 1, \ldots, K. \tag{2.19}$$

A decoding error at destination k occurs when $\hat{W}_k \neq W_k$. We consider that an error occurs unless all destinations decode their data sequences correctly. Thus, the error probability is given by the probability of the union of error events associated with incorrect detection on each of the different data sequences:

$$P_e = P\left[\bigcup_{k=1}^{K} \left[\hat{W}_k \neq W_k\right]\right]. \tag{2.20}$$

If the error probability can be made arbitrarily small for a code of sufficiently large n, the rates $(R_1, \ldots, R_K)$ will be simultaneously achievable in the considered network. More precisely, rates $(R_1, R_2, \ldots, R_K)$ are achievable if, for any $\epsilon > 0$, there exists, for sufficiently large n, an $(R_1, R_2, \ldots, R_K, n)$ code such that $P_e \leq \epsilon$. The capacity region is the closure of the set of all achievable rates $(R_1, R_2, \ldots, R_K)$, due to time-sharing between strategies associated with any set of points on the rate region.

2.3 Fundamental performance limits

The above formulation assumes a single destination for each data sequence. This definition can be extended to include multicasting to a set of destinations, broadcasting from one source to a set of destinations or multiple access from multiple sources to a single destination. The capacity region of a general wireless network is unknown. A general outer bound on the network performance is provided by the cut-set bound [2, 18, 23, 78], stated next.

Let S denote a subset of all network nodes and S^c be the complement of S. The pair (S, S^c) is a *cut* separating source s_k and destination d_k if source $s_k \in S$ and $d_k \in S^c$.

Cut-set outer bound
Any achievable $(R_1, \ldots, R_K)$ satisfies

$$\sum_{s_k \in S, d_k \in S^c} R_k \leq I(X(S); Y(S^c) \mid X(S^c)), \quad (2.21)$$

where mutual information is evaluated for some distribution $p(x_1, \ldots, x_K)$ for any S. R_k is the rate across the cut from source s_k to destination d_k. We observe that (2.21) bounds the sum rate going across a cut by the conditional mutual information between all sources in S and all destinations in S^c, given all sources in S^c.

As an example of the cut-set outer bound, consider K source–destination pairs with AWGN links of bandwidth W. The channel inputs and outputs are then vectors defined by

$$\mathbf{Y} = \mathbf{H}\mathbf{X} + \mathbf{Z}, \quad (2.22)$$

where $\mathbf{X}$ is the vector of channel inputs from all the sources with average power constraint $\mathcal{P}$, $\mathbf{Y}$ is the vector of all channel outputs, $\mathbf{H} \in R^{K \times K}$ is the channel gain matrix and $\mathbf{Z}$ is the vector of independent, unit-variance Gaussian noises at the destinations. The cut-set bound in (2.21) evaluates to [24]

$$\sum_{s_k \in S, d_k \in S^c} R_k \leq W \log_2 \det \left(\mathbf{I} + \mathbf{H}(S)\mathbf{K}(S)\mathbf{H}(S)^T \right), \quad (2.23)$$

where $\mathbf{I}$ is the identity matrix, $\mathbf{K}(S)$ is the covariance matrix of $\mathbf{X}(S)$ given $\mathbf{X}(S^c)$, and $\mathbf{H}(S)$ is determined such that

$$\mathbf{Y}(S^c) = \mathbf{H}(S)\mathbf{X}(S) + \mathbf{H}^T(S)\mathbf{X}(S^c) + \mathbf{Z}(S^c). \quad (2.24)$$

As another example of the cut-set outer bound, consider the three-node relay channel shown in Fig. 2.7. Assume that links are AWGN links of bandwidth W. Let X_s denote the symbol sent from the source, and X_r denote the symbol sent from the relay. Source and relay powers are denoted by $\mathcal{P}_s$ and $\mathcal{P}_r$. The received symbols at the destination and the relay, respectively, are Y_d and Y_r. There are two cuts in the network: in the first cut, only the source forms the set S, whereas in the second cut the source and the relay form the set S. The cut-set bound (2.21) evaluates to

$$R \leq \max_{p(x_s, x_r)} \min \{I(X_s; Y_r, Y_d | X_r), I(X_s, X_d; Y_r)\}. \quad (2.25)$$

In the AWGN relay channel this yields

$$R \leq \max_{0 \leq \rho \leq 1} \min \left\{ W \log_2 \left(1 + \mathcal{P}_s + \mathcal{P}_r + 2\rho \sqrt{\mathcal{P}_s \mathcal{P}_r}\right), \right.$$
$$\left. W \log_2 \left(1 + (\mathcal{P}_s + \mathcal{P}_r)(1 - \rho^2)\right) \right\}, \quad (2.26)$$

where we normalize the noise power to one. The parameter ρ determines the coherent combining gain of source and relay inputs X_s and X_r at the destination. A larger ρ corresponds to a larger gain.

The cut-set bound is a tight outer bound only for certain special scenarios. In general, the cut-set bound is loose relative to the network capacity region. Alternative outer bounds to the cut-set outer bound can be derived by using *genie-based* techniques in which the network is modified by assuming that additional information is known (a.k.a., given by a genie) to a subset of terminals. The goal of providing this information is to obtain a modified network for which capacity or an outer bound can be obtained. Due to the additional information, the modified network outperforms the original one. Consequently, its capacity (or any outer bound on it) yields a capacity outer bound for the original network. Outer bounds can also be obtained by the *theory of network equivalence* [50]. This approach provides conditions under which the capacity of a wireless network can be upper bounded by the performance of an equivalent noiseless network of bit pipes, for which the capacity can then be determined. Another technique to tighten the cut-set bound is by modification of the network connectivity graph [52, 53].

Obtaining the Shannon capacity region of a wireless network is generally intractable; in fact the capacity of several simple canonical topologies such as the relay channel and the interference channel have remained open problems for decades. As an alternative capacity metric, a landmark result by Gupta and Kumar [38] introduced the notion of scaling laws for noncognitive wireless network throughput as the number of nodes in the network K grows asymptotically large. They found that the throughput in terms of bits per second for each node in the network decreases with K at a rate between $1/\sqrt{K \log K}$ and $1/\sqrt{K}$. In other words the per-node rate of the network goes to zero, although the total network throughput, equal to the sum of rates, grows at a rate between $\sqrt{K/\log K}$ and $\sqrt{K}$. This surprising result indicates that when interference is treated as noise (as is typical in practical designs), even with optimal routing and scheduling, the per-node rate in a large ad hoc wireless network goes to zero. The reason is that in this relatively simple relaying scheme, intermediate nodes spend much of their resources forwarding packets for other nodes, so few resources are left to send their own data. There has been much follow-on work to this result, including the impact on wireless network scaling laws of mobility, multiple antennas, and cooperation [5, 36, 68]. In particular, [68] showed that more sophisticated cooperation schemes allow per-node rates in large networks to remain constant with network size rather than decrease. The tradeoff between throughput (in terms of scaling laws) and delay in asymptotically large networks was characterized in [22, 39, 85].

2.4 Interference channels without cognition

2.4.1 K-user interference channels

In cognitive radio networks, we would like to characterize capacity associated with communications between primary user pairs and between secondary user pairs. Although one could envision deployment of relays to improve the performance, these networks typically do not involve multihop routing of information, i.e., there is no forwarding of information through intermediate nodes. Without cognition, networks with K source–destination pairs can be modeled as a K-user interference channel, as shown in Fig. 2.10. Although the capacity region of this channel is in general unknown, there has been a lot of progress in understanding how to cope with interference in this model and, consequently, in developing spectrally-efficient transmission schemes for this channel. In some scenarios, these techniques lead to capacity. A cognitive radio network forms a two-tier K-user interference channel, due to the different capabilities and restrictions of primary and secondary users. Schemes that efficiently cope with interference can improve performance of both primary and secondary users in these networks. For that reason, some of the techniques developed for interference channels have been adopted for overlay cognitive networks as well. We next review these techniques, their performance and their known capacity results. In addition, cognition enables additional encoding/decoding techniques to improve the performance. Hence, performance of interference channels can serve as a lower bound to the capacity of cognitive networks.

In the K-user interference channel model, each of the K sources wishes to communicate with its corresponding destination over a shared wireless channel, as illustrated in Fig. 2.10. Source s_k encodes and sends a data sequence W_k at rate R_k to destination d_k. The K-user interference channel is a special case of a K-user wireless network, and hence we use the same definitions as in Section 2.3.1 for encoding, decoding, error probability, and capacity. However, the encoding function of the kth user in the interference channel is given by

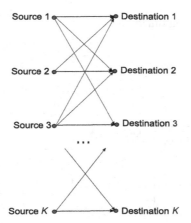

Figure 2.10 K-user interference channel. Sender k wishes to communicate to destination k.

$$X_k^n = f_k(W_k). \tag{2.27}$$

Thus, in this case a channel input at each source depends only on its own data sequence, which, in turn, is independent of data sequences from other sources. Hence, there is no cooperation (e.g., relaying) between sources in transmitting information about each other's data sequences. The cut-set bound (2.21) can then be tightened to become

$$\sum_{s_k \in \mathcal{S}_n, d_k \in \mathcal{S}_n^c} R_k \leq I(X(\mathcal{S}_n); Y(\mathcal{S}_n)|X(\mathcal{S}_n^c), V) \tag{2.28}$$

for any $p(v) \prod_k p(x_k|v)$, where V, referred to as a *time-sharing random variable*, has the property that the inputs $\{x_k\}$ are independent when conditioned on V.

Lower bounds to the capacity region are obtained by deploying specific communication techniques in the given network. We next present encoding schemes that achieve capacity for special cases of the two-user interference channel.

2.4.2 Two-user interference channel capacity

A two-user discrete-time memoryless interference channel is shown in Fig. 2.11. The interference channel contains only two communicating pairs ($K = 2$). Therefore, there are two channel inputs, X_1 and X_2, and two channel outputs, Y_1 and Y_2. The discrete-time channel is characterized by the conditional distribution $p(y_1, y_2|x_1, x_2)$. Each source s_k, $k = 1, 2$, wishes to send a data sequence $W_k \in \mathcal{W} = \{1, \ldots, 2^{nR_k}\}$ to destination d_k. Definitions for the channel code, the error probability and the capacity region follow the definitions for the K-user interference channel. In particular, an (R_1, R_2, n) code consists of two data sequence sets $\mathcal{W}_1, \mathcal{W}_2$, two encoding functions

$$X_1^n = f_1(W_1), \tag{2.29}$$
$$X_2^n = f_2(W_2) \tag{2.30}$$

and two decoding functions

$$\hat{W}_k = g_k(Y_k^n), \qquad k = 1, 2. \tag{2.31}$$

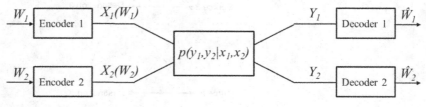

Figure 2.11 Two-user interference channel.

2.4 Interference channels without cognition

The error probability of the code is

$$P_e = P\left[\left[\hat{W}_1 \neq W_1\right] \bigcup \left[\hat{W}_2 \neq W_2\right]\right]. \tag{2.32}$$

A rate pair (R_1, R_2) is achievable if, for any $\epsilon > 0$, there exists for sufficiently large n an (R_1, R_2, n) code such that $P_e \leq \epsilon$. The capacity region of the interference channel is the closure of the set of all achievable rate pairs (R_1, R_2).

We will also consider the two-user AWGN interference channel, defined by

$$Y_1 = X_1 + h_{12}X_2 + Z_1,$$
$$Y_2 = h_{21}X_1 + X_2 + Z_2, \tag{2.33}$$

where h_{12} and h_{21} are the (real) cross-channel gains of the interfering links, $\mathbb{E}[X_k^2] \leq P_k$ are power constraints, and $Z_k \sim \mathcal{N}(0,1)$ for $k=1,2$, where $\mathcal{N}(0, \sigma^2)$ denotes the Gaussian distribution with variance σ^2.

The capacity of the interference channel is known in the case of *strong interference* [15]. In this regime, the interfering signal at each receiver is strong enough that the other user's data sequence carried by that signal can be decoded and hence removed. It is then optimal for each receiver to decode both data sequences. In the AWGN interference channel (2.33), the strong interference conditions are given by [40, 74]

$$|h_{12}| \geq 1,$$
$$|h_{21}| \geq 1, \tag{2.34}$$

implying that in this regime the cross-channel gains are larger than the direct link gains.

In general discrete memoryless channels, the strong interference conditions can be expressed in terms of the conditional mutual information inequalities. These conditions require that

$$I(X_1; Y_1|X_2) \leq I(X_1; Y_2|X_2), \tag{2.35}$$
$$I(X_2; Y_2|X_1) \leq I(X_2; Y_1|X_1) \tag{2.36}$$

are satisfied for every distribution $p(x_1)p(x_2)$. From (2.35) we observe that Y_2 contains more information about input X_1 than Y_1, given X_2. Hence, X_1 conveys more information to receiver 2 than to receiver 1. A similar interpretation can be made for (2.36).

The capacity region in strong interference is given by

$$\mathcal{C} = \bigcup \{(R_1, R_2) : R_1 \geq 0, R_2 \geq 0,$$
$$R_1 \leq I(X_1; Y_1 \mid X_2, V),$$
$$R_2 \leq I(X_2; Y_2 \mid X_1, V),$$
$$R_1 + R_2 \leq \min\{I(X_1, X_2; Y_1 \mid V), I(X_1, X_2; Y_2 \mid V)\}, \tag{2.37}$$

where the union is over distributions of the form $p(v)p(x_1|v)p(x_2|v)$ and V is once again a time-sharing random variable for which the inputs are independent when conditioned on it.

The capacity region (2.37) can alternatively be expressed as the intersection of two associated multiple-access channel (MAC) rate regions [18]:

$$\mathcal{C} = \bigcup \left\{ \mathcal{R}_{\text{MAC}_1} \bigcap \mathcal{R}_{\text{MAC}_2} \right\}. \qquad (2.38)$$

Recall that the multiple access channel consists of multiple transmitters simultaneously sending to one receiver. In Fig. 2.11, the channel MAC_1 consists of two encoders and decoder 1 and, similarly, the channel MAC_2 consists of two encoders and decoder 2. The capacity region of this network is in general a lower bound on the capacity of the interference channel because both receivers need to decode data sequences sent from both sources. In the interference channel, in contrast, each receiver decodes data sequences sent from only one source. In strong interference, however, each receiver can decode unwanted data sequences without reducing the capacity region of the interference channel, and the capacity region of the interference channel then coincides with (2.38).

Note that, in strong interference, the capacity is achieved by joint decoding of both data sequences at the decoders. In the special case of *very* strong interference, decoders do not need to perform joint decoding of the data sequences. Instead, interference cancellation can be performed successively, allowing for interference-free decoding of the desired data sequence [10].

We next consider the opposite interference regime in which the interference is weak. In this regime, the interfering data sequence cannot be decoded, but it is not strong enough to significantly degrade the rate of the impacted user. When the interference is weak, intuitively we expect that the optimal decoding strategy is to treat it as noise. However, it has been difficult to prove that this intuition is correct, and hence the capacity remains unknown. Conditions under which treating interference as noise leads to sum-rate capacity in two-user Gaussian channels were determined in [1, 63, 75]. In this regime (termed *noisy interference* in [75]), the cross-channel gains h_{12} and h_{21} in (2.33) as well as the transmit powers $\mathcal{P}_1$ and $\mathcal{P}_2$ are small; specifically, they satisfy

$$h_{12}(h_{21}^2 \mathcal{P}_1 + 1) + h_{21}(h_{12}^2 \mathcal{P}_2 + 1) \leq 1. \qquad (2.39)$$

The sum-rate capacity is then given by [75, Theorem 2]

$$C = W \log_2 \left(1 + \frac{\mathcal{P}_1}{1 + h_{12}^2 \mathcal{P}_2} \right) + W \log_2 \left(1 + \frac{\mathcal{P}_2}{1 + h_{21}^2 \mathcal{P}_1} \right). \qquad (2.40)$$

The proof of this result uses a genie-based outer bound for the Gaussian interference channel.

The above described regimes are two extremes with respect to the amount of interference that is being experienced and removed by receivers. Not surprisingly, in strong interference the highest-rate scheme is to decode the unwanted data sequences and subtract their corresponding signals, thus removing their interference from the received signal. In the other extreme of weak interference, the highest-rate strategy is to ignore the interference, that is, treat it as noise.

In regimes that are in between the two extremes, the interference is not strong enough so that decoding of the unwanted data sequence is optimal, nor is it weak enough to

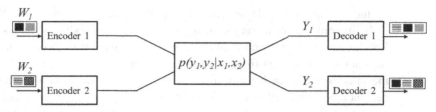

Figure 2.12 Rate splitting: each encoder splits its data sequence into two data subsequences of lower rate, and encodes them via superposition coding. Each decoder jointly decodes one data subsequence of the other user together with its desired data sequence.

be treated as noise without loss of optimality. In this scenario, decoding part of an interfering data sequence to partially remove interference from the received signal is beneficial. This idea is realized in the scheme developed by Carleial and subsequently improved by Han and Kobayashi, also referred to as *rate-splitting* [11, 40]. The rate-splitting concept is illustrated in Fig. 2.12. To perform rate-splitting, each encoder divides its data sequence into two data subsequences, each of lower rate than the original sequence, and encodes them via *superposition coding*, which was originally developed for the broadcast channel [16]. In superposition coding, the source encodes each of the two data sequences using separate but correlated codebooks. Specifically, it encodes the first data sequence based on the optimal input distribution, and the second data sequence based on the optimal conditional distribution between the two data sequences. Separate encoding enables each receiver to decode one of the data subsequences intended for the other user jointly with its own data sequence, while treating the signal carrying the other undesired data subsequence as noise. The communication rate for this user increases due to reduced interference, but the rate for the other communicating pair decreases due to an additional decoding constraint. Hence, there is a tradeoff between the amount of information sent only to the desired receiver and the amount of interference decoded at the other receiver.

In AWGN interference channels, this encoding tradeoff translates into optimizing the power allocated to each of the two parts of the encoder's data sequence. By choosing Gaussian codebooks, i.e., random codebooks generated according to a Gaussian distribution, and a specific power split, the Han–Kobayashi scheme achieves rates within one bit per dimension from the two-user interference channel capacity [26]. The power split is chosen so that the created interference at each receiver has the same power as the Gaussian noise at that receiver. Thus, the created interference is sufficiently weak so as not to significantly impair performance. At the same time, the undesired data sequence that is decoded at each destination allows for significant interference reduction. In Section 2.7 we will give more details on how rate-splitting can be used in overlay cognitive radio networks.

In a K-user interference channel, each receiver is exposed to interference arriving from multiple sources. A generalization of the Han–Kobayashi scheme would allow for partial decoding of each interfering signal. A receiver could then jointly decode its own data sequence along with some portion of the interfering data sequences dictated by the rate-splitting code design. While such a generalization is possible mathematically, it would result in very complex encoding and decoding schemes at each node. Specifically, this approach requires a receiver to *separately* decode parts of interfering data sequences sent from many interferers in order to reduce interference. Instead, the interference at each receiver can be treated *collectively* in a more efficient manner via *interference alignment* [7, 57] or via *structured codes* [65]. These approaches exploit the fact that a receiver is not interested in information associated with interfering data sequences and hence does not need to decode them (or parts of them). Interference alignment achieves the optimal capacity scaling law in the interference channel [69]. Lattice codes outperform the Han–Kobayashi scheme in the K-user interference channel [83].

The AWGN channel considered so far in this section assumes constant channel coefficients and hence does not capture flat or frequency-selective fading. Incorporating these channel characteristics leads in general to channel models that are more difficult to analyze. However, these characteristics open up possibilities for encoding and transmission strategies that exploit fading. In particular, frequency-selective and time-varying channels can be modeled as parallel interference channels [8]. Results in [8] demonstrate that parallel interference channels are optimized by joint encoding across subchannels. This is in contrast to point-to-point, multiple access, and broadcast parallel channels in which separate encoding over the subchannels is optimal. Further exploration of K-user interference channels has revealed that time-variations can be exploited to combat interference in the form of interference alignment. Interference alignment relies on channel time-variations to achieve half of the interference-free capacity for each user in the system [7, 66].

2.4.3 Interference channel techniques for cognitive radios

In interference networks with no cognition, a plausible transmission scheme to avoid interference is to split the bandwidth and assign orthogonal channels to each communicating pair, e.g., via a MAC protocol that divides channels orthogonally in time, frequency, or space. In cognitive radio networks, this corresponds to the interweave approach to avoid interference between cognitive and primary users. However, in general, dividing the bandwidth orthogonally reduces spectral efficiency in comparison with assigning the full bandwidth to all users and then coping with the introduced interference. One exception is in the *low-SNR regime* where the network is power-limited rather than interference-limited. In this regime, bandwidth is plentiful compared to power, and thus assigning orthogonal channels to users incurs no performance loss. In the interference-limited regime this is no longer true. Furthermore, as the number of network users increases, the rate per user of an orthogonal MAC scheme goes to zero. This reasoning emphasizes why the overlay approach may be a more spectrally-efficient technique for cognitive radios than the interweave approach. We analyze capacity regions of the interweave and the overlay cognitive radio in Sections 2.6 and 2.7, respectively.

Rate-splitting and superposition coding can also be applied to treat interference in overlay cognitive radio networks. An overlay cognitive radio shares its bandwidth with one or more primary users, thereby creating interference at the primary receivers. At the same time, the secondary receivers experience interference from the primary user transmissions. Efficiently coping with interference can thus improve the performance of both secondary and primary users in the system. We will see that in overlay cognitive radio systems there exists a regime of strong interference in which decoding of the unwanted data sequences is optimal. Similarly, there is a weak interference regime in which interference can be treated as noise without loss of optimality.

Furthermore, cognitive capabilities enable radios to deploy transmission strategies that cannot be deployed in interference channels without cognition. By listening to the channel, a secondary user can obtain information about primary user data sequences, assuming the security and privacy concerns of the primary system can be addressed. The secondary user can then cooperate with the primary user by relaying this information (or a part of it) to the primary receiver, thereby improving the primary user's performance. If the secondary user allocates part of its power to relay the primary data sequence and the rest of its power to transmit its own information, then the power split should be chosen such that the resulting SINR ratio at the primary receiver yields the same or better performance than if the secondary user was not present.

In summary, when compared to wireless networks with no cognition, a cognitive radio can cope with interference using techniques developed for interference channels. In addition, cognition enables the secondary encoder to protect its own information from the interference caused by the primary system by precoding against interference. These techniques will be described in more detail in Section 2.7.

2.5 Underlay cognitive radio networks

The underlay approach to cognitive radio allows for spectrum sharing between primary and secondary users, under the constraint that the interference caused by secondary users does not noticeably degrade the performance of the primary users. When the performance of a primary user is measured by its SINR in each signal dimension, the interference that secondary users cause to any primary user is typically constrained to be within a given spectral mask, i.e., a limit on the PSD of the interference over frequency and, for MIMO channels, over space. The interference constraint, in turn, imposes a constraint on the total PSD per dimension received from all secondary transmitters at any primary receiver. Alternatively, an average interference constraint may be defined whereby the interference power, averaged over all signal dimensions, is below a given threshold. In order to satisfy the interference constraint, several different conditions can be imposed on the secondary transmission. When primary and secondary users experience time-varying channels, the primary users' performance requirements may be based on meeting a given peak interference constraint at each time instant or constraining the average interference over time. An alternative underlay paradigm does not impose constraints on interference but rather on the minimum value of the primary user's capacity.

However, this is not a common paradigm since it is simpler for a secondary user to determine the interference it causes to a primary receiver than to determine the capacity degradation it causes. Thus, we will focus on the capacity of underlay systems given interference constraints imposed on the secondary users.

When multiple secondary and primary users coexist, their interference constraint needs to be satisfied at the primary user that is the most impaired by the interference from secondary users. As explained earlier, satisfying the interference condition requires that a secondary transmitter knows the channel to the primary receiver. Otherwise, in order to guarantee that the interference constraint is satisfied under unknown channel conditions, the secondary user's transmit power needs to be severely restricted. In the presence of multiple secondary users, determining the interference at the primary receiver becomes more demanding, as the interference depends on transmit powers and channel conditions from all secondary transmitters, as described in more detail later in this section.

2.5.1 Underlay capacity region

The capacity region of the underlay cognitive radio network can be defined following the capacity definition for wireless ad hoc networks, while taking into account the interference constraints at the primary users. Both capacity and the interference constraints will depend on the channel characteristics. To define the received power constraint, we consider first a narrowband system with one secondary and one primary user, as shown in Fig. 2.13. We extend this to networks with multiple primary and secondary users below. In this two-user network, an interference constraint translates to a received power constraint on the secondary user's signal at the primary user's receiver. Let X_i denote the channel input at time i by the secondary user and Y_i denote the corresponding channel output at the primary receiver. The channel between the secondary transmitter and the primary receiver is assumed to be memoryless and hence can be described by a conditional probability distribution $p(y|x)$ at each time i, as illustrated in Fig. 2.13 for the link between the secondary transmitter and primary receiver. With each input–output pair (x, y) we associate a cost function $c(x, y)$. The average cost function over n transmissions is then defined as the average of the per-symbol cost:

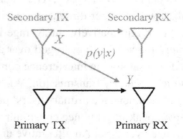

Figure 2.13 Underlay system with single primary and secondary pair.

2.5 Underlay cognitive radio networks

$$\mathbb{E}[c_n(X^n, Y^n)] = \frac{1}{n}\sum_{i=1}^{n}\mathbb{E}[c(X_i, Y_i)], \qquad (2.41)$$

where the expectation is with respect to X_i and Y_i. The interference constraint is based on imposing a maximum value for this average cost:

$$\mathbb{E}[c_n(X^n, Y^n)] \leq \eta. \qquad (2.42)$$

The average received power constraint is a special case of (2.42), where a constraint is imposed only on the received signal and the cost function is chosen to be $c(y) = |y|^2$:

$$\frac{1}{n}\sum_{i=1}^{n}\mathbb{E}\left[|Y_i|^2 \mid X_i\right] \leq \eta. \qquad (2.43)$$

Note that the constraint (2.43) is based on requiring that the average power of the received *interference plus noise* falls below η; to obtain the corresponding interference constraint, the noise power must be subtracted from η. The received peak power constraint, based on a per-symbol constraint, is given by

$$\mathbb{E}\left[|Y_i|^2 \mid X_i\right] \leq \eta_{\text{peak}}, \qquad i = 1, \ldots, n. \qquad (2.44)$$

When the channel is static, the channel and corresponding input distribution on X_i is the same for all i, so the average and peak power constraints are the same, given by (2.44) with $\eta = \eta_{\text{peak}}$. When the channel is wideband and the received power constraint is defined by the PSD over frequency, the average and peak power constraint definitions can be extended to constraints on the average or peak PSD. The capacity region can then be defined equivalently to the capacity region of the wireless network while taking into account the average or peak received power constraint, as we now describe in more detail. In this development we will assume a narrowband channel with average or peak power constraints given by (2.43) or (2.44), but the results can be easily extended to wideband channels with a spectral mask constraint by considering the received PSD over all frequencies.

Let us now consider a network with K source–destination pairs communicating over a common memoryless channel. The conditional distribution of the channel inputs and outputs is given by $p(y_1, \ldots, y_K | x_1, \ldots, x_K)$. We assume further that the network contains K_{p} primary and K_{s} secondary pairs such that $K_{\text{p}} + K_{\text{s}} = K$. Each source s_k wishes to communicate with a corresponding destination d_k at rate R_k, over n uses of the channel. The code for the network, $(R_1, \ldots, R_K, n)$, is defined as before by (2.18)–(2.19). The error probability of the code is given by (2.20). We denote the set of primary receivers as $\mathcal{S}_{\text{PR}}$, and the set of secondary transmitters as $\mathcal{S}_{\text{ST}}$.

Let $Y_{j,i}$ denote the interference at the jth primary receiver caused by all secondary transmitted symbols $X_{m,i}$ at time i, for $m \in \mathcal{S}_{\text{ST}}$. Under an average power constraint, we thus require that

$$\frac{1}{n}\sum_{i=1}^{n}\mathbb{E}\left[|Y_{j,i}|^2 \mid \{X_{m,i}, m \in \mathcal{S}_{\text{ST}}\}\right] \leq \eta, \quad \forall j \in \mathcal{S}_{\text{PR}}. \qquad (2.45)$$

Condition (2.45) thus extends condition (2.43) to a K-user setting by requiring that the average received interference power from all interfering secondary users is smaller than threshold η at all primary receivers. Alternatively, a received peak power constraint at each time i can be imposed as

$$\mathbb{E}\left[|Y_{j,i}|^2 \mid \{X_{m,i}, m \in \mathcal{S}_{\text{ST}}\}\right] \leq \eta_{\text{peak}}, \quad i = 1, \ldots, n, \quad \forall j \in \mathcal{S}_{\text{PR}}. \tag{2.46}$$

Rates $(R_1, \ldots, R_K)$ are achievable if for any $\epsilon > 0$ there exists an $(R_1, \ldots, R_K, n)$ code such that $P_e < \epsilon$ and the constraint (2.45) is satisfied for the set $\mathcal{S}_{\text{PR}}$. The capacity under the cost constraint is the closure of the set of the achievable rates since time-sharing between different strategies can be used.

The capacity region will characterize the set of achievable rates for both primary and secondary users. In underlay cognitive radio networks, the premise is that the primary users' performance is minimally affected by the presence of secondary users. This impact on capacity can be approximated by modeling the secondary users as an additional source of Gaussian noise with power given by the imposed received power constraint. The capacity for secondary users can then be considered separately, under the average or peak received power constraints, and treating the interference from the primary system to the secondary system as noise. The next section illustrates this approach for different network scenarios.

2.5.2 Capacity results for specific scenarios

In this section we present capacity results of the secondary user system for different channel models and numbers of secondary users. In each of these scenarios we consider only a single primary user. The case of multiple primary users is discussed at the end of this section.

Static scalar channels with one secondary user

Suppose a single secondary user transmits the symbol X_s over a point-to-point static channel. The channel between the secondary transmitter and the primary receiver has the input–output relationship

$$Y_p = h_{ps} X_s + Z_p \tag{2.47}$$

where h_{ps} is the cross-channel gain on the link from the secondary transmitter to the primary receiver, which is assumed known by the secondary user, and $Z_p \sim \mathcal{N}(0, \sigma^2)$ is the additive white Gaussian noise sample at the primary receiver. Since the channel is static, the expected power of each transmitted symbol is the same. Hence, the average received interference power constraint (2.43) for this channel is given by

$$\mathbb{E}\left[|Y_p|^2\right] = |h_{ps}|^2 \mathbb{E}\left[|X_s|^2\right] + \sigma^2 \leq \eta. \tag{2.48}$$

This can be translated to a transmit power constraint as

$$\mathbb{E}\left[|X_s|^2\right] \leq \frac{\eta - \sigma^2}{|h_{ps}|^2}. \tag{2.49}$$

2.5 Underlay cognitive radio networks

The secondary user's AWGN channel has the input–output relationship

$$Y_s = h_s X_s + Z_s, \qquad (2.50)$$

where $Z_s \sim \mathcal{N}(0, N_0/2)$. The capacity of the secondary user's channel is thus given by the capacity of the AWGN channel (2.9) with transmit power constraint (2.49).

Static MIMO channels with one secondary user
When the secondary transmitter and primary receiver have M_t and M_r antennas, respectively, the interfering link between them is a MIMO channel. Assuming this channel is static and Gaussian, its input–output relationship is given by

$$\mathbf{Y}_p = \mathbf{H}_{ps}\mathbf{X}_s + \mathbf{Z}_p, \qquad (2.51)$$

where $\mathbf{X}_s \in \mathbb{C}^{M_t}$, $\mathbf{Y}_p \in \mathbb{C}^{M_r}$, $\mathbf{H}_{ps} \in \mathbb{C}^{M_r \times M_t}$ is the channel gain matrix from the secondary transmit antennas to the primary receive antennas, and $\mathbf{Z}_p \in \mathbb{C}^{M_r}$ is the vector of AWGN noise components at each antenna of the primary receiver, where each component has variance σ^2. The underlay average received power constraint (2.43) corresponds to a received power constraint on the vector $\mathbf{Y}_p = (Y_1, \ldots, Y_N)$ received at the primary user, which is given by

$$\mathbb{E}\left[\|\mathbf{Y}_p\|^2 \mid \mathbf{X}_s, \mathbf{H}_{ps}\right] = \nu + M_r \sigma^2 \le \eta, \qquad (2.52)$$

where $\mathbf{Y}_p$ is the MIMO channel output defined in (2.51). We have separated the received power constraint into two components, ν and $M_r \sigma^2$, where the former is based on interference and the latter is based on noise. Thus, $\nu = \|\mathbf{H}_{ps}\mathbf{X}\|^2$. The secondary user's channel, assuming its receiver also has multiple antennas, is described by the input–output relationship

$$\mathbf{Y}_s = \mathbf{H}_s \mathbf{X}_s + \mathbf{Z}_s, \qquad (2.53)$$

where $\mathbf{Z}_s$ is the vector of AWGN noise components at each antenna of the secondary receiver. The received power constraint (2.52) changes the capacity of the secondary user's channel significantly, as it precludes the optimal spatial water-filling of power for the secondary user's MIMO channel. In particular, it was shown in [29, Sec. III] that the capacity of the secondary user MIMO channel $\mathbf{H}_s$ under constraint (2.52) is

$$C = \mathrm{rank}(\mathbf{H}_s) \log_2\left(1 + \frac{\nu}{\mathrm{rank}(\mathbf{H}_s)\sigma^2}\right). \qquad (2.54)$$

This capacity formula depends only on the rank of the secondary user's channel gain matrix $\mathbf{H}_s$, rather than on its singular values as in (2.12), although the code that achieves the capacity in (2.54) depends on the full matrix $\mathbf{H}_s$ and not just its rank.

Flat-fading channels with one secondary user
We next assume that the secondary user experiences flat-fading both to its own receiver and to the primary receiver and that these gains are known at the secondary transmitter. The Shannon capacity of the secondary user in this case is given by the ergodic capacity. As before, let h_{ps} denote the gain on the interfering link and h_s denote the gain on the

secondary link, where these parameters now vary over time. It is shown in [31] that the ergodic capacity of the secondary user channel under an average interference power constraint imposed by the primary system yields

$$C = \iint W \log_2 \left(\frac{|h_s|^2}{|h_{ps}|^2 W N_0 \lambda} \right) p(h_s) p(h_{ps}) dh_s dh_{ps}. \quad (2.55)$$

The optimal power allocation for the secondary user that achieves this capacity is given by

$$P^* = \left(\frac{1}{\lambda |h_{ps}|^2} - \frac{N_0 W}{|h_s|^2} \right)^+, \quad (2.56)$$

where λ is determined to satisfy the received power constraint as

$$\iint \left(\frac{1}{\lambda} - N_0 W \frac{|h_{ps}|^2}{|h_s|^2} \right)^+ p(h_s) p(h_{ps}) dh_s dh_{ps} = \eta. \quad (2.57)$$

This optimal power allocation is similar to the time water-filling given by (2.16) under a transmit power constraint. However, imposing a power constraint at the primary receiver results in a time-varying water level inversely proportional to $|h|^2$, the magnitude squared of the instantaneous time-varying channel gain. Moreover, the cutoff fade depth below which no data is transmitted by the secondary user depends on both their channel gain h_s and the gain h between the secondary transmitter and primary receiver. In particular, the secondary user increases his transmit power when either h_s increases (as their channel is better, i.e., conventional water-filling) or when h decreases, since less interference is caused to the primary. Note that, from (2.55), the capacity depends on channel statistics from the secondary transmitter to both primary and secondary receivers. Interestingly, it is shown in [31] that the capacity (2.55) when both h_s and h experience Rayleigh fading is larger than the capacity of the corresponding AWGN channel with the same average SNR. However, with this Rayleigh fading on both channels, the average SNR is low, and it is known that at low SNRs the ergodic capacity of a Rayleigh fading channel exceeds that of an AWGN channel with the same SNR [32, Figure 4.7]. Note, however, that this is in some sense an artifact of the infinitely long tail probability of the Rayleigh distribution, which results in a set of very high SNR channels with very low probability. These rare high-SNR channels can be exploited by allocating significant power and rate during these low probability events, which results in a slightly higher capacity than a static channel maintaining a fixed rate commensurate with the same average SNR. Note, also, that the propagation characteristics which induce fading also cause some reduction in average received power. Hence, comparing fading and nonfading channels with the same average SNR does not fully capture the impact of fading on performance.

Under the peak constraint (2.44), the optimum power strategy for the secondary user is simply to transmit at the highest power allowed by (2.44). The capacity is then given by

$$C = \iint W \log \left(\frac{|h_s|^2}{|h_{ps}| W N_0 \lambda} \right) p(h_s) p(h_{ps}) dh_s dh_{ps}. \quad (2.58)$$

Alternatively, the constraint imposed on the secondary user might be to meet an outage capacity constraint on the primary user. This condition is equivalent to a peak transmit power constraint that has to be satisfied at the secondary transmitter for some percentage of time P_{out}. The peak transmit power constraint will be chosen based on channel characteristics to guarantee that outage occurs at the primary with probability not exceeding P_{out}. Finally, imposing the ergodic capacity constraint at the primary receiver would require that the power allocation for the secondary user is determined for which the SINR at the primary receiver yields the desired ergodic rate. Further results on underlay cognitive radio in the presence of fading can be found in [31, 64].

In the presence of multiple secondary and primary users, a received signal at any primary user contains interference from all the secondary transmissions. Secondary transmitters then collectively have to choose their transmissions such that the desired interference constraints are satisfied at all primary users. Even in the presence of one primary user, secondary users form a wireless network and hence finding their capacity region is, in general, a difficult problem. In the case that there is no cooperation between secondary users, their transmitter/receiver pairs form an interference channel, and hence the capacity region is unknown. The performance will further be impacted by cooperation among terminals which allows for coherent combining in relaying of each other's information. Without cooperation, the received power from all secondary users simply sums incoherently. The capacity results with or without cooperation can be evaluated for some specific topologies [29]. We next present the capacity for one of these topologies, the Gaussian multiple access channel.

Static scalar channels with multiple secondary users
The static scalar AWGN multiple access channel consists of K secondary users communicating at rate R_k over an AWGN channel to a common secondary receiver. The rates that can be achieved are constrained by the constraint on the interference these users cause to primary receivers. Consider a single primary receiver where the channel between the kth secondary transmitter and the primary receiver has gain h_k. The received interference plus noise signal at the primary receiver is thus given by

$$Y_{\text{p}} = \sum_{k=1}^{K} h_k X_k + Z_{\text{p}}, \qquad (2.59)$$

where X_k is the kth secondary user's transmitted symbol and Z_{p} is the primary user's receiver noise, modeled as AWGN with variance σ^2. Since the channel is static, the average interference power constraint (2.45) corresponds to

$$\mathbb{E}\left[|Y|^2 \,|\, \{X_k, h_k\}, k=1,\ldots,K\right] = \nu + \sigma^2 \leq \eta, \qquad (2.60)$$

where ν represents the power of the received interference and σ^2 the noise power. It is shown in [29, Sec. IV] that under this constraint the capacity region of the secondary user multiple access channel is given by the set of rate vectors that satisfy

$$\sum_{k=1}^{K} R_k \leq \log\left(1 + \frac{\nu}{\sigma^2}\right). \qquad (2.61)$$

Secondary user broadcast channels
The secondary user broadcast channel consists of a secondary transmitter broadcasting to K receivers. Since there is only one secondary transmitter creating interference to the primary user, the interference constraint on the secondary system is similar to that of scalar channels. In particular, the received power constraint at the primary receiver translates to a transmit power constraint at the secondary transmitter. This holds for both static and fading channels. Hence, the capacity region of the secondary broadcast system is obtained using standard techniques for broadcast channels with an additional constraint on the transmit power imposed by the primary system.

Multiple primary users
Since the capacity of multiuser networks is in general unknown except in certain special cases, this will also be true for networks with multiple primary users. However, for some specific topologies capacity can be determined under both average and peak interference power constraints. For example, for a network with multiple primary users and a single secondary user, the received peak power constraint translates to multiple transmit peak power constraints at the secondary transmitter. The optimal power allocation is then to choose the lowest peak power allowed by these constraints. When there are multiple primary and secondary users, the power of all the secondary users must meet the interference power constraints at all the primary users. Optimization techniques can be used to find power allocation solutions that satisfy these constraints.

2.6 Interweave cognitive radio networks

The fundamental performance limits of interweave cognitive radios depend on assumptions about the overall system model as well as assumptions about the ability of the cognitive radio system to sense the interference it causes to primary users. In some cases these assumptions allow known capacity results or bounds to be applied. When detection is imperfect, the impact of missed detection and false alarms of primary user activity reduces capacity of both the secondary and primary users. In addition to capacity regions, scaling laws for interweave cognitive networks will be presented that indicate how capacity scales as the number of secondary users grows.

This section focuses on the rate limits based on information-theoretic capacity of interweave channels. Specifically, we discuss Shannon capacity, outage, and ergodic capacity, as well as scaling laws. In addition to these information-theoretic limits, performance of interweave cognitive systems can also be limited by detection and hardware constraints. In particular, interweave systems with poor sensing capability will have high probabilities of missed detection and false alarm. Similarly, the frequency agility of the secondary transmitter and receiver front ends drive performance limits, since systems without sufficient agility cannot exploit spectrum holes whose locations are changing rapidly. Thus, in addition to capacity, the performance limits of interweave systems are affected by limits of the system and hardware sensing capability as well as the front end capabilities of the radios, as discussed earlier in Section 1.7.

2.6.1 Shannon capacity

The Shannon capacity region of an interweave network dictates the maximum rates achievable for all secondary source–destination pairs, subject to the modeling assumptions and constraints imposed by the primary users. The modeling assumptions include the number of both primary and secondary users, the number of frequency bands available in the system, the statistics of the primary user traffic, and the topology of the network. When there are multiple primary or multiple secondary users, then a MAC protocol for how each type of user shares the available bandwidth with users of the same type must also be optimized. Alternatively, we can define a MAC protocol a priori and derive capacity based on the constraints associated with this MAC. In the interweave paradigm, the sharing of available bandwidth between primary and secondary users is subject to the constraint that the secondary user does not occupy any spectrum that it detects as currently occupied by primary users. It can still cause interference to primary users due to missed detection, when a spectrum hole is detected despite the presence of a primary user. The interweave system typically has a constraint on this missed detection probability imposed on it by the primary system.

We begin the discussion of capacity for interweave networks with the most basic model: one primary transmit–receive pair and one secondary transmit–receive pair. This system can be modeled by the four-node interference channel shown in Fig. 2.7, where source–destination pair 1 corresponds to the secondary user pair and source–destination pair 2 corresponds to the primary user pair. Suppose there is only one frequency band available to all users. We assume perfect detection of the primary user activity by the secondary user, and that the two users share the interference channel via a time-sharing medium access strategy whereby only one of the sources transmits at any given time (this will be the primary user if it has data to send, otherwise the secondary user). Let C_p denote the Shannon capacity of the primary user's channel assuming it transmits over all time and the secondary user is silent. Let α denote the fraction of time that the primary user occupies the channel. The primary user's channel capacity is then αC_p since that is the maximum rate it can achieve over all time. Let C_s denote the Shannon capacity of the secondary user's channel assuming it transmits all the time and the primary user is silent. Then the rate region (R_p, R_s) associated with the time-sharing strategy between the primary and secondary users is the triangle of Fig. 2.14 defined by $(\alpha C_p, (1-\alpha)C_s)$, with y-axis intercept C_p and x-axis intercept C_s. These ideas are easily extended to multiple primary and secondary users if there is perfect coordination among the primary users and among the secondary users, and perfect detection of primary user channel occupancy by the secondary users, as illustrated in Fig. 2.15. This figure shows the two primary users coordinating via time-sharing and the two secondary users coordinating via time-sharing to utilize the timeslots where primary users are absent. More generally, for a network with K_p primary users and K_s secondary users, $K_p + K_s = K$, assume a time-sharing strategy where the fraction of time allocated to primary user i on its channel of interference-free capacity C_i is α_i^p, with $\sum_{i=1}^{K_p} \alpha_i^p = \alpha$. Assume the jth secondary user with interference-free capacity C_j is allocated time fraction α_j^s of the remaining time fraction $1-\alpha$, such that $\sum_{j=1}^{K_s} \alpha_j^s = 1-\alpha$. Then the

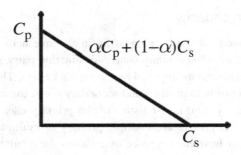

Figure 2.14 Two-user capacity region for an interweave channel with perfect spectrum hole detection.

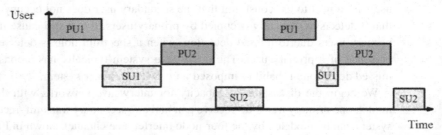

Figure 2.15 Channel occupancy under time-sharing: primary users PU1 and PU2 coordinate their channel time-sharing (alternate use in this case). Secondary users SU1 and SU2 perfectly detect the spectrum holes and coordinate sharing these holes (alternate use in this case).

capacity region for the K users, i.e., for the K_p primary users and the K_s secondary users, is

$$\bigcup (R_1 = \alpha_{p1}C_{p1}, R_2 = \alpha_{p2}C_{p2}, \ldots, R_{K_p} = \alpha_{pK_p}C_{pK_p},$$
$$R_{K_p+1} = \alpha_{s1}C_{s1}, \ldots, R_K = \alpha_{sK_s}C_{sK_s}), \quad (2.62)$$

where the union is taken over all $\{\alpha_{pi}\}$ and $\{\alpha_{sj}\}$ such that $\sum_{i=1}^{K_p} \alpha_{pi} + \sum_{j=1}^{K_s} \alpha_{sj} \leq 1$.

These ideas also extend to the case when there is more than one frequency band available. In this case the primary users coordinate to allocate the timeslots and frequency bands available via the primary user MAC protocol. The primary system uses spectral pooling, discussed in Section 1.8.1, to make unused time and frequency slots available to the secondary users. These resources are then shared either equally or unequally amongst secondary users via the secondary user MAC protocol. As discussed above, the capacity region is similar to that of a multiple user interference channel with the constraint that available resources (i.e., time and frequency slots) are first allocated to the primary users to support their traffic, then the remaining resources are allocated orthogonally among the secondary users. More sophisticated encoding and decoding strategies for medium access can be used such that the secondary and primary users simultaneously transmit over the same channel, with rate-splitting, superposition encoding, error-correction coding, and joint detection techniques used to ensure reliable data detection from all users. The capacity in this case is treated based on the more general capacity analysis of the unconstrained interference channel discussed previously in Section 2.4. These ideas can also be extended to the case where the secondary transmitters are part of a broadcast or multiple access channel within the cognitive network. Then, as before, available resources are first allocated to the primary users to support their traffic, e.g., over a time fraction α, then the remaining resources are allocated among the secondary users. If a secondary user is transmitting to multiple receivers over a time fraction $1-\alpha$, then its capacity over that time fraction will be the corresponding broadcast capacity region instead of a single rate. Similarly, if there are multiple secondary users transmitting to one receiver over a time fraction $1-\alpha$, then capacity over that time fraction will be the corresponding multiple access channel capacity region.

The capacity results discussed above all assume perfect detection of spectrum holes by the secondary users. There are two types of imperfect detection: missed detection and false alarm, which are defined formally in Section 4.3. A false alarm occurs when a primary user is detected but in fact none is present, i.e., a spectrum hole is undetected by the secondary user or users. For a single secondary user, assuming a stationary and ergodic system, a false alarm has no effect on the primary users (since no interference is generated). However, it reduces the secondary user's data rate by a fraction $(1-P_{\text{FA}})$ for P_{FA} the probability of false alarm since, under perfect detection, the missed spectrum holes would have been used for the secondary user's transmission. When there are multiple secondary users, the reduction in each of their data rates due to false alarm depends on how the spectrum holes are allocated. If the spectrum holes have the same length and are equally allocated among K_{s} secondary users, then each of the secondary user's data rates due to false alarm would be reduced by a fraction $(1-P_{\text{FA}})/K_{\text{s}}$.

In the case of missed detection, the secondary users fail to detect primary user activity when in fact these users are active, i.e., a spectrum hole is detected incorrectly. During missed detection, the channel becomes the interference channel of Fig. 2.7, since both primary and secondary users are transmitting simultaneously. However, in the interweave scenario, the encoding and decoding for both the primary and the secondary users do not take this interference into account since it occurs due to an unexpected detection error. Hence, interference from the secondary user to the primary user will degrade its capacity. The capacity reduction is typically derived by treating the interference as noise; capacity is computed based on the SINR associated with the interference generated by the secondary users at the primary users' receivers. Similarly, a secondary user transmitting during a spectrum hole obtained through missed detection experiences interference from the primary users also occupying that hole, and its capacity degradation can be determined in a similar manner by treating the interference from all primary users operating during that spectrum hole as noise. Alternatively, data transmitted when secondary and primary transmissions overlap can be declared as *erasures* at the secondary receiver, assuming the receiver detects the interference and hence does not attempt to decode the data. Erasures correspond to data that cannot be decoded except through channel codes that are designed to correct for erasures. An overview of erasure-codes for cognitive radio systems can be found in [54, 81]. Primary users can also treat data received while experiencing interference from secondary users as an erasure, but since the secondary user is typically at a much lower power than the primary user, this is rarely done in practice; instead the interference caused by secondary user transmissions is typically treated as noise. Since detection of spectrum holes is a dynamic and imperfect process, the capacity of both primary and secondary users under false alarm or missed detection is based on that of a randomly-varying channel, as discussed in the next subsection.

A model for the impact of false alarm and missed detection on the capacity of an interweave channel using cooperative detection between the secondary transmitter and receiver is developed in [45]. The distributed nature of this spectrum sensing is captured mathematically via a two-switch model, one switch at the secondary transmitter and one at the secondary receiver. When the secondary transmitter detects a spectrum hole then it moves its switch to the ON position, and similarly at the receiver. When both switches are ON, a hole is deemed to be present for utilization by the secondary user, and this information is disseminated to the secondary transmitter and receiver (perhaps via a separate cooperation channel). This model captures the fact that the farther apart the secondary transmitter and receiver, the less correlated the primary user activity will be that they each detect. Since the location of the primary receiver is unknown, detection of primary signals at more than one location via cooperative sensing will lead to more accurate estimation of spectrum holes, as discussed in more detail in Section 5.2.1. More general models for false alarm and missed detection can be mapped to this two-switch model via the switch probabilities of being ON or OFF. Inner and outer bounds on capacity of this two-switch model are developed in [45] based on information-theoretic results for the capacity of memoryless channels with causal and noncausal partial channel knowledge.

2.6.2 Random switch model for secondary channels

In the interweave paradigm, each secondary user's channel varies with time due to primary user activity and imperfect detection. Imperfect detection by the secondary system "breaks" the interweave paradigm as it introduces simultaneous primary and secondary transmissions during the times when the primary user transmissions are missed. Moreover, false detection of primary user activity reduces the capacity of the secondary system as transmission opportunities are missed. Specifically, the secondary user has an interference-free channel when a spectrum hole is correctly detected, a channel with interference when a spectrum hole is incorrectly detected, and a zero-capacity channel when primary user activity is (correctly or incorrectly) sensed. In this section we capture these time-varying channel characteristics caused by primary user activity and its imperfect detection through a random switch model. Note that in this model the underlying channels of both the primary and secondary users are assumed static, although the random switch model can be extended to incorporate fading.

Under the random switch model for secondary channels, at any given time primary users may be transmitting or not. When they transmit, their activity can be detected correctly as the lack of a spectrum hole, or not; when they do not transmit, their lack

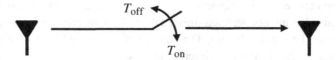

Figure 2.16 Channel model for interweave secondary user: random switch position indicates detected presence (OFF) or absence (ON) of a primary user.

of activity can be detected correctly as a spectrum hole, or not. We first consider the random switch model for a single secondary user under perfect detection of spectrum holes, as shown in Fig. 2.16. When primary user activity is detected, the switch is OFF and the channel is unavailable. When no primary activity is detected, the switch is ON, indicating that the channel is available for the secondary user. The channel is randomly varying since within any time interval $[0, T]$ the amount of time that the switch is ON or OFF, T_{off} and T_{on}, respectively, is random. Assume that when the switch is ON, the capacity of the secondary user channel is C_s, e.g., for an AWGN channel $C_s = W \log_2(1 + \mathcal{P}_s/(N_0 W))$ bps where W is the channel bandwidth associated with the spectrum hole, $\mathcal{P}_s$ is the received signal power at the secondary receiver, and $N_0/2$ is the PSD of the secondary receiver noise. Note that this channel is interference-free since detection is assumed to be perfect.

In general the capacity of the random-switch channel depends on what is known about the switch position at the transmitter and receiver. Under the interweave paradigm the switch position is assumed known at both the secondary transmitter and receiver: the secondary transmitter must know the switch position so that it does not transmit during a primary user's transmission, and the secondary receiver must know the switch position in order to decode the secondary user's transmission. Since the switch position is known and detection is perfect, at each time instant the channel is known at both the transmitter and receiver, and hence the appropriate capacity metric is ergodic or outage capacity. In particular, when the switch is ON, the secondary user can reliably transmit at any rate $R < C_s$. Over a time interval $[0, T]$, the fraction of time that the secondary user transmits is T_{on}/T. For stationary and ergodic primary user activity, in the limit as T approaches infinity, this equals the probability that the switch is ON, P_{on}. The ergodic capacity of this two-state channel is then $C_{\text{erg}} = P_{\text{on}} C_s$, since the secondary user obtains zero rate when the switch is OFF, which occurs with probability $P_{\text{off}} = 1 - P_{\text{on}}$. The outage capacity of this channel is also $P_{\text{on}} C_s$, since the secondary user reliably transmits at any rate $R < C_s$ when the switch is ON and is in outage when the switch is OFF. Note that, as in [45], the details of detection that lead to a switch being ON or OFF are not relevant for capacity analysis; only the probability P_{on} is needed, which can be determined based on the primary user activity characteristics and the given detection strategy.

The same switch model also applies when detection is imperfect, but the capacity metric for both the secondary and primary users in this case is no longer ergodic or outage capacity, since the channel is no longer known at the transmitter. Consider the simplest case of one primary user and one secondary user where both have fixed transmit powers and static channels to their respective receivers. When the primary user's transmissions are not detected (missed detection), its channel is degraded due to interference from the secondary user. The channel of the secondary user during these transmissions is also degraded due to interference from the primary user. Let C_p equal the capacity of the primary user's channel when the secondary user is inactive, and let C_p^I equal the capacity of the primary user's channel when the secondary user is active. For AWGN channels of bandwidth W, noise PSD $N_0/2$, and interference treated as noise, these correspond to $C_p = W \log_2(1 + \mathcal{P}_p/(N_0 W))$ and $C_p^I = W \log_2(1 + \mathcal{P}_p/(N_0 W + \mathcal{I}_s))$, respectively, for $\mathcal{P}_p$ the received power at the primary user's receiver and $\mathcal{I}_s$ the interference power caused by the secondary transmitter to the primary user's receiver.

Let P_{MD} denote the probability of missed detection, i.e., the probability that the primary user's activity is not detected. We will first examine the impact of missed detection on the primary user's capacity *during the time it is transmitting*; the primary user has zero rate when it does not transmit. Assuming the primary user is not aware of the secondary user's transmission, the correct metric for its performance while it is transmitting is capacity versus outage. This is given by C_p^I for $P_{\text{out}} < P_{\text{MD}}$, since the primary user experiences secondary user interference with probability P_{MD}, and C_p for outage probability $P_{\text{out}} \geq P_{\text{MD}}$. Note that the primary user in general does not know the secondary user's interference, and thus it does not know C_p^I. Hence, it must choose a *fixed* transmission rate $R \leq C_p$ without this knowledge. If an outage probability of $P_{\text{out}} = P_{\text{MD}}$ is acceptable, then the primary user can transmit at any rate $R < C_p$ and accept that a fraction P_{MD} of the data will be decoded in error. If that is an unacceptable outage probability then, assuming it can bound the maximum interference of the secondary user as $\mathcal{I}_{s_{\max}}$, it can reliably transmit at any rate $R < W \log_2(1 + \mathcal{P}_p/(N_0 W + \mathcal{I}_{s_{\max}}))$. As long as the bound is accurate, data sent at this rate will always be correctly decoded. Similarly, if it does know C_p^I then it can send at that rate with no outage. The primary user is active when the switch is OFF due to correct detection of primary user activity, or when the switch is ON due to missed detection of primary user activity. Recalling that P_{FA} is the probability of false detection and letting P_{MD} denote the probability of missed detection, we have that the primary user is active with probability $(1 - P_{\text{FA}})P_{\text{off}} + P_{\text{MD}}P_{\text{on}}$. Hence, the primary user's throughput averaged over all time, given a fixed transmission rate of R when it is active, is $((1 - P_{\text{FA}})P_{\text{off}} + P_{\text{MD}}P_{\text{on}})R$.

A primary user adapting to the behavior of a secondary user is atypical for an interweave channel, where primary users are meant to be oblivious to the presence of secondary transmissions. However, some systems have been proposed where primary users can react to secondary users. Suppose the primary user can measure the interference caused by the secondary user due to missed detection. Then it can adapt its transmission rate during the time that it transmits to $R < C_p$ in the absence of secondary interference (i.e., when the switch is OFF) and to $R = C_p^I$ in the presence of secondary interference (i.e., when the switch is ON due to missed detection). This yields an ergodic rate for the primary user during the time it transmits of

$$C_{p,\text{erg}} = P_{\text{MD}} C_p + (1 - P_{\text{MD}}) C_p^I. \qquad (2.63)$$

Note that for typical wireless channels this rate can be achieved even when the primary user's transmitter is not adapting its rate to the presence or absence of interference, as long as the primary user's receiver is aware of the interference from the secondary user [32, Section 4.2].

The effect of missed detection on the secondary user is similar to that of the primary user. In particular, the secondary user is unaware of primary user transmissions during missed detection, which happens with probability P_{MD}. Hence the two-state switch channel of the secondary user becomes a three-state channel: when the switch is OFF capacity is zero since the secondary user does not transmit; when the switch is ON the capacity of the channel equals the interference-free capacity when the spectrum hole is correctly detected, and it equals capacity with the interference from the primary user when the hole is falsely detected. Let C_s equal the capacity of the secondary user's channel during transmission when the primary user is inactive, and let C_s^I equal the capacity of the secondary user's channel during transmission when the primary user is active (missed detection). For AWGN channels with bandwidth W, noise power $N_0 W$ and interference treated as noise, these correspond to $C_s = W \log_2(1 + \mathcal{P}_s/(N_0 W))$ and $C_s^I = W \log_2(1 + \mathcal{P}_s/(N_0 W + \mathcal{I}_p))$, respectively, for $\mathcal{P}_s$ the received signal power at the secondary user's receiver, and $\mathcal{I}_p$ the interference caused by the primary transmitter to the secondary user's receiver. Since the channel is unknown, the correct metric for secondary user capacity when it is transmitting is capacity versus outage. Similarly to the case of the primary user under missed detection, this function is given by C_s^I for $P_{\text{out}} \leq P_{\text{MD}}$ and C_s for $P_{\text{out}} > P_{\text{MD}}$. Hence, if the outage probability P_{MD} is acceptable, then the secondary user can transmit at any rate $R < C_s$ and accept that a fraction P_{MD} of the data will be decoded in error. If that is an unacceptable outage probability then, assuming it can bound the maximum interference power of the primary user as $\mathcal{I}_{p_{\max}}$, it can transmit at any rate $R < W \log_2(1 + \mathcal{P}_s/(N_0 W + \mathcal{I}_{p_{\max}}))$ such that, as long as the bound is accurate, data sent at this rate will be correctly decoded. Similarly, if it does know C_s^I then it can send at that rate with no outage. Since the secondary user is active with probability P_{on}, the secondary user's throughput averaged over all time, given a fixed transmission rate of R when it is active, is $P_{\text{on}} R$.

In addition to incorrectly detecting a spectrum hole, the secondary user detects the presence of a primary user when none is present with probability P_{FA}. Hence, if the secondary user transmits at rate R when the switch is ON, the capacity of the secondary user is reduced by $P_{\text{FA}} R$ relative to the case of no false alarm due to the missed transmission opportunities. Note that the exact detection strategy is not needed for capacity calculations under missed detection and false alarms. Only the probability of missed detection and false alarm is used in these capacity calculations, and this can be determined for any given detection strategy, as discussed in Chapters 4 and 5.

2.6.3 Scaling laws for interweave networks

In Section 2.3.3 we discussed scaling laws for K-user wireless networks. We now discuss scaling laws in interweave networks with one or more primary and secondary users. Not much is known about scaling laws for this type of network, although scaling laws have been derived for a single-hop interweave network where multiple secondary users transmit in the presence of a single primary user in [92]. This work defined the notion of a *primary exclusive region*, or PER, around the primary transmitter. Secondary transmitters cannot operate within this PER, and the distance between a secondary user's transmitter and receiver must be less than some specified distance. These two constraints restrict the amount of interference a secondary transmitter can cause to the primary receiver, and hence its outage probability, thereby enabling a closed-form derivation of scaling laws. The capacity was found to scale *linearly* with the number of secondary users, in contrast to the classical sublinear scaling in Gupta and Kumar's result, where multihop routing was assumed. These interweave scaling law results were extended in [46] to a network with multiple primary as well as multiple secondary users, along with multihop routing. This work assumes that the locations of the primary user transmitters and receivers are known by the secondary network, and also that the primary network is less dense than the secondary network. The multihop routing protocols assume a *preservation region* around each primary node such that the secondary users route their traffic around these regions. This routing protocol is shown to achieve almost the same scaling law as if the primary network was absent, while the primary network throughput is subject to only a fractional loss that decreases asymptotically. Specifically, a K_p-user primary network achieves a throughput scaling of $\sqrt{K_p} = K_p^{0.5}$, while a K_s-user secondary network achieves a throughput scaling of $K_s^{0.5-\delta}$ for any $\delta > 0$ with an arbitrarily small probability of outage. Similar scaling laws under the same model, constraints, and assumptions were obtained in [100], however in this work the location of the primary receivers was unknown. Hence preservation regions were only formed around the primary transmitters. This paper also showed that the same throughput-delay tradeoff as was obtained in [22] for a single asymptotically large network can be achieved by both the primary and secondary networks as the number of each type of user grows asymptotically. These results assume that the relative density of the primary users relative to the secondary users remains the same as the number of users grows.

2.7 Overlay cognitive radio networks

The motivation for overlay cognitive radio networks is to exploit intelligent radio capabilities to their fullest: (1) from the information-theoretic point of view, the interweave constraint of orthogonal transmissions used by secondary and primary communicating pairs is unnecessarily restrictive and, as such, reduces capacity; (2) cognitive capabilities can be exploited more broadly than just for spectrum hole detection.

For these reasons, and in contrast to interweave networks, overlay cognitive radio networks allow for concurrent secondary and primary transmissions over the same dimensions. Furthermore, in contrast to both interweave and underlay networks, in overlay networks a secondary transmitter may improve the communication of a primary user. In this setting, relaying and techniques for coping with interference become key tools to maximize the performance of both primary and secondary users.

2.7 Overlay cognitive radio networks

Modeling cognition

In the analysis of overlay cognitive radio models, the cognitive capability is typically captured by assuming that the secondary user's encoder, called the *cognitive encoder*, knows the data sequence to be sent by the primary encoder in the next transmission block. This assumption is often too idealistic for practical systems. However, it is reasonable when the secondary and primary transmitters are close to each other, or the primary data sequence is being retransmitted after an initial failure and the secondary decoder was able to successfully decode it in the first transmission. This assumption is also applicable when the primary transmitter sends its data sequence in advance to a secondary transmitter, which might be done in a separate frequency band. Although idealistic, once the encoding and decoding strategies under this assumption are fully understood, this assumption can be relaxed to assume: (1) that primary data sequences are conveyed to the secondary user over orthogonal links of finite capacities; (2) that only partial data sequence knowledge is available, and this partial knowledge is used for partial decoding, or (3) the primary data sequence is learned causally, to account for the time needed for decoding. These relaxations were investigated in [9, 58, 60].

2.7.1 Cognitive encoder for the two-user overlay channel

The simplest overlay cognitive network consists of one secondary and one primary pair, as shown in Fig. 2.17. The cognitive encoder corresponding to the secondary user is assumed to have full knowledge of data sequence W_p communicated by the primary encoder to the primary decoder. The encoding strategies and channel gains are assumed known to all users in the channel. Overlay encoding techniques have mostly been investigated for this two-user channel. Without cognition, this network reduces to the two-user interference channel of Fig. 2.11 with the secondary user corresponding to user 1 and the primary user corresponding to user 2. Due to the similarities between these two models, in the information theory literature, this network is referred to as the *interference channel with one cognitive encoder*. It is also referred to as the cognitive radio channel, the interference channel with asymmetric message knowledge, the interference channel with degraded message sets, or the cognitive interference channel. Another special case of this channel model is obtained when the primary user does not transmit.

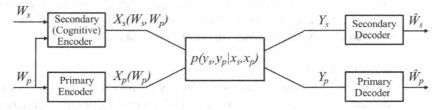

Figure 2.17 Two-user overlay channel.

Since the secondary transmitter knows the messages intended for both receivers, the channel reduces to the broadcast channel [16]. The two-user overlay channel thus contains elements of both interference and broadcast channels. The encoding strategies that have been developed for these canonical channels, or their combinations, are capacity-achieving for the overlay channel in certain conditions. We next review the encoding strategies that have been proposed for the two-user overlay channel, and discuss scenarios in which these schemes or their combinations achieve capacity.

Formally, the two-user overlay channel, with one secondary (cognitive) user and one primary user, consists of two input alphabets $\mathcal{X}_s, \mathcal{X}_p$, two output alphabets $\mathcal{Y}_s, \mathcal{Y}_p$, and a conditional probability distribution $p(y_s, y_p | x_s, x_p)$, where $(x_s, x_p) \in \mathcal{X}_s \times \mathcal{X}_p$ are channel inputs and $(y_s, y_p) \in \mathcal{Y}_s \times \mathcal{Y}_p$ are channel outputs. The secondary source wishes to send a message $W_s \in \mathcal{W}_s = \{1, \ldots, 2^{nR_s}\}$ to its destination, and the primary source wishes to send a message $W_p \in \mathcal{W}_p = \{1, \ldots, 2^{nR_p}\}$ to its destination. Data sequence W_p is also known at the secondary (cognitive) encoder. An (R_s, R_p, n) code consists of two data sequence sets $\mathcal{W}_s, \mathcal{W}_p$, encoding functions

$$X_s^n = f_s(W_s, W_p), \qquad (2.64)$$
$$X_p^n = f_p(W_p), \qquad (2.65)$$

and two decoding functions

$$\hat{W}_s = g_s(Y_s^n), \hat{W}_p = g_p(Y_p^n). \qquad (2.66)$$

The error probability of the code is

$$P_e = P\left[\left[\hat{W}_s \neq W_s\right] \bigcup \left[\hat{W}_p \neq W_p\right]\right]. \qquad (2.67)$$

A rate pair (R_s, R_p) is achievable if, for any $\epsilon > 0$, there exists, for sufficiently large n, an (R_s, R_p, n) code such that $P_e \leq \epsilon$. The capacity region of the two-user overlay channel is the closure of the set of all achievable rate pairs (R_s, R_p).

We observe that the only difference between this definition and the corresponding definitions for the interference channel is in the encoding function of the secondary cognitive encoder (2.64). Unlike in the interference channel, in this case the encoder knows both data sequences W_s and W_p and can thus form encoded sequences that depend on both of them. We will also consider the AWGN interference channel given by (2.33), augmented with cognitive encoding by the secondary user.

The additional information allows the cognitive encoder to deploy cooperation to increase the rate of the primary pair and precoding against interference to increase its own rate. Before summarizing encoding strategies used by the cognitive encoder, we give a brief overview of the Gelfand–Pinsker encoding technique. This technique, also referred to as *binning*, is used for the precoding against interference.

Gelfand–Pinsker coding

For overlay networks, transmission by any primary transmitter causes interference at the secondary receiver. Since the cognitive encoder knows this interference, it deploys an encoding scheme that mitigates the effect of the interference at its receiver, thereby

2.7 Overlay cognitive radio networks

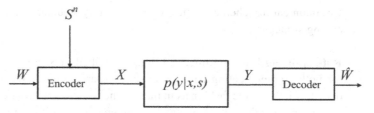

Figure 2.18 A channel with a random state. The state S^n is known noncausally at the encoder.

increasing its rate. In the overlay channel, since the cognitive encoder has perfect knowledge of the data sequence and the encoding strategy of the primary user, the secondary transmitter–receiver pair can view this situation as communication in a channel with a random state noncausally known at the encoder, also known as the Gelfand–Pinsker problem [30], shown in Fig. 2.18. In this figure, a single source communicates a data sequence $W \in \{1, \ldots, 2^{nR}\}$ to the destination over a channel given by $p(y|x, s)$. The channel state sequence S^n is assumed to be a random iid sequence generated from the distribution $p(s)$ over the random state. S^n is noncausally known at the encoder.

We next consider an AWGN channel where the random state is an additional interference term S. Specifically, an AWGN channel with a random interference state has input–output relationship given by

$$Y = X + S + Z, \qquad (2.68)$$

where Z is zero-mean Gaussian noise at the receiver with power $\mathcal{N}$ and S is zero-mean Gaussian noise at the receiver with power $\mathcal{N}_s$. The transmit power constraint is $\mathbb{E}[X^2] \leq \mathcal{P}$. The Gelfand–Pinsker encoding in this channel model reduces to DPC, described above for the interference channel, whereby precoding is used to cancel the effect of the inteference term S. In this setting DPC achieves the capacity of the AWGN channel as if there were no interference, i.e., it completely cancels the effect of interference S, yielding capacity

$$C = W \log_2 \left(1 + \frac{\mathcal{P}}{\mathcal{N}}\right). \qquad (2.69)$$

The capacity of the general discrete-time memoryless channel with random state is given by

$$C = \max_{p(u|s), f(\cdot)} I(U; Y) - I(U; S), \qquad (2.70)$$

where $U - (X, S) - Y$ form a Markov chain, and is achieved by Gelfand–Pinsker encoding [30]. The random variable U is used to generate a codebook at the encoder such that, in each block transmission, the codeword depends both on the data sequence and on the random state S^n. Gelfand–Pinsker encoding can play a crucial role in overlay cognitive radio, as will be described next.

Communication schemes deployed in the overlay channel contain the following three encoding strategies.

1. **Rate-splitting** Rate-splitting, as briefly described in Section 2.4, can improve rates for both communicating pairs by enabling (partial) interference cancellation at the decoders in the same fashion as in interference channels without cognition. This technique can potentially be deployed by both encoders, as it requires no cognition about other user's data sequences as the encoder is only splitting its own rate. For the rate-splitting to be exploited by the primary user, this approach requires that the decoding at the primary receiver be modified to decode and cancel part of the secondary user's transmission. This approach differs from the approach used in interweave and underlay cognitive radio where the communication technique of the primary users is not adapted in the presence of secondary users. If this constraint were to be imposed in overlay cognitive networks as well, the rate-splitting would be no longer applicable since it requires that the primary receiver decodes a part of the message of the secondary user.

2. **Cooperation** Full, partial, or delayed knowledge of the primary user's data sequence allows the cognitive encoder to relay this information to the primary user, thereby increasing the rate of the primary pair. The cognitive encoder uses a fraction of its power to transmit message W_p to the primary user and thus increases the primary rate R_p. When the primary user's data sequence (or a part of it) is known to the cognitive encoder without the delay, i.e., at the beginning of the transmission block, relaying is performed via superposition coding, where secondary and primary messages are superimposed and sent by the cognitive encoder. Due to the noncausal knowledge, there is no need for Markov block encoding otherwise used for relaying [17].

3. **Gelfand–Pinsker encoding (binning)** Any signal carrying all or part of W_p is interference at the secondary receiver. Hence, both the primary encoder and the cognitive encoder cause interference at the secondary receiver. However, if the cognitive encoder knows W_p at the beginning of the transmission block (as well as the codebooks and channel state information), it can infer the interference created at the secondary receiver. Knowing the interference, it can apply Gelfand–Pinsker encoding and, specifically in AWGN channels, DPC in order to precancel this interference. Another motivation for deploying Gelfand–Pinsker encoding is found in the relationship between the overlay channel and the broadcast channel whereby when the primary transmitter is silent, the overlay channel reduces to a two-user broadcast channel. For the discrete-time memoryless broadcast channel, the encoding scheme that yields the largest known achievable rate region is based on Gelfand–Pinsker encoding [61]. In the Gaussian case the two-user overlay channel is related also to the MIMO Gaussian broadcast channel. Specifically, a MIMO broadcast channel with two transmit antennas is equivalent to a two-user overlay channel when *both* encoders are cognitive, i.e., both encoders know the message intended for each receiver, since in fact these encoders are part of the same broadcasting transmission strategy. For MIMO Gaussian broadcast channels, DPC is the optimal encoding strategy [91, 94]. Motivated by these observations, binning and DPC have been applied to the two-user

overlay channel, resulting in the highest known achievable rates, including capacity, in some cases, as will be presented later in this section.

2.7.2 Capacity results

Determining the capacity region for the general overlay channel remains an open problem. However, as for the interference channel without cognition, the capacity has been determined for some regimes. The capacity in strong interference has been determined in [60, Thm. 5]. The capacity for the Gaussian channel in weak interference has been derived in [49, 95]. For a more general case that unifies the results under these two regimes, the capacity was determined in [70]. For a class of discrete memoryless cognitive Z-channels (i.e., channels where there is no interference at the primary receiver) in which only the receiver of the secondary pair suffers interference the capacity has been determined in [56]. Capacity of a class of Gaussian cognitive Z-channels for the opposite case, when the interference is at the primary receiver, has been found in [47]. In each case, some combination of rate-splitting, cooperation, and/or binning achieves capacity. These regimes and their capacity-achieving encoding techniques are as follows.

1 **Strong interference** [60]

As in the interference channel, this is the regime in which both decoders can decode each other's messages without rate penalty. In AWGN overlay channels, described by (2.33) with $X_1 = X_s$, $X_2 = X_p$, $h_{12} = h_{sp}$, $h_{21} = h_{ps}$ and the primary user's message known at the secondary user's transmitter, the following interference conditions are sufficient for strong interference to be satisfied:

$$|h_{ps}| \geq 1,$$
$$|h_{sp}| \geq \frac{h_{ps}}{\alpha} + \frac{|\alpha - 1|}{\alpha} \quad \text{if } h_{sp}h_{ps} > 0,$$
$$|h_{sp}| \geq \frac{h_{ps}}{\alpha} + \frac{\alpha + 1}{\alpha} \quad \text{if } h_{sp}h_{ps} < 0, \qquad (2.71)$$

where $\alpha = \sqrt{\mathcal{P}_s/\mathcal{P}_p}$, for $\mathcal{P}_s$ the received secondary transmit power at the secondary receiver and $\mathcal{P}_p$ the received primary transmit power at the primary receiver. For $\mathcal{P}_s = \mathcal{P}_p$ and $h_{sp}, h_{ps} \geq 0$ these conditions simplify to

$$h_{ps} \geq 1, h_{sp} \geq h_{ps}. \qquad (2.72)$$

In a general discrete-time memoryless channel, these conditions can be expressed in terms of the conditional mutual information as

$$I(X_s; Y_s | X_p) \leq I(X_s; Y_p | X_p),$$
$$I(X_s, X_p; Y_s) \leq I(X_s, X_p; Y_p), \qquad (2.73)$$

and need to be satisfied for all input distributions $p(x_s, x_p)$. We observe that the first inequality is the same condition as in the interference channel (2.35). In this regime, cooperation via superposition coding achieves capacity. The capacity region in an AWGN channel is

$$\mathcal{C} = \bigcup_{|\rho| \leq 1} \Big\{ (R_\text{s}, R_\text{p}) : R_\text{s} \geq 0, R_\text{p} \geq 0,$$

$$R_\text{s} \leq \frac{1}{2} \log \left(1 + (1 - \rho^2)\mathcal{P}_\text{s} \right), \tag{2.74}$$

$$R_\text{s} + R_\text{p} \leq \frac{1}{2} \log \left(1 + \mathcal{P}_\text{s} + \mathcal{P}_\text{p} + h_\text{ps}^2 \mathcal{P}_\text{s} + 2\rho h_\text{ps} \sqrt{\mathcal{P}_\text{s} \mathcal{P}_\text{p}} \right) \Big\}, \tag{2.75}$$

where ρ is the correlation between inputs X_s and X_p. Note that a larger ρ results in a larger coherent combining gain at the primary receiver of the secondary and primary transmissions of the primary user's data. The general capacity region is

$$\mathcal{C} = \bigcup \Big\{ (R_\text{s}, R_\text{p}) : R_\text{s} \geq 0, R_\text{p} \geq 0,$$

$$R_\text{s} \leq I(X_\text{s}; Y_\text{s}|X_\text{p}), \tag{2.76}$$

$$R_\text{s} + R_\text{p} \leq I(X_\text{s}, X_\text{p}; Y_\text{p}) \Big\}, \tag{2.77}$$

where the union is over all input distributions $p(x_\text{s}, x_\text{p})$.

2 **Weak interference at the primary receiver** [49, 95]

In the AWGN overlay channel (2.33), weak interference at the primary receiver corresponds to $|h_\text{ps}| \leq 1$, i.e., the cross-channel gain from the secondary to the primary user is smaller than the direct link gain. Because the interference is weak, the primary receiver does not attempt to decode the unwanted data sequence and instead treats it as noise. Interference at the secondary receiver can be eliminated by DPC at the cognitive encoder, allowing for the interference-free, single-user rate to be achieved. The optimum coding strategy at the cognitive encoder consists of encoding message W_s via DPC while treating the primary user's input sequence X_p^n as interference, and superposition coding to help convey W_p to the primary receiver. Thus, there is no need for rate-splitting; DPC and cooperation via superposition coding achieve capacity. The capacity region is

$$R_\text{s} \leq \frac{1}{2} \log(1 + (1 - \alpha)\mathcal{P}_\text{s}), \tag{2.78}$$

$$R_\text{p} \leq \frac{1}{2} \log \left(1 + \frac{(h_\text{ps}\sqrt{\alpha \mathcal{P}_\text{s}} + \sqrt{\mathcal{P}_\text{p}})^2}{1 + h_\text{ps}^2 (1 - \alpha)\mathcal{P}_\text{s}} \right), \tag{2.79}$$

where α is the fraction of power that the cognitive encoder uses for cooperation. The rest of the power, $(1 - \alpha)\mathcal{P}_\text{s}$, the cognitive encoder uses to transmit the signal carrying its own data. That signal is the interference at the primary receiver.

In the case of the discrete-time memoryless channel, the weak interference conditions are given by:

$$I(U; Y_\text{p}|X_\text{p}) \leq I(U; Y_\text{s}|X_\text{p}),$$
$$I(X_\text{p}; Y_\text{p}) \leq I(X_\text{p}; Y_\text{s}) \tag{2.80}$$

for all distributions $p(u, x_\text{s}, x_\text{p})$. The random variable U has the same role as in the Gelfand–Pinsker encoding, i.e., it is used for precanceling the interference via binning. The capacity region is

2.7 Overlay cognitive radio networks

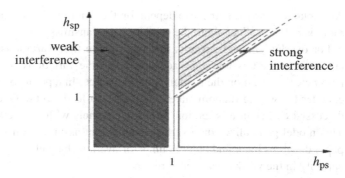

Figure 2.19 Values of (h_{sp}, h_{ps}) for which the channel is in weak or strong interference and capacity is known.

$$\mathcal{C} = \bigcup \Big\{ (R_s, R_p) : \; R_s \geq 0, R_p \geq 0,$$
$$R_s \leq I(X_s; Y_s | U, X_p), \tag{2.81}$$
$$R_p \leq I(U, X_p; Y_p) \Big\}, \tag{2.82}$$

where the union is over all distributions $p(u, x_s, x_p)$.

Regimes of strong and weak interference for Gaussian channels with $\mathcal{P}_s = \mathcal{P}_p$ for $h_{sp}, h_{ps} \geq 0$ are shown in Fig. 2.19.

3 **Better cognitive decoding regime** [70]

This is the regime for which the condition

$$I(U, X_p; Y_p) \leq I(U, X_p; Y_s) \tag{2.83}$$

holds for any $p(u, x_s, x_p)$.

The capacity region is given by

$$R_s \leq I(U, X_s; Y_s | X_p),$$
$$R_p \leq I(U, X_p; Y_p),$$
$$R_s + R_p \leq I(U, X_p; Y_p) + I(X_s; Y_s | X_p, U),$$
$$R_s + R_p \leq I(U, X_s, X_p; Y_p). \tag{2.84}$$

Again, the random variable U has the same role as in the Gelfand–Pinsker encoding, i.e., it is used for binning. In addition to binning and superposition coding, the encoding scheme requires rate-splitting at the cognitive encoder.

4 **Common information**

If the cognitive decoder wishes to decode both data sequences, there is no interference at that decoder and hence there is no need for binning. Then, rate-splitting and superposition coding achieve capacity [48, 55].

The capacity results presented above indicate that in order to maximize performance of cognitive radio networks, its users should be allowed to share the bandwidth and cope with the introduced interference via (partial) decoding, relaying or precoding. Gains from the various techniques depend on channel conditions and the network topology.

As explained earlier, gains also depend on the amount of information that the cognitive radio is able to collect about the primary user transmissions.

For overlay channels, results that incorporate time variations in the channel model are scarce. One possible approach to model time variations is to assume that channel parameters depend on the random state sequence. In a point-to-point channel when the encoder knows the random state, this approach leads to the Gelfand–Pinsker problem discussed earlier in this section. Overlay channels with a state were analyzed in [82]. This model generalizes the overlay channel that assumes constant channel gains. The paper develops inner and outer bounds for overlay channels with a state, and establishes capacity in the weak interference regime.

Another important aspect of the overlay channel capacity is the impact of imperfect channel state information (CSI) at the cognitive encoder. The lack of channel knowledge will affect all cognitive encoding schemes. With phase uncertainty at the cognitive encoder, DPC cannot be applied directly. Hence this uncertainty precludes canceling interference and results in significant performance loss [37]. In this case, feedback about channel state information to the cognitive encoders could be very beneficial. Relaying strategies can be realized without CSI at the transmitter, but would require CSI at the receiver [51]. However, CSI at the transmitter could allow for additional relaying approaches (e.g., beamforming) and thus further improve the network performance.

2.7.3 K-user overlay networks

Analysis of the two-user overlay channel focuses on a single secondary pair and a single primary pair. Yet, as in the interweave networks, multiple secondary and multiple primary users in the overlay network can simultaneously share the same spectrum. As with the two-user overlay case, in this more general setting we also seek efficient ways to minimize interference among users.

We have seen in Section 2.4 that for the K-user interference channel, generalizing the rate-splitting approach as a way to cope with interference from multiple senders is too complex. Moreover, this approach can be outperformed by collective treatment of interference via interference alignment or structured codes. Not surprisingly, interference alignment as well as structured codes are promising approaches for the K-user overlay cognitive radio network as well. These techniques could be deployed by cognitive nodes to reduce the interference at both the primary and the cognitive receivers. But in contrast to the encoders in interference channels, secondary users can also perform relaying of primary messages, and precoding against interference. The interplay between these techniques is not yet well understood and presents an important and interesting research topic.

Another interesting problem is the impact of cognition when present at different points in the network. In particular, not only encoders but also (or only) decoders can be cognitive. This will significantly impact the available encoding and decoding techniques. No cognition at the encoder will preclude both cooperation and precoding techniques. On the other hand, it will enable decoders to directly cancel interference. The impact of different points of cognition in the case of a single primary and secondary pair has been investigated in [44].

Once the capacity results and achievable rates, along with their encoding and decoding strategies, are obtained, indicating the most promising encoding approaches, protocols for coexistence among many secondary users will need to be developed. As a final remark we note that in principle the presence of cognitive radios in overlay systems can improve not only spectrum utilization and rates of both primary and secondary users, but also the robustness, the delay, and the energy consumption of the primary system. The presence of cognitive transmitters that have the ability to learn primary users' information enables this information to be sent over multiple paths in the network, thereby adding diversity to the primary system, and deploying more energy and delay efficient paths for information flow in the network. However, the implementation challenges associated with overlay systems may preclude many of these potential benefits being realized in practice.

2.8 Summary

Determining the fundamental performance limits of general cognitive radio networks is a daunting challenge. Fundamental performance bounds are available for conventional wireless networks only in a few special cases. Adding a second tier of cognitive users makes the capacity even more difficult to obtain. The most common metric for performance of wireless systems is Shannon capacity. Although there are few capacity results for general wireless networks, this chapter has described how existing Shannon-theoretic tools can be used directly or extended to obtain some results or bounds on the fundamental capacity limits of cognitive networks.

The building block for all known cognitive networks is the two-user interference channel, where one primary transmit–receive pair and one secondary transmit–receive pair share a common channel. This simple four-node network models all three cognitive radio paradigms – underlay, interweave, and overlay – with one secondary user and one primary user. Specifically, in the underlay paradigm the cognitive transmitter is restricted such that its interference to the primary receiver is below some threshold, in the interweave paradigm the cognitive transmitter may only transmit when it detects a spectrum hole, and in the overlay paradigm both users transmit simultaneously, with the secondary user improving the performance of the primary user while obtaining some spectral resources for its own communication. Based on this four-node model, we have shown that the capacity region of an underlay network can be determined based on Shannon capacity analysis with the interference from the primary users to the secondary users, and the interference from the secondary users to the primary users, treated as AWGN. Hence, capacity of an underlay single-user channel, with fading and multiple antennas, as well as that of a multiple access and broadcast underlay channel, is the same as in the absence of primary users, except that the transmit power of the secondary user(s) is constrained relative to the interference caused to the primary users. In the case of multiple secondary users, a MAC protocol may also be used when the sum of secondary interference exceeds the primary users' threshold, so that some of the secondary users stop transmitting to reduce the total interference below the required level.

Treating interference as noise is also used in the capacity analysis of interweave networks. However, unlike in underlay networks, interweave capacity analysis requires that the dynamics associated with detection of primary user activity, as well as the probability of missed detection and false alarm, be incorporated into the capacity analysis. To capture these dynamics, the secondary user's interweave channel in the four-node case is modeled as a switch. When the switch is ON, the secondary user has detected a spectrum hole (lack of primary user activity) and transmits. When the switch is OFF, the secondary transmitter is silent, and the primary user experiences no interference. If a spectrum hole is correctly detected, then the secondary user experiences no interference; when the spectrum hole is incorrectly detected, both the secondary and primary users transmit simultaneously and interfere with each other's transmissions. Given the time variations of the switch, under perfect detection the secondary user's channel can be modeled as a two-state channel corresponding to the switch being ON or OFF. Using this model, the ergodic and outage capacities are derived. Under imperfect detection both the primary and secondary user channels have three states. For the secondary user's channel, the switch can be OFF or ON; when the switch is OFF there is no transmission, when the switch is ON, the spectrum hole can be correctly (resulting in no interference) or incorrectly (primary user interference) detected. Since the secondary user is not aware of false detections, the correct capacity metric is capacity versus outage, which can be computed based on the probability of missed detection and the SINR that results due to primary user interference. Similarly, the primary user experiences interference from the secondary user due to missed detection, and its capacity versus outage can be computed based on this missed detection probability and the SINR resulting from secondary user interference. These capacity calculations can be generalized to multiple primary and secondary users by generalizing the models for missed detection and false alarm along with the resulting interference.

Overlay networks do not treat interference as noise. Instead, novel encoding and decoding strategies are used by the secondary users to enhance the performance of the primary users while obtaining some spectral resources for their own transmissions. These encoding techniques include rate-splitting, whereby one user splits its data sequence and encodes it via superposition coding such that a part of this sequence can be decoded and removed by the receiver of the other user. In the overlay network, this would allow the secondary transmitter's interference to be removed by the primary receiver. However, this interference removal requires a modified decoder for the primary user, so it is not applicable when the primary system is oblivious to the secondary system in terms of its design or operation. In addition to rate-splitting, the secondary encoder uses its knowledge of the primary user's encoded data sequence to enhance the transmission of the sequence to the primary receiver via a cooperative protocol. Finally, knowledge of the primary user's data sequence allows the secondary transmitter to employ Gelfand–Pinsker encoding, or binning, to precancel the interference that the primary user causes to the cognitive receiver. Similarly, if the primary user knows the

2.8 Summary

secondary user's sequence, it can use binning in a similar fashion. These information-theoretic strategies have not yet made their way into commercial systems, where underlay and interweave currently dominate. Moreover, extensions of these ideas to networks of more than a few secondary and primary users remain open problems. However, as this chapter has shown, the overlay paradigm holds significant promise for increasing the capacity of both secondary and primary users above that of the other cognitive radio paradigms. If these information-theoretic results can be translated to practice, then perhaps at some point in the future overlay systems may dominate some segments of the cognitive radio landscape.

There are many open problems in determining fundamental performance limits of cognitive networks. In the context of underlay networks, capacity when nodes have multiple antennas that can direct secondary transmissions away from primary users has been investigated in [93]. Capacity versus outage based on incorrect information about the interference caused between primary and secondary users has not yet been explored. For interweave networks, there is much active research to determine optimal medium access protocols that achieve capacity. In addition, the scaling laws to date assume single-hop routing and an exclusion region around the primary users – perhaps these scaling laws can be improved upon by considering relaying as well as hierarchical cooperation, which have proved fruitful in developing improved scaling laws for noncognitive networks. Capacity results for overlay networks to date mainly assume one secondary and one primary user. Since the capacity regions for conventional ad hoc networks with more than a few users have proved elusive, we expect the same will be true with overlay networks, unless the intelligence associated with cognition proves to simplify the problem and facilitate a solution. There are few fundamental performance results for metrics other than capacity such as energy, delay, and complexity, nor have asymptotic scaling laws been developed for any of these networks. In particular, while capacity strategies for overlay networks with more than a few nodes may be difficult to obtain, perhaps scaling laws that exploit previous work on scaling in the absence of primary users can be extended to take the primary network into account. Performance bounds with respect to energy require models for energy consumption that includes both transmit and circuit (analog and processor) energy, which have proved difficult to obtain. This complicates determining tradeoffs between performance (which improves via sophisticated signal processing and encoding/decoding) and energy consumption. Indeed, the development of fundamental performance metrics for cognitive networks is a rich and active area of research. Results in this area will prove essential to obtain insights and performance bounds for these emerging systems.

The next chapter develops the models associated with signal propagation in cognitive networks. These models are critical to determine the nature of interference at primary receivers and of the desired received signals at all receivers. In particular, different radio bands will have very different interference characteristics since propagation is highly dependent on frequency. Moreover, the time-variations of these channels dictate how well the information required under the different cognitive user paradigms can be obtained, and how fast such systems need to adapt.

2.9 Further reading

The four-node cognitive radio channel is a special case of the four-node interference channel, whose capacity was first investigated in the late 1970s by Carleial and by Sato [11, 73]. The model where one encoder in the interference channel is a cognitive encoder was first proposed and its capacity region analyzed in [19, 59]. Since then there has been tremendous activity determining achievable rate regions and capacity outer bounds for this channel, as well as special cases such as strong, very strong, and weak interference regimes where these bounds meet. Tutorial papers [20, 35] describe the large body of work on the capacity of interference channels with cognitive encoders and the resulting capacity regions and encoding/decoding strategies for these special regimes. Scaling laws for interweave networks were first investigated in [92] using the idea of a primary exclusion region. The first work to analyze capacity of the interweave channel based on a switch model for primary user detection was [45].

References

[1] V. S. Annapureddy and V. Veeravalli, "Gaussian interference networks: sum capacity in the low interference regime and new outer bounds on the capacity region," *IEEE Trans. Inf. Theory*, **55**, no. 7, 3032–3050, July 2009.

[2] M. R. Aref, *Information Flow in Relay Networks*, Ph.D. Dissertation, Stanford University, Stanford, CA, 1980.

[3] C. Bae and W. E. Stark, "End-to-end energy/bandwidth tradeoff in multihop wireless networks," *IEEE Trans. Inf. Theory*, **55**, no. 9, 4051–4066, Sep. 2009.

[4] E. Biglieri, J. Proakis, and S. Shamai (Shitz), "Fading channels: information theoretic and communication aspects," *IEEE Trans. Inf. Theory*, **44**, no. 6, 2619–2692, Oct. 1998.

[5] H. Bölcskei, R. U. Nabar, O. Oyman, and A. J. Paulraj, "Capacity scaling laws in MIMO relay networks," *IEEE Trans. Wireless Commun.*, **5**, no. 6, 1433–1444, June 2006.

[6] R. Brodersen, A. Wolisz, D. Cabric, S. M. Mishra, and D. Willkomm, "CORVUS: A cognitive radio approach for usage of virtual unlicensed spectrum," *White Paper: Berkeley Wireless Research Center*. bwrc.eecs.berkeley.edu/research/mcma/CR_White_paper_final1.pdf.

[7] V. R. Cadambe and S. A. Jafar, "Interference alignment and degrees of freedom of the K-user interference channel," *IEEE Trans. Inf. Theory*, **54**, no. 8, 3425–3441, Aug. 2008.

[8] V. R. Cadambe and S. A. Jafar, "Parallel Gaussian interference channels are not always separable," *IEEE Trans. Inf. Theory*, **55**, no. 9, 3983–3990, Sep. 2009.

[9] Y. Cao and B. Chen, "Interference channel with one cognitive transmitter," *Proc. of the Asilomar Conference on Signals, Systems, and Computers*, Pacific Grove, CA, Oct. 26–29, 2008.

[10] A. B. Carleial, "A case where interference does not reduce capacity," *IEEE Trans. Inf. Theory*, **21**, no. 5, 569–570, Sep. 1975.

[11] A. B. Carleial, "Interference channels," *IEEE Trans. Inf. Theory*, **24**, no. 1, 60–70, Jan. 1978.

[12] T. C. Clancy, "Achievable capacity under the interference temperature model," *Proc. of the IEEE International Conference on Computer Communications (INFOCOM)*, Anchorage, AK, May 6–12, 2007, pp. 794–802.

[13] C. B. Cormio and K. R. A. Chodhury, "A survey on MAC protocols for cognitive radio networks," *Elsevier J. Ad Hoc Netw.*, **7**, no. 7, 1315–1329, July 2009.

[14] M. H. M. Costa, "Writing on dirty paper," *IEEE Trans. Inf. Theory*, **29**, no. 3, 439–441, May 1983.
[15] M. H. M. Costa and A. El Gamal, "The capacity region of the discrete memoryless interference channel with strong interference," *IEEE Trans. Inf. Theory*, **33**, no. 5, 710–711, Sep. 1987.
[16] T. Cover, "Broadcast channels," *IEEE Trans. Inf. Theory*, **18**, no. 1, 2–14, Jan. 1972.
[17] T. Cover and A. El Gamal, "Capacity theorems for the relay channel," *IEEE Trans. Inf. Theory*, **25**, no. 5, 572–584, Sep. 1979.
[18] T. Cover and J. Thomas, *Elements of Information Theory* (2nd edn.). J. Wiley & Sons, 2006.
[19] N. Devroye, P. Mitran, and V. Tarokh, "Achievable rates in cognitive radio channels," *IEEE Trans. Inf. Theory*, **52**, no. 5, 1813–1827, May 2006.
[20] N. Devroye, M. Vu, and V. Tarokh, "Cognitive radio networks," *IEEE Signal Processing Mag.*, **25**, no. 6, 12–23, Nov. 2008.
[21] M. Effros, A. Goldsmith, and Y. Liang, "Generalizing capacity: new definitions and capacity theorems for composite channels," *IEEE Trans. Inf. Theory*, **56**, no. 7, July 2010.
[22] A. El Gamal, J. Mammen, B. Prabhakar, and D. Shah, "Optimal throughput-delay scaling in wireless networks – part I: the fluid model," *IEEE Trans. Inf. Theory*, **52**, no. 6, 2568–2592, June 2006.
[23] A. El Gamal, "On information flow in relay networks," *Proc. of the IEEE National Telecommunications Conference*, **2**, Nov. 1981, pp. D4.1.1–D4.1.4.
[24] A. El Gamal and Y.-H. Kim, *Network Information Theory*, Cambridge University Press, 2012.
[25] H. El Gamal, G. Caire, and M. O. Damen, "The MIMO ARQ channel: diversity-multiplexing-delay tradeoff," *IEEE Trans. Inf. Theory*, **52**, no. 8, 3601–3621, Aug. 2006.
[26] R. Etkin, D. N. C. Tse, and H. Wang, "Gaussian interference channel capacity to within one bit," *IEEE Trans. Inf. Theory.*, **54**, no. 12, 5534–5562, May 2008.
[27] Federal Communications Commission Spectrum Policy Task Force, Report of the Spectrum Efficiency Working Group, *Technical Report 02-135*, (Nov. 2002). www.fcc.gov/sptf/files/SEWGFinalReport_1.pdf.
[28] R. G. Gallager, *Information Theory and Reliable Communication*, J. Wiley & Sons, 1968.
[29] M. Gastpar, "On capacity under receive and spatial spectrum-sharing constraints," *IEEE Trans. Inf. Theory*, **53**, no. 2, 471–487, Feb 2007.
[30] S. I. Gelfand and M. S. Pinsker, "Coding for channel with random parameters," *Probl. Peredachi Informatsii*, **9**, no. 1, 19–31, Jan. 1980.
[31] A. Ghasemi and E. Sousa, "Fundamental limits of spectrum-sharing in fading environments," *IEEE Trans. Wireless Commun.*, **6**, no. 2, 649–658, Feb 2007.
[32] A. J. Goldsmith, *Wireless Communications*, Cambridge University Press, 2005.
[33] A. J. Goldsmith, S. A. Jafar, N. Jindal, and S. Vishwanath, "Capacity limits of MIMO channels," *IEEE J. Sel. Areas Commun.*, **21**, no. 5, 684–702, June 2003.
[34] A. Goldsmith, M. Effros, R. Koetter, M. Médard, A. Ozdaglar, and L. Zheng, "Beyond Shannon: the quest for fundamental performance limits of wireless ad hoc networks," *IEEE Commun. Mag.*, **49**, no. 5, 195–205, May 2011.
[35] A. Goldsmith, S. Jafar, I. Marić, and S. Srinivasa, "Breaking spectrum gridlock with cognitive radios: an information theoretic perspective," *Proc. IEEE*, **97**, no. 5, 894–914, May 2009.
[36] M. Grossglauser and D. N. C. Tse, "Mobility increases the capacity of ad hoc wireless networks," *IEEE/ACM Trans. Netw.*, **10**, no. 4, 477–486, Aug. 2002.
[37] P. Grover and A. Sahai, "What is needed to exploit knowledge of primary transmissions?," *Proc. of the IEEE International Symposium on New Frontiers in Dynamic Spectrum Access Networks (DySPAN)*, Dublin, Ireland, Apr. 17–20, 2007, pp. 462–471.

[38] P. Gupta and P. R. Kumar, "The capacity of wireless networks," *IEEE Trans. Inf. Theory*, **42**, no. 2, 388–404, March 2000.

[39] P. Gupta and P. R. Kumar, "Towards an information theory of large networks: an achievable rate region," *IEEE Trans. Inf. Theory*, **49**, no. 8, 1877–1894, Aug. 2003.

[40] T. Han and K. Kobayashi, "A new achievable rate region for the interference channel," *IEEE Trans. Inf. Theory*, **27**, no. 1, 49–60, Jan. 1981.

[41] S. Haykin, "Cognitive radio: brain-empowered wireless communications," *IEEE J. Sel. Areas Commun.*, **23**, no. 2, 201–220, February 2005.

[42] W. Hirt and J. L. Massey, "Capacity of the discrete-time Gaussian channel with intersymbol interference," *IEEE Trans. Inf. Theory*, **34**, no. 3, 380–388, May 1998.

[43] J. Huang, R. A. Berry and M. L. Honig, "Spectrum sharing with distributed interference compensation," *Proc. of the IEEE International Symposium on New Frontiers in Dynamic Spectrum Access Networks (DySPAN)*, Baltimore, MD, Nov. 8–11, 2005, pp. 88–93.

[44] S. A. Jafar and S. Shamai (Shitz), "Degrees of freedom region for the MIMO X channel," *IEEE Trans. Inf. Theory*, **54**, no. 1, 151–170, Jan. 2008.

[45] S. A. Jafar and S. Srinivasa, "Capacity limits of cognitive radio with distributed and dynamic spectral activity," *IEEE J. Sel. Areas Commun.*, **25**, no. 3, 529–537, Apr. 2007.

[46] S.-W. Jeon, N. Devroye, M. Vu, S.-Y. Chung, and V. Tarokh, "Cognitive networks achieve throughput scaling of a homogeneous network," *IEEE Trans. Inf. Theory*, **57**, no. 8, 5103–5115, Aug. 2011.

[47] J. Jiang, I. Marić, A. Goldsmith, S. Shamai (Shitz), and S. Cui, "On the capacity of a class of cognitive Z-interference channels," *Proc. of the IEEE International Conference on Communications (ICC)*, Kyoto, Japan, June 5–9, 2011.

[48] J. Jiang, Y. Xin, and H. Garg, "Interference channels with common information," *IEEE Trans. Inf. Theory*, **54**, no. 1, 171–187, Jan. 2008.

[49] A. Jovičić and P. Viswanath, "Cognitive radio: an information-theoretic perspective," *IEEE Trans. Inf. Theory*, **55**, no. 9, 3945–3958, Sep. 2009.

[50] R. Koetter, M. Effros, and M. Médard, "On a theory of network equivalence," *Proc. of the IEEE Information Theory Workshop (ITW)*, Volos, Greece, June 10–12 2009. arxiv.org/abs/1007.1033

[51] G. Kramer, I. Marić, and R. D. Yates, "Cooperative communications," *NOW J. Found. Trends Netw.*, **1**, no. 3-4, 271–425, 2006.

[52] G. Kramer and S. Savari, "Capacity bounds for relay networks," *Proc. of the UCSD Workshop on Information Theory and Applications*, La Jolla, CA, February 6–10, 2006.

[53] G. Kramer and S. Savari, "Edge-cut bounds on network coding rates," *J. Netw. Syst. Management*, **14**, no. 1, 49–67, March 2006.

[54] H. Kushwaha, R. Chandramouli, and K. P. Subbalakshmi, "Cognitive networks: towards self-aware networks," in Q. H. Mahmoud (ed.), *Cognitive Networks – Towards Self-Aware Networks*. J. Wiley & Sons, 2007.

[55] Y. Liang, A. Somekh-Baruch, V. Poor, S. Shamai (Shitz), and S. Verdú, "Cognitive interference channels with and without secrecy," *IEEE Trans. Inf. Theory*, **55**, no. 2, 604–619, February 2009.

[56] N. Liu, I. Marić, A. Goldsmith, and S. Shamai (Shitz), "Bounds and capacity results for the cognitive Z-interference channel," *Proc. of the IEEE International Symposium on Information Theory*, Seoul, Korea, June 28-July 3, 2009.

[57] M. A. Maddah Ali, S. A. Motahari, and A. K. Khandani, "Communication over MIMO X channels: interference alignment, decomposition, and performance analysis," *IEEE Trans. Inf. Theory*, **54**, no. 8, 3457–3470, Aug. 2008.

[58] I. Marić, A. Goldsmith, G. Kramer, and S. Shamai (Shitz), "On the capacity of interference channels with a partially-cognitive transmitter," *Proc. of the IEEE International Symposium on Information Theory*, Nice, France, June 24–29, 2007.

[59] I. Marić, R. D. Yates, and G. Kramer, "Strong interference channel with unidirectional cooperation," *Proc. of the UCSD Workshop on Information Theory and Applications*, La Jolla, CA, February 6–10, 2006.

[60] I. Marić, R. D. Yates, and G. Kramer, "Capacity of interference channels with partial transmitter cooperation," *IEEE Trans. Inf. Theory*, **53**, no. 10, 3536–3548, Oct. 2007.

[61] K. Marton, "A coding theorem for the discrete memoryless broadcast channel," *IEEE Trans. Inf. Theory*, **25**, no. 3, 306–311, May 1979.

[62] J. Mitola III, *Cognitive Radio: Model-Based Competence for Software Radio*, Ph.D. Dissertation, KTH Royal Institute of Technology, Stockholm, Sweden, Aug. 1999.

[63] A. S. Motahari and A. K. Khandani, "Capacity bounds for the Gaussian interference channel," *Proc. of the IEEE International Symposium on Information Theory*, Toronto, Canada, July 6–11, 2008.

[64] L. Musavian and S. Aissa, "Capacity and power allocation for spectrum-sharing communications in fading channels," *IEEE Trans. Wireless Commun.*, **8**, no. 1, 148–156, Jan. 2009.

[65] B. Nazer and M. Gastpar, "Computation over multiple-access channels," *IEEE Trans. Inf. Theory*, **53**, no. 10, Oct. 2007.

[66] B. Nazer, M. Gastpar, S. Jafar, and S. Vishwanath, "Ergodic interference alignment," *Proc. IEEE International Symposium Information Theory*, Seoul, S. Korea, June 28–July 3, 2009, pp. 1769–1773.

[67] A. Ozgur and O. Leveque, "Throughput–delay tradeoff for hierarchical cooperation in ad hoc wireless networks," *IEEE Trans. Inf. Theory*, **56**, no. 3, 1369–1377, March 2010.

[68] A. Ozgur, O. Leveque, and D. N. C. Tse. "Hierarchical cooperation achieves optimal capacity scaling in ad hoc networks," *IEEE Trans. Inf. Theory*, **53**, no. 10, 3549–3572, Oct. 2007.

[69] A. Ozgur and D. N. C. Tse, "Achieving linear scaling with interference alignment," *Proc. of the IEEE International Symposium on Information Theory*, Seoul, Korea, June 28–July 3, 2009, pp. 1754–1758.

[70] S. Rini, D. Tuninetti, and N. Devroye, "New inner and outer bounds for the memoryless cognitive interference channel and some capacity results," *IEEE Trans. Inf. Theory*, **57**, no. 7, 4087–4109, July 2011.

[71] V. Rodoplu and T. H. Meng, "Bits-per-joule capacity of energy-limited wireless networks," *IEEE Trans. Wireless Commun.*, **6**, no. 3, 857–865, March 2007.

[72] J. Sachs, I. Marić, and A. J. Goldsmith, "Cognitive cellular systems within the TV spectrum," *Proc. of the IEEE International Symposium on New Frontiers in Dynamic Spectrum Access Networks (DySPAN)*, Singapore, Apr. 6–9, 2010, pp. 1–12.

[73] H. Sato, "Two user communication channels," *IEEE Trans. Inf. Theory*, **23**, no. 3, 295–304, May 1977.

[74] H. Sato, "The capacity of the Gaussian interference channel under strong interference," *IEEE Trans. Inf. Theory*, **27**, no. 6, 786–788, Nov. 1981.

[75] X. Shang, G. Kramer, and B. Chen, "A new outer bound and the noisy-interference sumrate capacity for Gaussian interference channels," *IEEE Trans. Inf. Theory*, **55**, no. 2, 689–699, February 2009.

[76] C. E. Shannon, "A mathematical theory of communication," *Bell Sys. Tech. Journal*, 379–423, 623–656, 1948.

[77] C. E. Shannon, "Communications in the presence of noise," *Proc. IRE*, **37**, 10–21, 1949.

[78] C. E. Shannon, "Two-way communication channels," *Proc. Berkeley Symposium on Math, Statistics, and Probability*, **1**, 1961, 611–644.

[79] C. E. Shannon and W. Weaver, *The Mathematical Theory of Communication*, University of Illinois Press, 1949.

[80] Shared Spectrum Company, "Comprehensive spectrum occupancy measurements over six different locations," Aug. 2005. www.sharedspectrum.com/?section=nsf_summary.

[81] F. Shayegh and M. R. Soleymani, "Rateless codes for cognitive radio in a virtual unlicensed spectrum," *Proc. IEEE Sarnoff Symposium,*, Princeton, NJ, May 2–4, 2011.

[82] A. Somekh-Baruch, S. Shamai (Shitz), and S. Verdú, "Cognitive interference channels with state information," *Proc. IEEE International Symposium on Information Theory*, Toronto, Canada, July 6–11, 2008.

[83] S. Sridharan, A. Jafarian, S. Vishwanath, S. A. Jafar, and S. Shamai (Shitz), "A layered lattice coding scheme for a class of three user Gaussan interference channels," *Proc. Allerton Conference on Communication, Control and Computing*, Monticello, IL, Sep. 24–26, 2008, pp. 531–538.

[84] E. Telatar, "Capacity of multi-antenna Gaussian channels," *European Trans. Telecommun.*, **10**, no. 6, 585–596, Nov. 1999.

[85] S. Toumpis and A. J. Goldsmith, "Capacity regions for wireless ad hoc networks," *IEEE Trans. Wireless Commun.*, **2**, no. 4, 736–748, July 2003.

[86] D. N. C. Tse and P. Viswanath, *Fundamentals of Wireless Communication*, Cambridge University Press, 2005.

[87] Ultra-wideband wireless communications – theory and applications, *IEEE J. Sel. Areas Commun.*, **24**, no. 4, Apr. 2006.

[88] J. Unnikrishnan and V. V. Veeravalli, "Algorithms for dynamic spectrum access with learning for cognitive radio," *IEEE Trans. Signal Process.*, **58**, no. 2, 750–760, Feb. 2010.

[89] S. Verdú, "On channel capacity per unit cost," *IEEE Trans. Inf. Theory*, **36**, no. 5, 1019–1030, Sep. 1990.

[90] S. Verdú and T. S. Han, "A general formula for channel capacity," *IEEE Trans. Inf. Theory*, **40**, no. 4, 1147–1157, July 1994.

[91] S. Vishwanath, N. Jindal, and A. Goldsmith, "Duality, achievable rates, and sum-rate capacity of Gaussian MIMO broadcast channels," *IEEE Trans. Inf. Theory*, **49**, no. 10, 2658–2668, Oct. 2003.

[92] M. Vu, N. Devroye, M. Sharif, and V. Tarokh, "Scaling laws of cognitive networks," *Proc. of the IEEE International Conference on Cognitive Radio Oriented Wireless Networks and Communications (CrownCom)*, Orlando, FL, July 31–Aug 3, 2007, pp. 2–8.

[93] H. Wang, J. Lee, S. Kim, and D. Hong, "Capacity enhancement of secondary links through spatial diversity in spectrum sharing," *IEEE Trans. Wireleless Commun.*, **9**, no. 2, 494–499, Feb 2010.

[94] H. Weingarten, Y. Steinberg, and S. Shamai, "The capacity region of the Gaussian multiple-input multiple-output broadcast channel," *IEEE Trans. Inf. Theory*, **52**, no. 9, 3936–3964, Sep. 2006.

[95] W. Wu, S. Vishwanath, and A. Arapostathis, "Capacity of a class of cognitive radio channels: interference channels with degraded message sets," *IEEE Trans. Inf. Theory*, **53**, no. 11, 4391–4399, Nov. 2007.

[96] Z. Wu and B. Natarajan, "Interference tolerant agile cognitive radio: maximize channel capacity of cognitive radio," *Proc. IEEE Consumer Communications and Networking Conference (CCNC)*, Las Vegas, NV, Jan. 11–13, 2007, pp. 1027–1031.

[97] Y. Xie and A. J. Goldsmith, "Diversity-multiplexing-delay tradeoffs in MIMO multihop networks with ARQ," *Proc. IEEE International Symposium on Information Theory*, Austin, TX, June 13–18, 2010, pp. 2208–2212.

[98] Y. Xie, D. Gündüz, and A. J. Goldsmith, "Multihop MIMO relay networks with ARQ," *Proc. IEEE Global Telecommunications Conference (GLOBECOM)*, Honolulu, HI, Nov. 30–Dec. 4, 2009, pp. 1–6.

[99] F. Xue and P. R. Kumar, "Scaling laws for ad-hoc wireless networks: an information theoretic approach," *NOW J. Found. Trends Netw.*, **1**, no. 2, 145–270, 2006.

[100] C. Yin, L. Gao, and S. Cui, "Scaling laws for overlaid wireless networks: a cognitive radio network versus a primary network," *IEEE/ACM Trans. Netw.*, **18**, no. 4, 1317–1329, Aug. 2010.

[101] R. Zhang and Y.-C. Liang, "Exploiting multi-antennas for opportunistic spectrum sharing in cognitive radio networks," *IEEE J. Sel. Topics Signal Process.*, **2**, no. 1, 88–102, Feb. 2008.

[102] Q. Zhao, L. Tong and A. Swami, "Decentralized cognitive MAC for opportunistic spectrum access in ad hoc networks: A POMDP framework," *IEEE J. Sel. Areas Commun.*, **25**, no. 3, 224–232, Apr. 2007.

[103] L. Zheng and D. N. C. Tse, "Diversity and multiplexing: a fundamental trade-off in multiple antenna channels," *IEEE Trans. Inf. Theory*, **49**, no. 5, 1073–1096, May 2003.

3 Propagation issues for cognitive radio

Larry J. Greenstein, Andreas F. Molisch, and Mansoor Shafi

3.1 Introduction

The fundamental principal of cognitive radio (CR) is to detect other radios in the environment that are using the same spectral resources, and to then deploy transmission and reception strategies that permit secondary users to communicate, while minimizing interference to and from those radios. For the design, analysis, and implementation of such transmission and reception strategies, it is essential to understand the relevant propagation channels [80]. The power emitted by a transmitter (TX) might be determined by the system designer, but it is the channel that dictates how much of it arrives as useful power at the intended receiver (RX), and also how much interference it creates at unintended receivers. Similarly, the signal sensing process of the CR system might be determined by the system designer, but it is the channel that dictates the intervals in time, frequency, and space at which samples should be taken. For example, the temporal variation rate of the channel response dictates how often samples are needed and, thus, how often transmit or receive strategies might have to be adapted. As we will see, many other properties of the channel influence CR design and analysis as well.

3.1.1 Propagation in the cognitive radio bands

CRs may be deployed over a wide range of the frequency spectrum. The bands below about 3.5 GHz have relatively low propagation loss and are sought after by all services. These bands are therefore ideal candidates for the deployment of CR. They have different incumbent (primary) systems, each with its own mix of service type, architecture, bandwidth, and resilience to interference. A partial description of these bands is in Chapter 1.

In CR bands, as in most other bands, transmitted signals are affected by the propagation in a number of ways. They undergo attenuation, the local spatial average of which – expressed in decibels – is called the path loss. This is a large-scale property of the channel, in that it changes mildly over distances of a great many wavelengths. There is also a small-scale variation of signal level, for a given value of path loss, which is due to the presence of multiple signal paths; this is called multipath fading and is characterized by rapid fluctuations in signal level over distances of a radio wavelength. The multiple paths arise from scattering, reflection, and diffraction related to objects and structures in the physical environment and can lead to serious signal distortion. The properties of wireless channels can vary with carrier frequency and signal bandwidth. For example, large-scale properties like the path loss tend to increase linearly with the logarithm of the carrier frequency; by contrast, small-scale properties like the signal distortion are relatively insensitive to carrier frequency but their effect can depend a lot on the service bandwidth. All of this pertains to both primary and secondary (i.e., cognitive) radios, and so both radio types require good channel models to enable effective designs.

Here, we provide an overview of the propagation models that might be used by CR system designers and researchers. These models can help to determine, via analysis or simulation, potential CR interference to primary systems; and they can dictate the design of transmission formats, routing algorithms, and control schemes for mitigating interferences from CRs. Moreover, because the spectrum acquired for CR use may span a wide and fragmented range, propagation models are needed for a wide variety of frequencies and bandwidths. We stress that the physical propagation channel is independent of whether the radio is cognitive or not. However, the issues of what aspects of the channel are important, and thus have to be modeled especially carefully, are different. For example, a special feature of CR is the strong potential of interference to primary users, which can occur under all CR paradigms (interweave, underlay, and overlay, Chapters 1 and 2).

3.1.2 Impact of propagation on sensing

The energy-sensing process is influenced in a myriad of ways by the propagation channel. For example:

- The signal level at a sensor (part of the secondary system) is determined by the path loss of the link between this sensor and the primary transmitter, as well as by the large-scale and small-scale fading of this link. The signal level, in turn, influences the tradeoff between the false-alarm probability (spectrum is declared occupied even though it is free) and the missed-detection probability (spectrum is declared empty even though it is occupied by a primary user), as discussed in Chapter 4.
- When wideband sensing is used, the signal levels are different at different frequencies, due to the frequency-selectivity of small-scale fading and, possibly, the frequency dependence of the path loss as well. Knowledge of these effects is essential for building appropriate wideband sensors.
- When distributed sensing is used (Chapter 5), knowledge of the correlation of the channel characteristics from the primary transmitter to the sensors is required. Such knowledge includes the path loss laws and the correlation statistics of the shadowing.
- When antenna arrays are used for sensing, the angular spread determines the correlation of the signals at the different antenna elements, and thus helps determine the optimum detection process.
- Further improvements of the detection process can be achieved when sensing data are averaged over time. In order to perform proper averaging, the secondary system needs to know both the transmission statistics of the primary system and the temporal variability of the channel.

As noted, a sensor usually can only detect the presence of a primary transmitter, not of a receiver. It is nontrivial to draw conclusions from these observations about whether transmission from secondary devices create excessive interference to a primary receiver. Imagine as a simple scenario a primary system using frequency-domain duplexing. Then, the two ends of a primary link transmit in separate bands. It is thus essential to understand the correlation of the signal levels in the transmit and receive bands of the primary system. This is related to the frequency selectivity of the small-scale fading and the frequency dependence of the path loss. Propagation channels also have a key influence on the "hidden node problem" [90], where a secondary transceiver is outside the "listening range" for a primary transmitter but close enough to the primary receiver to create interference. The likelihood for hidden nodes to occur increases with increasing path loss and shadowing variance.

3.1.3 Impact of propagation on transmission

After sensing, the next step for the secondary system is to find a good transmission strategy. In the case of a spectrum-sensing radio (i.e., involving just changing the operating frequency), this entails decisions about which secondary transmitter can transmit in which frequency band, and for how long. In other words, the "free" spectrum has to be distributed among the different secondary devices that want to transmit information. Optimum assignment strategies can often be obtained from game-theoretic considerations (Chapter 1). As an input to such computations, knowledge of the propagation channels is essential. The actual performance, as well as the relative effectiveness of different schemes, depends on the observed signal levels. For an overlay approach, the secondary system needs accurate knowledge of the propagation channels within and between the primary and secondary systems at the time of actual transmission. In order to develop proper transmission strategies, it is vital that the secondary transmitter be able to make channel predictions (based on the past and present observations) and that it understands the achievable accuracy of those predictions. This, in turn, requires good channel models.

The necessity of updates in the optimum transmission scheme is intimately tied to the temporal properties of the channel, including the Doppler spectrum for mobile users and temporal variations of the Rice factor for fixed users. This also has an impact on whether cooperative games or noncooperative games can be used to develop a transmission strategy. Cooperative games require exchange of information between the different secondary users, but if the channel changes rapidly (large Doppler spread) and/or too strongly (small Rice factor), the overhead for the required information exchange becomes large.

The channel characteristics also have an impact on the design of transmission and reception strategies for CRs. For example, in CR systems using orthogonal frequency division multiplexing (OFDM), the optimal subcarrier spacing depends on the Doppler spectrum and also the power delay profile. Thus, a CR design using OFDM should choose a subcarrier spacing that is well-matched to the expected channel properties.

Similarly, the CR design should permit the modulation alphabet size to be adapted to the signal level, which depends on path loss and small-scale fading.

Finally, we stress the importance of propagation models in analyzing interference in CR systems and in designing ways to minimize or avoid it. In overlay and underlay, for example, a CR transmitter is operating co-channel with the primary receiver(s); in interweave, it may be operating in adjacent channels. Therefore, both co-channel and adjacent-channel interferences are important, and their impact depends on numerous factors: transmission spectrum, including the out-of-band energy; receive filter characteristics, including the out-of-band response; duplex plan and parameters (e.g., the spacing of bands in frequency-division duplexing); and such propagation aspects as path loss (to estimate total power at victim receivers) and the temporal, spatial, and spectral variations of small-scale fading (to relate sensing measurements at one time–place–frequency to conditions at another). As a general rule, we can say that every operating parameter and strategy in a CR design depends in some way on the propagation channel.

3.1.4 Outline of the chapter

We will strive, throughout this chapter, to point out the CR-specific impact of the propagation issues we describe, whether the impact is on overlay, underlay, or interweave scenarios. At the same time, one cannot appreciate the CR-specific aspects of propagation without understanding the fundamentals of radio channels *per se*. The reader will therefore note that we discuss propagation phenomena and channel models in a general way, which will hopefully be an added benefit to those not already familiar with this subject. Section 3.2 introduces the generic channel response, showing it to be a product of a large-scale term (related to path loss), which incorporates the effects of distance and shadowing; and a small-scale term, which incorporates the effects of multipath. The next three sections (3.3 to 3.5) deal, in order, with the ideal free-space path loss, the path loss in wireless systems, and the path loss in tower-based systems. The next four sections (3.6 to 3.9) deal with the small-scale phenomena caused by multipath. These include, in order, the spatial variability of field strength, temporal variability (Doppler spectra) for mobile and fixed terminals, delay dispersion in wideband channels, and angle dispersion. Finally, we introduce the topic of polarization, which can have an impact on CR designs using multiple antennas (Section 3.10); discuss some new, less thoroughly modeled propagation environments of importance to CR (Section 3.11); and summarize key propagation parameters of the radio channels we have described (Section 3.12). We end with a summary of the chapter (Section 3.13) and some suggestions for further reading (Section 3.14).

3.2 Generic channel response

It is not our purpose here to discuss the physics of radio propagation in CR-relevant bands, but to describe the impact of propagation on the behavior of the medium as seen by a given CR system. This leads us to the notion of the channel response on a particular path in a particular frequency band. The response of any linear time-invariant medium can be characterized by the *impulse response*, $h(\tau)$, or the *frequency response*, $H(f)$, the two functions being related by the Fourier transform. We will focus on the impulse response and how it is affected by the interaction of the propagation medium, system geometry, and surrounding environment.

A distinguishing feature of any particular radio path to be analyzed is the joint positions of the transmitter (TX) and receiver (RX). We denote the combination of the TX and RX location coordinates by $\mathbf{x}$, which uniquely specifies the path (or *link*) geometry in space. We further note that the impulse response of interest is just the response within the radio bandwidth of interest, e.g., between $f_c - W/2$ and $f_c + W/2$, where f_c is the center frequency of the signal band and W is the signal bandwidth or, more generally, the width of any frequency interval for which one chooses to measure or model the channel response. Strict confinement to this f-interval, which we denote by $\mathbf{F}$, is generally enforced by the TX and RX filters, so the $h(\tau)$ we seek to characterize is just the response within $\mathbf{F}$. We therefore invoke the notation $h(\tau; \mathbf{x}, \mathbf{F})$ to signify the spatial and spectral specificity of the impulse response we are discussing. The impulse and frequency responses are, in all cases, low-pass functions referenced to the center frequency f_c.

We now introduce the *power gain*, G, of the path, which is the average of $|H(f)|^2$ over both the signal bandwidth (i.e., over the spectral domain $\mathbf{F}$) and the local spatial domain of whichever end(s) of the path are potentially mobile. Thus, we can write

$$G = \mathbb{E}_\mathbf{x}\left[\frac{1}{W}\int_\mathbf{F} |H(f)|^2 df\right] = \mathbb{E}_\mathbf{x}\left[\frac{1}{W}\int_\tau |h(\tau, \mathbf{x}, \mathbf{F})|^2 d\tau\right], \qquad (3.1)$$

the second equality being a consequence of Parseval's theorem.[1] We can therefore write the impulse response as

$$h(\tau, \mathbf{x}, \mathbf{F}) = [G]^{\frac{1}{2}} \hat{h}(\tau, \mathbf{x}, \mathbf{F}), \qquad (3.2)$$

where $\hat{h}(\tau; \mathbf{x}, \mathbf{F})$ is normalized such that its squared magnitude, averaged over $\mathbf{x}$, integrated over τ and normalized by W, is 1. For concreteness the spatial average can be thought of as taken over a circle centered on the mobile's nominal location, with a radius of about 10 wavelengths.

The relevance of this representation is that the two factors in $h(\tau, \mathbf{x}, \mathbf{F})$ above are important in different ways, and they change on different scales of distance and frequency. The power gain, G, is important in determining interference levels from and into CR networks, and the signal-to-noise ratios of links within the CR network. The normalized impulse response, $\hat{h}(\tau, \mathbf{x}, \mathbf{F})$, is important in determining CR signal distortion and in devising techniques for combating it. Moreover, G varies in a *large-scale* way, i.e., it changes over location separations of tens of meters or more and frequency separations of hundreds of MHz or more. By contrast, $\hat{h}(\tau, \mathbf{x}, \mathbf{F})$ varies in a *small-scale* way, i.e., it can change significantly over distances as small as a fraction of the radio

---

[1] The spatial averaging denoted by $\mathbb{E}_\mathbf{x}$ is over a local area about the mobile's nominal position. It is equivalent to averaging over small changes in the mobile position coordinates, which are part of $\mathbf{x}$.

wavelength (e.g., 30 cm at 1 GHz) and over frequencies well within the bandwidth of a typical communications signal (20 MHz or smaller).

Finally, we note how G relates to the very important radio quantity called path loss, PL. Specifically, path loss is the negative decibel value of G,[2]

$$PL = -10\log(G). \quad (3.3)$$

We shall treat the large-scale path loss and the small-scale impulse response separately, dropping the adjective "normalized" in the latter case for convenience.

3.3 Introduction to path loss

3.3.1 Free-space path loss

If a radio wave at a given frequency f is radiated from a transmitter, and there are no sources of reflection or scattering in its path, it spreads outward with no loss of total power. Therefore, its power density, say in watts/m^2, decreases as the inverse square of distance, d. A receiver that detects the wave will therefore see a power proportional to d^{-2}. Assuming no antenna gain at either end of the TX-RX path, the relationship between receive power $\mathcal{P}_\mathrm{r}$ and transmit power $\mathcal{P}_\mathrm{t}$ is given by

$$\mathcal{P}_\mathrm{r} = \mathcal{P}_\mathrm{t} \left(\frac{c}{4\pi f d}\right)^2, \quad (3.4)$$

where c is the speed of radio wave propagation in the medium and c, f, and d are expressed in consistent units. This is the famous Friis formula for free-space propagation [34]. Noting that the power gain is $\mathcal{P}_\mathrm{r}/\mathcal{P}_\mathrm{t}$, we see that G for the free-space case is just the squared term in (3.4). Or, from (3.3), we can say that the free-space path loss is

$$FSPL = 20\log \frac{4\pi d}{\lambda}, \quad (3.5)$$

where $\lambda = c/f$ is the wavelength. For example, on a path of length 1 km at $f = 1$ GHz, $FSPL = 92.4$ dB.

The physical world being what it is, there are few radio services in the terrestrial domain where (3.5) might apply. Though the formula for free-space path loss may thus seem irrelevant to CR, we introduce it nonetheless, not only for its historic importance, but because it provides a baseline for other path loss formulas we will discuss.

3.3.2 Path loss in CR scenarios

There are potential CR applications where propagation between towers can be important. For example, one use of CR is as a backbone network for carrying Internet traffic, replacing fiber optic transport [36]. In such cases, the CR towers might be lower and embedded in urban areas, with antennas that are compact and thus of limited directivity.

---

[2] Logarithms in this chapter are $\log_{10}$ unless otherwise specified.

The careful engineering that permits the use of (3.5) for the path loss is not possible here. Hence, other path loss formulas are needed to quantify the desired signal levels received within a CR backbone network; the interference levels from primary transmitters (especially TV towers) to CR receivers; and those from CR transmitters to primary receivers.

More typically, CR systems are *wireless* networks, defined in this chapter as applications wherein one or both link ends are in the clutter of the environment; and neither antenna is more than 80 m above ground. For the outdoors, this means that transmission is significantly affected by reflections from the ground, mountains, buildings, trees, vehicles, etc. For indoor environments, the reflections are from walls, floors, ceilings, furniture, people, etc. In all these cases, the path loss is important to know, again, because of its impact on the received levels of both desired signals and interference.

3.4   Path loss models for wireless channels

3.4.1   General formulation

For the most part, radio paths in wireless systems do not meet free-space conditions. There is often no unobstructed LOS path, and the non-LOS paths are numerous, due to multiple obstructions, reflections, and scattering that make up the received signal (see Fig. 3.1).

Over the years, there have been extensive measurement campaigns leading to empirical models for wireless system path loss in different bands and environments. The range of the many reported studies spans outdoor mobile environments, including large cells (macrocells) and small cells (microcells); indoor environments with many walls and with large open spaces; and outdoor fixed wireless environments, all at frequencies ranging from ~300 MHz to ~6 GHz.

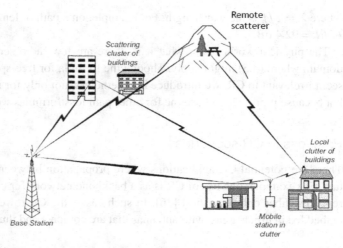

Figure 3.1 Multipath propagation in a wireless system (from [80]).

There is a generic formula for approximating the power gain, G, that stands up well for most measured bands and wireless environments. It is

$$G = \frac{\langle \mathcal{P}_r \rangle}{\mathcal{P}_t} \sim a \left(\frac{d}{d_0}\right)^{-\gamma}, \qquad (3.6)$$

where $\mathcal{P}_t$ and $\langle \mathcal{P}_r \rangle$ are the transmit and average received power, respectively, in consistent units, with the average of $\mathcal{P}_r$ being taken over the signal bandwidth and local area; d_0 is a reference distance, in the same units as d; and a and γ are dimensionless model parameters. Typical values for d_0 are 1 m in indoor environments and 100 m or 1 km in outdoor environments. It is easy to see that the power gain in free-space propagation, given by the squared quantity in (3.4), is a special case of (3.6), where $a = (\frac{4\pi d_0}{\lambda})^2$ and $\gamma = 2$. In most wireless environments, *the path loss exponent*, γ, is greater than 2 because, if there are obstructions along the path, power falls off more quickly with distance than predicted by spreading loss alone. However, in some indoor scenarios such as transmission down hallways, γ can be less than 2 because of waveguiding effects [62].

The above formula cannot be precise because, given the vagaries of man-made structures and terrain from one TX-RX path to another, no one function of distance d can represent G everywhere in a given environment. It is customary, therefore, to regard the bracketed quantity as the median power gain, G_{med}, for a given distance, d; and to express G as the product of $G_{\text{med}}(d)$ times a variable factor ζ that is different for every TX-RX path and has a median value over the environment of 1. Thus,

$$G = \left[a \left(\frac{d}{d_0}\right)^{-\gamma}\right] \zeta = G_{\text{med}}(d)\zeta. \qquad (3.7)$$

Since $PL = -10\log(G)$, we can write the generic path loss formula as

$$PL(d) = \left[A + 10\gamma \log\left(\frac{d}{d_0}\right)\right] + S, \qquad (3.8)$$

where A and S are the negative dB values of a and ζ, respectively. The bracketed term is the median path loss, $PL_{\text{med}}(d)$, over all TX-RX paths with distance d; and S represents a random-like variation about the median, different for every TX-RX path. It has been observed empirically that the linear variation of median path loss with log-distance, as implied by (3.8), is remarkably consistent over all bands and most environments; and that the variable quantity S can be accurately modeled in most cases as a Gaussian random variable with zero mean and a standard deviation, σ, whose value depends on the type of environment. Thus,

$$S = \mathcal{N}(0, \sigma^2) \qquad (3.9)$$

and wireless systems can be analyzed with PL having the form (3.8) and (3.9), with empirically derived values (or formulas) for the model parameters A, γ, and σ.

This mathematical description is illustrated in Fig. 3.2 for cellular measurements in the 2.6 GHz band for a number of suburban macrocellular environments. The circles in the upper figure represent a scatter plot of PL vs. d over about 1500 paths, and the straight line is a least-squares fit over all paths. The distances of the data points

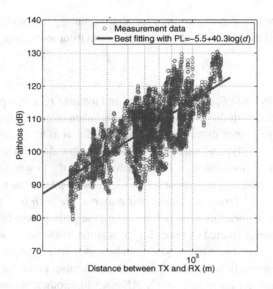

(a) Scatter plot of path loss vs. distance and a linear fit

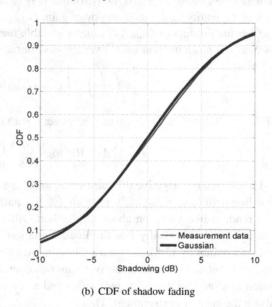

(b) CDF of shadow fading

Figure 3.2 Path loss and shadow fading for cellular signals in a suburban environment.

(circles) from the straight line were used to form a population of deviations, whose CDF is plotted in the lower figure. Also shown is a Gaussian CDF with the same mean (namely, 0) and standard deviation (≈ 5.8 dB). The agreement between the two CDFs is excellent.

The variable quantity S is called *shadow fading*, for historical reasons. In the early days of cellular radio, the fluctuation of path loss about a linear growth with log-distance was attributed to the shadowing effect of obstructions (e.g., buildings) between the base station and the mobiles. It has since been learned, however, that the fluctuations occur even on paths with a clear line-of-sight but with reflections along the path, e.g., in LOS microcells [41]. We will continue to refer to this random-like process by its popular name. We note that, since S is Gaussian, the factor ζ in (3.7) is log-normal; hence, the popular terminology in the literature for the variation about the median is *lognormal shadow fading*.

A popular convention among wireless engineers is to regard the bracketed term in (3.8) as the "path loss," to which shadow fading (S) is added, as if the former is a natural attribute of the distance and environment and the latter is a perturbation. In reality, the separation into these parts is merely an artifact of the measurement program (which locations are measured and how many) and the data processing used (least-squares fitting of a linear function of log-distance). We will use here the more precise convention that the bracketed term represents a *median* path loss for distance d in a given environment; and the path loss at distance d is the sum of both this median and a zero-mean, random-like component called shadow fading. The most widely used formulation for the median path loss, as discussed later, is in [53].

3.4.2 Shadow fading, S

The standard deviation (σ) of S can vary from 3 to 12 dB, depending on the environment. In a typical cellular environment, the most commonly used value is 8 dB; for indoor environments, it is $\sim$ 4 dB. This description summarizes the first-order statistics of shadow fading and is well supported by measurements. Less well documented are the second-order statistics of shadow fading, e.g., the autocorrelation function of its 2-dimensional distribution over space for a specified TX location. This information is not always needed in system studies, but one can envision CR scenarios where it would be useful. For example, in an environment where n CR nodes are receiving and one source (either primary or CR) is transmitting, simulating the path losses to the n CR nodes would require knowing the $n \times n$ correlation matrix among the n source-to-node shadow fading components. It is easy to appreciate the complexity of the measurement programs needed to derive such information, which explains the sparse literature on the subject.

Nevertheless, we can cite a few useful results pertaining to the spatial distribution of shadow fading. One is due to Gudmundson [47], who reported that, along a path moving towards or away from the TX in the x direction, the correlation between S-values for points at x and $x + \Delta x$ can be approximated by

$$\mathbb{E}_x\{S(x)S(x+\Delta x)\} = \sigma^2 \exp(-|\Delta x|/X_c) = \rho(\Delta x), \qquad (3.10)$$

where X_c is the "correlation distance" of shadow fading. Typical values for X_c range from 10 m (i.e., in urban microcells) to 500 m (i.e., in some rural macrocells). Roughly speaking, X_c is a measure of how far a mobile must travel before it encounters a different set of obstructing or scattering objects. In CR environments where sensor networks are used to obtain and update radio maps of the environment, the correlation distance is an important parameter for determining the necessary density of sensors. This correlation function for shadow fading along a path has been widely used, in part because it was among the first to be reported; in addition, it is simple to describe, and is readily simulated via an AR(1) process [54]. However, other empirically-derived functions have been proposed that work better in some environments. For example, see [116], which expresses $\rho(\Delta x)$ – when averaged over many routes – as a sum of *two* decaying exponentials in $|\Delta x|$; and expresses $\rho(\Delta x)$ for individual routes as an exponential combined with a sinusoid.

Besides the need to estimate shadow fading correlations along a single path, there can be a need to correlate the shadow fading values on two paths, as when one radio is communicating with two other radios. In such cases, the angle between the two paths becomes important in addition to distances, see [2, 42, 112]. A simple approach often used in cellular system simulations is to assume that the shadow fading, S_1 and S_2, from any two bases to the same mobile are correlated Gaussian variables, with the correlation coefficient being ~ 0.5 [107].

Other measurements and models relating to the properties of S have also been reported. Here, we briefly summarize those of major import.

- In the model of PL vs. d, (3.8), the standard deviation of S can vary with distance along the TX-RX path. Specifically, it tends to be smaller at close-in distances, where the path loss is close to that of free space, with little variation; and it can grow with distance, leveling off to a final value. To capture this effect, a model is given in [55] which we rewrite here as

$$\sigma(d) = 1.5 + \sigma_\mu \left[1 - \exp\left(-\frac{|PL_{\text{med}}(d) - FSPL(d)|}{4} \right) \right], \quad (3.11)$$

where σ_μ is a data-derived fitting parameter.

- We also know that shadow fading is a "semi-local" phenomenon, in the sense that, from one local environment to another, the urban or natural terrain can change character (e.g., from built-up neighborhoods to parks), leading to a different value for σ. This effect is partially quantified in [27], which reports results at 1.9 GHz from drive measurements in 95 environments located around the United States. The outcome is a probability distribution for σ for each of three representative kinds of terrain. For each terrain category, the variability among cells is shown to be describable by a truncated Gaussian distribution.

- Finally, σ changes very slowly with frequency. Though there are few if any explicit measurements tracking its value across bands, there is anecdotal evidence that σ increases slightly with frequency. This is likely due to the greater diffraction losses at higher frequencies, so that the presence/absence of obstructions magnifies the spread of path loss values in a cell.

3.4.3 Median path loss, PL_{med}

Many papers, reports and standards documents have been published on the median path loss, PL_{med} (e.g., [27, 53]). The vast majority use the bracketed term in (3.8) as the

3.4 Path loss models for wireless channels

generic form. We note, however, that some formulas contain "break points" in distance, d, where the value of γ changes; and some formulas for indoor environments add one or more terms to capture the effects of transmission through floors and walls [92]. Primarily, though, what distinguishes one set of authors from another is how they characterize A and γ. We will describe here two median path loss models in some depth. In so doing, we hope to illustrate, by concrete examples, the nature of such models for wireless systems.

For cellular systems in particular, the most popular model is the one developed by Hata [53] based on extensive measurements made by Okumura in the Tokyo area in the 1960s [85]. The model covers distances from 1 to 20 km; frequencies from 150 MHz to 1.5 GHz; base antenna heights from 30 to 200 m; and mobile antenna heights from 1 to 10 m. The environments covered are urban areas (large cities and medium-small cities); suburban areas; and open areas. In the 1990s, as personal communications services (PCS) opened up worldwide near 2 GHz, the Hata–Okumura model was expanded to include frequencies up to 2 GHz, and small urban cells (diameters below 1 km). This expansion, developed by the European Cooperative for Scientific and Technical Research, is the COST 231–Hata model [111].

The original Hata–Okumura model is applicable to wide-area scenarios wherein the user terminals are mobile. A situation like this is suitable for cognitive radio scenarios where the delay tolerance of the service is high. For urban environments and user terminals at height 1.5 m, the formula for PL_{med} is identical to the bracketed quantity in (3.8), where $d_0 = 1$ km;

$$A = [69.55 + 26.16 \log_{10} f_c - 13.82 \log_{10} h_b]; \qquad (3.12)$$

and

$$\gamma = [4.49 - 0.655 \log h_b]. \qquad (3.13)$$

In these formulas, f_c is the carrier frequency in MHz and h_b is the base antenna height in meters. The model provides additive terms to A that adjust it for terminal heights different from 1.5 m and environments different from urban. For similar CR applications over a smaller area (radii less than 1 km) and frequencies closer to 2 GHz, COST 231–Hata is more applicable. The term $26.16 \log_{10} f_c$ in A is largely due to the antenna aperture effect in all path loss equations, including the one for free-space propagation, which accounts for $20 \log_{10} f_c$; the remaining part, $6.16 \log_{10} f_c$, is due to radio wave diffraction around obstructing objects along the radio path, causing an incremental loss that increases with frequency.

The Erceg path loss model, devised for fixed wireless applications [27], was cited in Section 3.4.2 in connection with shadow fading. A CR context for this model might be transmission to and from fixed sensors (e.g., as used for radio mapping) distributed over a wide area. The model was derived from 1.9 GHz data collected from 95 suburban/residential areas, all with radii of 2 km or less and all downlink terminals at 2-m height. The data span a variety of base antenna heights and terrain types, and the model delineates their influences. Extensions to other carrier frequencies and terminal heights can be made via simple scaling laws, e.g., see [25].

The Erceg model for PL_{med} in fixed wireless scenarios, as presented in [27], is as follows.

- Up to the reference distance d_0, specified in this case to be 100 m, the median path loss follows the free-space path loss formula, i.e., $A = 20\log_{10}(\frac{4\pi d_0}{\lambda})$ and $\gamma = 2$. For a carrier frequency of 1.9 GHz, $\lambda = 15.8$ cm and $A = 78.0$ dB.
- For $d > d_0$, A remains the same, but γ becomes a quantity that varies in a seemingly random way from one area (or "cell") to another. Over all cells, it has a mean value, μ_γ, that depends on the base antenna height, h_b, and on the terrain type; and a zero-mean variable part whose standard deviation, σ_γ, depends on the terrain type only.
- Specifically, $\mu_\gamma = (a - bh_b + \frac{c}{h_b})$, where h_b is in meters; (a, b, c) are constants that depend on terrain category, in units that make μ_γ dimensionless; and this applies for h_b from 10 m to 80 m.
- The terrain categories, A, B, and C, are, respectively, hilly/moderate-to-heavy tree density; hilly/light tree density or flat/moderate-to-heavy tree density; and flat/light tree density.
- The variable part of γ is represented using a truncated Gaussian distribution, the truncation being chosen to preclude γ from going negative.

The Erceg model combines this description of median path loss with the shadow fading description in Section 3.4.2. Numerical values of all model parameters are in Table I of [27] for each of the three terrain categories. The model thus incorporates the randomness of γ and σ from one area to another, the influence of base antenna height, and the influence of terrain type.

3.4.4 Antenna gain and the gain reduction factor

We have, up to this point, written the relationship between transmit power and received power under the implicit assumption that there is no antenna gain, i.e., that the antennas transmit or receive equally in all directions. In truth, this is never quite the case. In cellular base stations, for example, the antennas have gain in the elevation plane in order to focus transmission and reception at the heights containing most user terminals; and they may have moderate azimuth gains (e.g., azimuthal beamwidths between 60° and 120°) to improve frequency reuse efficiency through sectoring. If the gain of a transmit antenna beam pattern towards a receiver is G_t dB, and the gain of the receive antenna beam pattern toward the transmitter is G_r dB, then the dB power gain will be augmented, in theory, by an amount $(G_t + G_r)$ dB. This composite gain is in fact realized in engineered routes like those for point-to-point radio relay.

The notion of an antenna with a highly directive beam, and correspondingly low sidelobes, suggests a potential benefit for CR receivers: by pointing a receive antenna beam at the intended transmitter, it can both improve the received signal level against noise *and* suppress interfering signals coming from off-beam directions. This approach is most feasible to contemplate when the link ends are not moving. It thus has relevance to fixed wireless applications, such as a CR-based network of fixed sensors. Nonetheless, propagation measurements conducted for fixed wireless systems in multipath environments have shown that the full pattern gains of highly directive receive antennas will not necessarily be realized [43]. The reason is that directive beams reduce wide-angle scatter that would otherwise add to the received power, and this loss offsets part of the gain acquired from the antenna.

The extensive measurements reported in [43] were conducted at 1.9 GHz in suburban residential environments, both in summer and winter to assess the impact of trees. The data were used to quantify the amount by which receiver gain, G_r, is reduced by multipath scatter. The amount of this reduction – whose dB value is the *gain reduction factor (GRF)* – was shown to vary from one path to another. A statistical model for *GRF*, as reported in [43] and standardized in [25], is as follows.

- In suburban residential environments, *GRF* is a Gaussian random variable from path to path, with a mean μ_{grf} and a standard deviation σ_{grf}.
- For antenna heights between 3 m and 10 m, there are but minor changes in μ_{grf} and σ_{grf}, so the model averages the values of these parameters for the two heights.
- $\mu_{\text{grf}} = -(0.53 + 0.1I)\ln(\frac{\beta}{360}) + (0.5 + 0.04I)(\ln(\frac{\beta}{360}))^2$, where $I = 1$ for winter and -1 for summer; and β is the directive antenna beamwidth, in degrees.
- $\sigma_{\text{grf}} = -(0.93 + 0.02I)\ln(\frac{\beta}{360})$.

The practical effect of this model, in the kinds of environment where CR systems often operate, is this: a directive receive antenna can suppress some of the received power in the presence of wide-angle scatter, thereby diminishing the hoped-for benefit of using high-gain antennas. For example, a link whose receive antenna has a 30° beamwidth, with a corresponding gain of $\sim$ 11dB, loses all but 4 dB of that gain with 50% probability. For antennas with higher directivity (and correspondingly higher gain), the reduction is even greater.

3.5 Path loss models for tower-based scenarios

We have defined the wireless channel to be one where at least one end is in clutter and the other end has an antenna height of up to 80 m. Here we treat channels in which either one end is higher than 80 m (as in TV towers) or both ends are above the clutter (as in roof-mounted antennas). We will give the possible CR contexts for these cases and provide relevant path loss models.

3.5.1 Transmissions from TV towers

A good starting point for modeling path loss in this case is the ITU-R recommendation P.1546 [94], which covers a wide range of antenna and terrain conditions. Formulas and graphs are given in terms of median received field strength, E_{dB}, in dB-μV/m, as a function of distance, for different frequencies and tower heights and with an effective radiated power of 1 kW. This can be translated into path loss using the formula

$$PL = 139.3 - E_{\text{dB}} + 20\log(f_{\text{MHz}}). \tag{3.14}$$

The field strength vs. distance data in [94] are given as functions of two antenna height parameters, h_1 and h_2, the latter being the actual height of the receiving antenna. The height h_1 pertains to the transmitter antenna and the path geometry; specifically, it is a function of the actual antenna height above the ground, h_a, and the *effective* antenna height, h_{eff}, defined for land paths as the antenna height above the average ground level between distances of 3 and 15 km in the direction of the receiver. The value of h_1 is calculated from h_a and h_{eff} as follows:

$$h_1 = \begin{cases} h_a, & d < 3 \text{ km}; \\ h_a + (h_{eff} - h_a)(d-3)/12, & 3 \text{ km} \leq d < 15 \text{ km}; \\ h_{eff}, & d \geq 15 \text{ km}, \end{cases} \quad (3.15)$$

where d is the TX-RX distance in km.

Some example curves are shown in Fig. 3.3 for a frequency of $f_c = 600$ MHz, some typical TV tower heights, and a fixed receiver height, h_2, of 20 m. In these curves, the difference between h_{eff} and h_a is fixed, i.e., $h_{eff} - h_a = 50$ m. These kinds of curve can be used for a TV tower radiating interference into a CR tower. The curves shown can be adjusted for receiver heights other than 20 m via additive correction terms given in Annex 5 of [94].

It should be emphasized that the curves in Fig. 3.3 are median values over space, based on measurements carried out in Europe and North America. Also, since broadcast signals are subject to diurnal variations, the median values over space have been derived for different percentiles of the temporal variations. The values plotted in Fig. 3.3 represent the path loss values exceeded 50% of the time. It is also implicitly assumed that the spatial variations of path loss are Gauss-distributed.

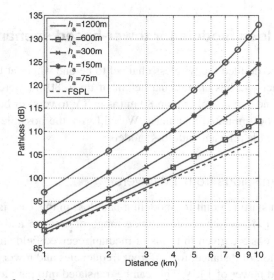

Figure 3.3 Path loss from high towers vs. distance using the P.1546 model. For all curves, $h_2 = 20$ m, $h_{eff} - h_a = 50$ m, and $f_c = 600$ MHz.

3.5.2 Tower-to-tower paths at low-to-moderate heights

These kinds of scenario can occur, for example, in CR systems using roof-mounted towers, and include cases where one end of the path is a primary station (e.g., a cellular BS) and one is a CR station. We cite two models, one for suburban areas and one for urban areas. Both are based on IEEE 802.16j, as reported in [55].

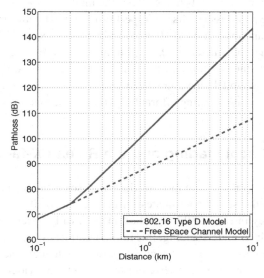

(a) Type-D model

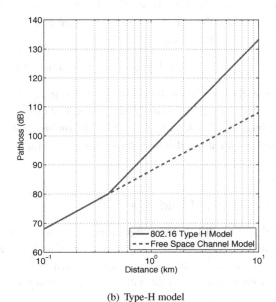

(b) Type-H model

Figure 3.4 Tower-to-tower path loss vs. distance using the IEEE 802.16j model.

For suburban environments, the Type-D path category is appropriate, and the path loss for this case is described in Section 2.1.2.2 of [55] as a suburban macrocell model for above rooftop (ART) propagation. The model is similar to the Erceg model [27], as standardized in [25]. For urban environments, the Type-H path category is appropriate, as described in Section 2.1.2.6 of [55]. Here, the underlying model is a modified form of COST 231–Walfisch–Ikegami [111], which captures the over-the-rooftop transmissions of urban environments but excludes the final rooftop-to-street loss of the original model.

Some path loss results for Type-D and Type-H paths are given in Fig. 3.4 for a transmitting antenna height of 30 m, a receiving antenna height of 10 m, and $f_c = 600$ MHz.[3] Additional parameters $h_{\text{roof}} = 15$ m and $b = 30$ m are given in the urban model for the height of the rooftop and the distance between buildings, respectively. We note that these two models have not been empirically validated, although they build on prior measurement/modeling efforts for geometric scenarios with similar features.

3.6 Small-scale fading and the Ricean K-factor

We now turn our attention to the small-scale fading term, $\hat{h}(\tau; \mathbf{x}, \mathbf{F})$. Recall that $\mathbf{x}$ denotes the set of coordinates on the terrain for the TX and RX; and $\mathbf{F}$ denotes the frequency interval of the signal of interest, with center frequency f_c and bandwidth W. To facilitate understanding, this initial treatment of small-scale fading will consider the extreme case where W shrinks to 0, i.e., small-scale fading for a single frequency, f_c. Later sections will generalize our treatment to the wideband case.

It is obvious that, for a single frequency, $\hat{h}(\tau; \mathbf{x}, \mathbf{F})$ becomes just a constant over τ, dependent on $\mathbf{x}$ and f_c. In short, it becomes the (normalized, complex) frequency response at f_c for the path between the TX and RX whose coordinates are given by $\mathbf{x}$. Thus, we can use the notation

$$\hat{h}(\tau; \mathbf{x}, \mathbf{F}) \to \hat{H}(f_c; \mathbf{x}); \text{ single-frequency case.} \tag{3.16}$$

Here, we will elaborate on the statistical variation of this complex amplitude over space in the vicinity of the location coordinates $\mathbf{x}$. For concreteness, we assume that the TX is at the base and fixed, and the RX is at the mobile and may be moving in the vicinity of its nominal location. If it does move, then the complex amplitude will vary with time as $\mathbf{x}$ does.

3.6.1 Spatial variation of field strength

As noted, the electric field strength fluctuates rapidly over space if there is multipath scatter, as in typical wireless environments. This is because the field at a particular point is the sum of several components, as seen in the example of Fig. 3.5, where the mobile is moving to the right with velocity v. The receiving antenna sees the sum of

---

[3] Fig. 3.4b takes into account the additional constraint that the path loss should be greater than $FSPL$, (3.5).

3.6 Small-scale fading and the Ricean K-factor

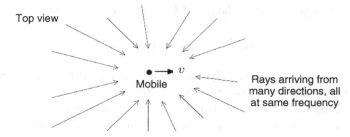

Figure 3.5 Multipath echoes arriving at a mobile receiver from all azimuth angles.

the echoes or *multipath components* (MPCs), which are attenuated and phase shifted on their way from the transmitter to the receiver. Depending on the differences among the phase shifts, the summation of the MPCs can be constructive or destructive, and these differences change with mobile position.

This effect can also be visualized as follows. The channel gain for each MPC can be represented by a complex vector. The combining of the MPCs at the receiving antenna is then equivalent to the addition of complex vectors. As the location of the receiver changes, the phase shift of each MPC changes, namely by $(2\pi/\lambda)d\cos(\phi)$, where ϕ is the angle between the direction of the arrival of the MPC and the direction of movement. This fact further implies that the *relative* phases among the MPCs change, since they are arriving at the receiver from different directions. Thus, MPCs that interfere destructively at one location can interfere constructively at a different location. This variation over space is what we call either small-scale fading or multipath fading, as discussed in Section 3.3.

If the net field, i.e., the superposition of waves, were sampled at many points within a radius of, say, 10λ, we would obtain a population of many (complex) samples. These can be interpreted as a statistical ensemble, whose values are describable using a *probability density function* (pdf). The most common and widely understood situation is the one where all the multipath components are of similar magnitude so that, in the limit of a great many components, their sum has a variation over space that resembles a complex Gaussian process. It has been shown that the phase of a complex Gaussian random variable (rv) has a uniform pdf over $(-\pi, \pi]$; and the magnitude r has a *Rayleigh* pdf [56, 76], i.e.,

$$p_r(r) = \frac{2r}{\bar{\mathcal{P}}_r} \cdot \exp\left[-\frac{r^2}{\bar{\mathcal{P}}_r}\right], \qquad r \geq 0, \tag{3.17}$$

where $\bar{\mathcal{P}}_r$ is the second moment of the distribution, i.e., the statistical average of r^2. In reality, the received power $\mathcal{P}_r = r^2$ has an average value defined by its local spatial average $\langle \mathcal{P}_r \rangle$. Thus, to the extent that (3.17) accurately models the distribution of r, the statistical average, $\bar{\mathcal{P}}_r$, and the local spatial average $\langle \mathcal{P}_r \rangle$ can be regarded as virtually the same. This case is referred to as *Rayleigh fading*.

It is easy to derive the pdf of $\mathcal{P}_r = r^2$ from (3.17) and the result is

$$p_{\mathrm{P}}(\mathcal{P}_r) = \frac{1}{\bar{\mathcal{P}}_r} \exp\left[-\frac{\mathcal{P}_r}{\bar{\mathcal{P}}_r}\right], \qquad 0 \leq \mathcal{P}_r \leq \infty. \qquad (3.18)$$

Using this equation the dynamic range between the 1st and 99th percentiles of Rayleigh fading can readily be shown to be 26.6 dB. This signifies a great deal of small-scale fading within the local area of a receive antenna. Adding to this the large-scale spatial variation in path loss, *PL*, we see that sensing devices in CR networks must be capable of measuring over a very large dynamic range.

It is clear from the above description that the received amplitude or power does not change abruptly, but rather continuously as the receiver moves along a trajectory. If all MPCs are coming from similar directions, then the rate of change as a function of distance along the trajectory is slow (because the relative phase changes among the MPCs are small); if the MPCs are coming from all directions, then the change is faster. A mathematical description of these facts is given by the spatial covariance function of the envelope

$$\frac{\mathbb{E}_x\{(r(x) - \langle r \rangle)(r(x + \Delta) - \langle r \rangle)\}}{\mathbb{E}_x\{[r(x) - \langle r \rangle]^2\}}. \qquad (3.19)$$

If MPCs are arriving at the receiver uniformly from all directions, the "correlation distance," i.e., the distance Δ for which the correlation function is 0.5, is a fraction of a wavelength.

In some cases, there is a dominant term among the many arriving MPCs, i.e., one whose magnitude is much higher than all others. This happens, for example, when there is a direct ray (LOS path) from the transmitter to the receiver. In this case, the vector sum of all arriving component magnitudes has a pdf called *Ricean*, which is characterized by two parameters: (i) the power of the direct component, $\mathcal{P}_d$, and (ii) the local spatial average of the power sum of all the weaker (scatter) terms, $\bar{\mathcal{P}}_s$. A more popular but equivalent parameter pair is the total spatial average of received power, $\mathcal{P}_d + \bar{\mathcal{P}}_s$; and the ratio $\mathcal{P}_d/\bar{\mathcal{P}}_s$. The former parameter can be recognized as the average received power, $\bar{\mathcal{P}}_r$, discussed in the previous section. The latter parameter is called the *Ricean K-factor*, or *Rice factor*, or just K. It is a measure of how much local fading the signal undergoes over space. The pdf of the received amplitude can then be written as

$$p_r(r) = 2r \frac{\exp(-K)}{\bar{\mathcal{P}}_s} \cdot \exp\left[-\frac{r^2}{\bar{\mathcal{P}}_s}\right] \cdot I_0\left(\sqrt{\frac{4r^2 K}{\bar{\mathcal{P}}_s}}\right); \qquad 0 \leq r < \infty. \qquad (3.20)$$

$I_0(x)$ is the modified Bessel function of the first kind and zero order [1]. The associated pdf of the received power, $\mathcal{P}_r = r^2$, is then

$$p_{\mathrm{P}}(\mathcal{P}_r) = \frac{\exp(-K)}{\bar{\mathcal{P}}_s} \exp\left[-\frac{\mathcal{P}_r}{\bar{\mathcal{P}}_s}\right] \cdot I_0\left(\sqrt{\frac{4\mathcal{P}_r K}{\bar{\mathcal{P}}_s}}\right), \qquad 0 \leq \mathcal{P}_r < \infty. \qquad (3.21)$$

Note that by replacing $\bar{\mathcal{P}}_s$ with $\bar{\mathcal{P}}_r/(K+1)$, each of the above two pdfs can be characterized in terms of the parameters $\bar{\mathcal{P}}_r$ and K.

3.6 Small-scale fading and the Ricean K-factor

When K is infinite, there is no fading (only a single component, e.g., an LOS term, as in free space) and when $K = 0$ ($-\infty$ dB), there is Rayleigh fading as described above (no dominant component, only scatter). It is also noteworthy that in the general Ricean case, the phase is not independent of the amplitude anymore; in other words, the joint amplitude–phase distribution does not factor into a pdf of r times a pdf of the phase. We also note that, for an LOS component, the magnitude does not change rapidly over space but the phase does, while the scatter component changes rapidly in both magnitude and phase.

There are little published data on K-factors for different environments and distances. It is conventional in mobile system studies to assume the fading is Rayleigh ($K = 0$), since this assumption generally leads to performance estimates on the conservative side. In Section 3.6.3, we will comment on links where both ends are fixed; and in Section 3.11, we will cite fading distributions that are even "worse than Rayleigh," as reported for links in which both terminals are near the ground.

Another amplitude distribution that has found widespread use for describing the effects of sums of MPCs is the Nakagami distribution [84], even though its physical interpretation is not as clear-cut as for the case of Rayleigh/Rice fading. The amplitude pdf in this case is

$$p_r(r) = \frac{2}{\Gamma(m)} \left(\frac{m}{\bar{\mathcal{P}}_r}\right)^m r^{2m-1} \exp\left(-\frac{m}{\bar{\mathcal{P}}_r} r^2\right); \qquad 0 \leq r < \infty \qquad (3.22)$$

for $m \geq 1/2$, where $\Gamma(m)$ is Euler's Gamma function [1], the parameter $\bar{\mathcal{P}}_r$ is the mean square value $\bar{\mathcal{P}}_r = \langle r^2 \rangle$, and the parameter m is

$$m = \frac{\bar{\mathcal{P}}_r^2}{\langle (r^2 - \bar{\mathcal{P}}_r)^2 \rangle}. \qquad (3.23)$$

It is straightforward to extract those parameters from measured values. If the amplitude is Nakagami fading, then the received power follows a Gamma distribution,

$$p_P(\mathcal{P}_r) = \frac{m}{\bar{\mathcal{P}}_r \Gamma(m)} \left(\frac{m\mathcal{P}_r}{\bar{\mathcal{P}}_r}\right)^{m-1} \exp\left(-\frac{m\mathcal{P}_r}{\bar{\mathcal{P}}_r}\right); \qquad 0 \leq \mathcal{P}_r < \infty. \qquad (3.24)$$

The Nakagami distribution is in widespread use because it can be handled in a rather straightforward way analytically, and also because it has shown agreement with a considerable number of measurements. In the range where amplitude values occur with high probability, i.e., near the center of the pdf, Nakagami and Rice distributions have similar shapes, and one can be converted into the other using relationships between the Rician K-factor and the Nakagami m-factor [15]. However, this equivalence does not hold in the tails (high and low powers) of the distribution, which can be important regions for CR and other applications. In particular, in the high-power tails, the Nakagami distribution decays faster with increasing power than does the Rice distribution; and in the small-power tails, Nakagami distributions decay faster with decreasing power.

Finally, we emphasize that the local average of received power is closely related to the path loss, PL. Assuming omni-directional antennas at both ends of the wireless link, we can write $\bar{\mathcal{P}}_r$ as simply $\mathcal{P}_t 10^{-0.1 PL}$, so that knowing path loss and K (for the Ricean case) or path loss and m-factor (for the Nakagami case) is sufficient for characterizing the pdf of small-scale fading.

3.6.2 Temporal fading on mobile radio links

So far we have discussed signal fading as a random-like variation over space, i.e., over potential mobile locations on the ground. What translates this spatial process into a *temporal* process is the fact that a mobile terminal moves over the ground, and as it does it will experience fading, in time, of its received signal strength. Thus, the Rayleigh/Ricean fading descriptions given above materialize as temporal processes in mobile operation. The higher the speed of the mobile, the more rapid is the fading. The correlation distance can be mapped into a correlation time; to a first approximation, we can say that the receive amplitude remains constant within a correlation time. This is captured in the Doppler spectrum and autocorrelation function, elaborated upon in Section 3.7.

3.6.3 Temporal fading on fixed wireless links

The notion of temporal fading on a fixed wireless path may seem strange at first, because we have been explaining it as a random time variation of the signal level received by a mobile terminal. We recall that fading arises from the presence of many received MPCs, each with a phase that is highly sensitive to both antenna position and signal frequency (or wavelength). The result is that the net received signal magnitude is a random-like function of both position and frequency. In a mobile terminal, it is the variation over space that produces the temporal fading of the received signal. In fixed wireless, the two ends of a link are nonmoving, so whereas the fading over frequency can be discerned, one would not expect to see temporal fading. That such fading exists nonetheless leads us to a different way of thinking about and modeling the Ricean K-factor.

In a mobile link, a mobile traces out the spatial fading as it moves, the scaling factor between distance traveled and time elapsed being the mobile speed, v. Thus, if $E(x)$ is the field variation along the mobile direction x, then the temporal variation is just $E(vt)$. In a fixed wireless link, the variation is due instead to the fact that some scattering objects are moving, a prime example being wind-blown leaves and limbs of trees. In fixed wireless systems operating in suburban residential areas, trees can be a large factor in producing temporal fading of received signal strength. Moving cars and people can have a similar impact in wireless ad hoc networks and wireless LANs.

This also implies that the K-factor needs to be interpreted in a different light. It is *not* the ratio of the LOS power to that of the scatterer component anymore. Rather, it is the ratio of the powers of the temporally fixed MPCs to that of the temporally varying MPCs. In a completely static environment, this K-factor is infinity, independent of whether a line-of-sight exists or not. In order to distinguish this K-factor from the one described in Section 3.6.1, it is often called the *temporal* Rice factor.

Extensive measurements reported in [45], covering many frequencies, observation times, and paths, and several locales within the United States, produced the following findings: (1) temporal fading on fixed wireless links can be characterized by Ricean distributions; (2) the Ricean K-factor shows a significant variation from one observation time, frequency, and path to another; and (3) the rate of fading is generally related to the prevailing wind speed. In this case, the fixed scatterers produce the constant term in the complex magnitude and the moving scatterers produce the variable term. The ability to represent temporal fading as Ricean is fortunate, again, because it involves only two parameters (path loss and K-factor), and also because it permits the use of analysis methods similar to those developed for mobile radio.

The model for Ricean K-factor on fixed wireless paths has been standardized in [25], based largely on [45] but also corroborated by data in [6]. It can be described as follows:

- the K-factor is expressed as a random variable

$$K = F_s F_h F_b K_0 d^\gamma u, \qquad (3.25)$$

where $K_0 = 10$; $\gamma = -0.5$; d is the transmitter–receiver distance in km; u is a log-normal variable whose dB value has a mean of 0 and standard deviation of 8.0 dB; and F_s, F_h, and F_b are factors that account for season, user terminal height, and base antenna beamwidth, respectively;
- $F_s = 1.0$ in summer (leaves on trees) and $F_s = 2.5$ in winter (no leaves);
- $F_h = (\frac{h}{3})^{0.46}$, where h is the height of the user antenna, in m;
- $F_b = (\frac{b}{17})^{-0.62}$, where b is the base antenna beamwidth, in degrees.

This model applies to paths between fixed primary TXs and fixed secondary RXs and helps to quantify the amount of variation of fixed wireless interference power. Low K-factors (which, the model predicts, occur quite often) can lead to periods of very low interference power (deep fades). Whether such fades can be exploited by CR receivers depends on the fade durations, which again relates to the Doppler spectrum of fading.

3.7 Small-scale fading and the Doppler spectrum

3.7.1 Doppler frequency

We now put the notion of Doppler frequency and spectra on a formal basis. Figure 3.6 indicates a mobile (large dot) moving in an arbitrary direction with constant velocity v; with no loss in generality, the x-axis is aligned with the direction of motion. The gray screen shown represents the wavefront of a propagating wave at frequency f_c, moving in a direction that makes an angle ϕ with the direction of motion of the mobile. We can regard this wave as an MPC propagating toward the mobile receiver.[4] From simple

---

[4] We will make the reasonable approximation that all arriving rays have elevation angles close to 0°, so the direction vector of each MPC is parallel to the x–y plane.

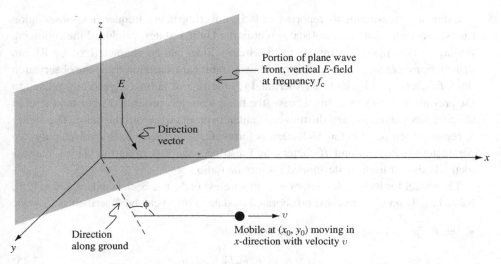

Figure 3.6 A schematic for explaining Doppler frequency at a moving terminal.

geometry, we can see that the complex magnitude of the received E-field is proportional to $\exp(j\Phi(t))$, where

$$\Phi(t) = 2\pi f_c t - \frac{2\pi}{\lambda}(-x\cos\phi + y\sin\phi), \qquad (3.26)$$

and where (x, y) is the mobile position at a given time. Assume that the mobile is at the location (x_0, y_0) at time 0, so that at time t, $x = x_0 + vt$ and $y = y_0$. Inserting this into (3.26) and rearranging, we get

$$\Phi(t) = 2\pi \left(f_c + \frac{v}{\lambda}\cos\phi \right) t - \frac{2\pi}{\lambda}(-x_o \cos\phi + y_o \sin\phi). \qquad (3.27)$$

Finally, the frequency, in Hz, is $\frac{\Phi'(t)}{2\pi} = f_c + \frac{v}{\lambda}\cos\phi$, where the second term is the frequency change due to motion. This is the *Doppler frequency*,

$$\nu = \frac{v}{\lambda}\cos\phi. \qquad (3.28)$$

3.7.2 The angle-of-arrival and Doppler spectra

In a typical wireless environment, there is not just one MPC; there are many. Figure 3.5 shows a top view of the mobile position and motion, with MPCs arriving from all directions. Assume there are N such components caused by local scattering, and that the set of azimuth arrival angles is $\{\phi_1, \ldots, \phi_n, \ldots, \phi_N\}$. The corresponding Doppler frequencies will be

$$\{\nu_n\} = \{\frac{v}{\lambda}\cos\phi_1, \ldots \frac{v}{\lambda}\cos\phi_n, \ldots \frac{v}{\lambda}\cos\phi_N\}. \qquad (3.29)$$

3.7 Small-scale fading and the Doppler spectrum

Thus, the received signal is the sum of N MPCs at slightly different frequencies. The result is a random-like variation over time of the envelope of the sum. This is just another (more mathematical) way of explaining the small-scale spatial fading described in Section 3.6. Note that, in light of reciprocity, this fading occurs in both directions, i.e., including transmissions from the mobile to the base. For each MPC, with arrival angle ϕ_n, there is a received power, P_n. The pairing of the set P_n with the set $\{\phi_n\}$ is called the *angle-of-arrival (AOA) spectrum*,

$$S_\phi(\phi) = \sum_n \rho_n \delta(\phi - \phi_n); \qquad -\pi < \phi \leq \pi; \qquad n = 1, N, \qquad (3.30)$$

where the ρ_n are the P_n normalized so that they sum to 1. Corresponding to the set $\{\phi_n\}$ is the set of Doppler frequencies, $\{\nu_n\}$, as given above. Pairing the two sets ρ_n and $\{\nu_n\}$ gives us the Doppler spectrum, $S_d(\nu)$, where ν ranges from $-v/\lambda$ to v/λ. For future reference, we designate v/λ as ν_{max}, the magnitude of the largest possible Doppler frequency.

Now consider the case where the AOA spectrum is defined over a *continuum* of angles. We can then write the Doppler spectrum as

$$S_d(\nu) = [S_\phi(\phi^+) + S_\phi(\phi^-)]/|\frac{d\nu}{d\phi}|, \qquad (3.31)$$

where, from (3.28),

$$\phi^+ = \left|\cos^{-1}\left(\frac{\nu}{\nu_{max}}\right)\right|; \quad \phi^- = -\left|\cos^{-1}\left(\frac{\nu}{\nu_{max}}\right)\right|; \qquad (3.32)$$

and

$$|d\nu/d\phi| = -\nu_{max} \sin\phi = \nu_{max}\left[1 - \left(\frac{\nu}{\nu_{max}}\right)^2\right]^{\frac{1}{2}}, \qquad (3.33)$$

for both $\phi = \phi^+$ and $\phi = \phi^-$. The two parts of (3.31) are due to the symmetric nature of the cosine function, resulting in the fact that each ν corresponds to two angles (equal and opposite) on the interval $(-\pi, \pi]$. As an example of the Doppler spectrum, assume that $S_\phi(\phi)$ is uniform over all ϕ. The Doppler spectrum in this case is

$$S_d(\nu) = \frac{1}{\pi \nu_{max}} \left[1 - \left(\frac{\nu}{\nu_{max}}\right)^2\right]^{-\frac{1}{2}}; \qquad |\nu| < \nu_{max}. \qquad (3.34)$$

This is the well-known Doppler spectrum of Clarke [15] and Jakes [56]. It has the shape shown in Fig. 3.7a. For other angular distributions (i.e., other AOA spectra), the more general formula (3.31) can be used to obtain the Doppler spectrum.

3.7.3 The autocorrelation function, $A(\Delta t)$

The random-like variation of the signal envelope as a terminal moves over the local terrain at constant velocity v has the same first-order statistics in time as in space, as the

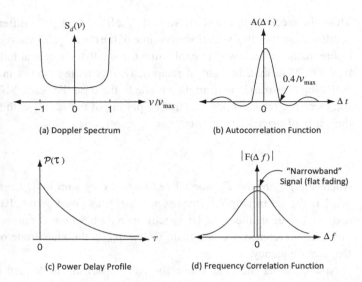

Figure 3.7 Time and frequency functions for characterizing dispersion in a wireless channel.

former variation is just a manifestation of the latter. Thus, for example, if the probability law for the spatial variation is Ricean, the same is true for the temporal variation, including the same K-factor.

As for the second-order statistics of the temporal variation, we can derive them from the Fourier relationship that exists between the Doppler spectrum and the autocorrelation function (ACF). To be clear, however, the first-order statistics we refer to most often concern the signal *envelope* (r in the treatment of Section 3.6) or its square $\mathcal{P}_r$, while the ACF referred to here is that for the complex amplitude of the path response. The first-order distribution for that quantity is complex Gaussian. For the Clarke–Jakes Doppler spectrum, (3.34), the Fourier transform is

$$A(\Delta t) = J_0(2\pi \nu_{\max} \Delta t) = J_0\left(\frac{2\pi v \Delta t}{\lambda}\right), \quad (3.35)$$

where $J_0(\cdot)$ is the Bessel function of zero order [1], and Δt is the time separation between two points on the complex envelope's temporal variation. This function is plotted in Fig. 3.7b.

Since $v\Delta t$ is the distance traveled in time Δt, we can say that, for MPCs arriving uniformly from all directions, the small-scale fading is totally decorrelated for spatial positions separated by $\sim 0.4\lambda$. The correlation distance for multipath fading (different from that for shadow fading, Section 3.4.2) is the value of $v\Delta t$ for which $A(\Delta t) = 0.5$.

3.7.4 The Doppler spectrum for fixed terminals

The notion of time variations on fixed wireless links has been introduced in Section 3.6.3. Here we discuss the associated Doppler spectrum. It is important to note that the phenomenology of Doppler dispersion is subtly different from the case of mobile terminals. In the latter case, the angular spectrum of the MPCs as seen from the terminal, together with the velocity vector of the terminal, determines the Doppler spectrum. A change of the magnitude of the velocity scales (compresses or dilates) the Doppler spectrum, but does not change its shape. For fixed terminals, the Doppler shift of the MPCs is created by moving scatterers, each of which has not only a different angle of the associated MPC at the terminal, but also a different velocity. The random time variation of the return for a sinusoidal transmitted signal is largely caused by the back-and-forth motion of wind-blown tree branches and leaves (outdoor environments) or by the motion of people (indoor environments). In these cases, the power density spectrum of the received signal is found to be concentrated in the center of its frequency range, which is quite different from the Jakes spectrum (see Fig. 3.7a).

The amount of hard data on Doppler spectra for fixed terminals is sparse. What is established is that the width of these spectra tends to be related to wind (or people) speed; is inversely related to wavelength; and is of the order of 1 Hz in microwave bands. A useful and simple model, derived for wireless LANs and reported in [26], can be written in the following compact way:

$$S_d(\nu) = [1 + (\nu/\nu_0)^2]^{-1}; \qquad \nu_0 = 1/9\lambda, \tag{3.36}$$

where the wavelength λ is in meters. A similar shape and width apply to outdoor fixed wireless systems, as reported in [25]. For the spectrum in (3.36), $A(\Delta t)$ is a two-sided decaying exponential, $\exp(-2\pi\nu_0|\Delta t|)$.

3.7.5 Dispersion

The Doppler spectrum describes the spreading of a signal from its transmission as a single sinewave to its reception: it starts with zero bandwidth (i.e., its frequency spectrum is an impulse function) and is received with a nonzero bandwidth (specifically, $2\nu_{\max}$). This dispersion of the spectrum from an impulse function to a spread in frequency is called *frequency dispersion*.

Another form of dispersion relevant to wireless channels is the spread in angle-of-arrival due to multipath scatter: on an ideal LOS path, there is only one component reaching the receiver, so the AOA spectrum is an impulse function. The reality in wireless environments, however, is a *spread* of arriving angles, which constitutes *angle dispersion*. (As we have shown, it is the angle dispersion that produces frequency dispersion, so these are not separate phenomena.) The topic of angle dispersion, and how to model it based on empirical data, is elaborated upon in Section 3.9.

The next section deals with *delay dispersion*, whereby a radio impulse sent into the channel emerges as a set of echoes with a nonzero range of delays, as in the case of multipath scatter. In the process, we will provide a formal distinction between "wideband" channels and the "narrowband" channels we have discussed in this section and the previous one.

3.8 Delay dispersion

3.8.1 "Narrowband" vs. "wideband"

We return now to the normalized impulse response introduced in Section 3.2, $\hat{h}(\tau; \mathbf{x}, \mathbf{F})$, where the average of the squared magnitude over the spatial domain $\mathbf{x}$ and the spectral domain $\mathbf{F}$ is unity. The spectral domain $\mathbf{F}$ is the frequency interval from $f_c - W/2$ to $f_c + W/2$. If a radio impulse is passed through the radio channel in this spectral space, a multipath component arriving at the receiver with complex channel gain a_1 at a delay τ_1 will appear as $a_1 \xi(\tau - \tau_1)$, where $\xi(\tau) = W \sin(\pi W \tau)/(\pi W \tau)$. Thus, the impulse response to an MPC within a channel of bandwidth W is a sinc pulse over τ of null-to-null width $2/W$.[5] For present purposes, we can ignore the fact that this response is noncausal.

Now assume there are N multipath components, with delays $\tau_1, \ldots, \tau_n, \ldots, \tau_N$ and complex amplitudes $a_1, \ldots, a_n, \ldots, a_N$. Then the impulse response can be written as

$$\hat{h}(\tau) = \sum_n a_n \xi(\tau - \tau_n). \tag{3.37}$$

As noted, we arbitrarily specify that the delay of the first MPC is at the τ-origin, so that $\tau_1 = 0$. The total delay spread (i.e., the support) of the peaks of the summed sinc functions is therefore τ_N. The spread is of minimal impact in cases where τ_N is very small compared to the nominal width of the individual sinc pulses, which we can take to be $\approx 2/W$. By "minimal impact" we mean that a signal transmitted in the specified band arrives with little distortion. This is because, given the extensive temporal overlap of the MPC terms in (3.37), the impulse response deviates little from a sinc function. The channel in this case, $W\tau_N \ll 2$, is what we can refer to as "narrowband." It corresponds to a frequency response that is virtually flat over the bandwidth W. As $W\tau_N$ grows from very small values, the temporal dispersion can no longer be ignored or approximated away; we will refer to the channel in that case as "wideband," in the sense that the detailed nature of the delay dispersion must be accounted for in analyzing signal transmission.

3.8.2 Wideband channels

We will refer to the pulse $\xi(\tau)$ as the *bandlimited impulse function*, noting that, in the limit of infinite bandwidth, $\xi(\tau)$ converges to the familiar Dirac impulse function, $\delta(\tau)$. We now rewrite (3.37) under different regimes of $\xi(\tau)$. We have already discussed the "narrowband" case, $W\tau_N \ll 2$, wherein the different terms in (3.37) overlap considerably and the channel appears as flat fading. At the other extreme is the situation where the nominal τ-spacing between MPC arrivals is larger than $1/W$ (equivalently, the nominal arrival rate is greater than W). Consequently, the different MPCs can be resolved by the receiver and processed individually. This condition applies to certain contexts, such

---

[5] The sinc pulse is defined by $\text{sinc}(x) = \sin(\pi x)/\pi x$.

as ultra-wideband (UWB) systems, discussed later. The more typical "wideband" condition occurs when $W\tau_N > 1$ *and* there are several MPC arrivals within each τ-interval of width $1/W$. We next develop a useful and time-honored model for this case, the *tapped delay line* (TDL) filter.

The representation (3.37) is quite general, but not measurable or useful when there is overlap among the individual terms. Another general form that can be quite useful in that case is the sampled-and-interpolated version, using Nyquist theory to express $\hat{h}(\tau)$ via uniformly spaced sinc functions:

$$\hat{h}(\tau) = \sum_m \operatorname{sinc}((1/\Delta)(\tau - m\Delta)) \sum_n a_n \xi(m\Delta - \tau_n), \qquad (3.38)$$

where Δ is the sampling interval and can be any value that does not exceed $1/W$. We will assume here that equality holds, $\Delta = 1/W$. We can thus rewrite (3.38) more simply as

$$\hat{h}(\tau) = \sum_m c_m \xi(\tau - m/W), \qquad (3.39)$$

where c_m is the inner sum over n in (3.38), normalized by W. Whereas all of the MPCs contribute to this mth amplitude, we can see that the major contributors are those for which τ_n lies in the time-of-arrival interval $[(m-1/2)/W, (m+1/2)/W]$. We call this interval the "resolvable delay bin" about $\tau = m/W$ and note that a bin with multiple MPCs will exhibit fading as the mobile moves over space. Under our assumption that the number of MPCs falling into a delay bin is large, the fading statistics tend towards complex Gaussian, as discussed in Section 3.6. Otherwise, different fading statistics, like Nakagami or lognormal fading, might be more suitable. If a line-of-sight exists between the TX and RX, then the amplitude fading in the first delay bin is typically Rice-distributed.

As we can see from (3.37), the pulses corresponding to *actual* MPCs are spaced in time at irregular intervals, corresponding to the placement of the scatterers in the environments. Representing these by regularly spaced taps is exact only if the tap spacing is at most $1/W$. Furthermore, according to the interpolation function, one actual MPC can impact many regularly spaced taps, thus possibly introducing correlation into the fading of the taps. Lastly, we note that the regular spacing of the taps leads to a periodicity of the frequency transfer function, with the period being the inverse of the tap spacing. When the tap spacing is $1/W$ or less, the period is W or more. Since the signal band is of width W, there is no overlap of the repeated spectra.

3.8.3 Time-variant impulse response

If the TX or RX moves, the spatial variability of c_m we have described is converted into a time variation. In the following, we thus invoke the commonly used "time-variant impulse response," $\hat{h}(t, \tau)$, under conditions of mobility instead of the equivalent $\hat{h}(\tau; (\mathbf{x}, \mathbf{F}))$. Thus, (3.39) becomes

$$\hat{h}(t,\tau) = \sum_m c_m(t)\xi(\tau - m/W). \qquad (3.40)$$

To write an expression for $c_m(t)$, we refer to the inner sum of (3.38) and consider the time variation of the nth MPC, with complex gain a_n and arrival time τ_n. For the normal geometry of a TX–RX path and its nth scatterer, the amplitude $|a_n|$ and arrival time τ_n change very little in the neighborhood of a mobile's position, i.e., over a travel distance of many wavelengths. By contrast, the *phase* of a_n changes rapidly, i.e., by $360°$ for every travel distance of λ relative the scatterer. Thus, over a travel distance of many wavelengths, corresponding in time to a great many digital symbols,[6] the mth tap gain can be written as

$$c_m(t) = \sum_n |a_n|\exp(j\phi_n(t))\xi(m\Delta - \tau_n). \qquad (3.41)$$

Assuming as before that the dominant terms in this sum (those peaking in an interval of width Δ about $m\Delta$) are numerous, $c_m(t)$ can be regarded as a complex Gaussian temporal process. In addition, the mth tap has its own angular power spectrum (based on the set of arrival angles for the MPCs in the mth bin) and, under mobility, its own associated Doppler spectrum. Even so, in most system studies, it is conventional to assign the same Doppler spectrum to all taps of the TDL representation, usually the one for uniform angle-of-arrival distribution, (3.34).

3.8.4 The power delay profile, $\mathcal{P}(\tau)$

For the assumed complex Gaussian statistics, the MPC phases are uniformly random over the small-scale fading, and the mean of $c_m(t)$ is zero. Then, the only parameter needed to characterize the first-order statistics of $c_m(t)$ is its mean-square value. This raises the topic of *power delay profile* (PDP), which plays a prominent role in analyzing and simulating wireless channels.

A formal mathematical definition of the PDP is

$$\mathcal{P}(\tau) = (1/W)\mathbb{E}_t\{|\hat{h}(t,\tau)|^2\}, \qquad (3.42)$$

where $\mathbb{E}_t$ denotes expectation over a time interval where small-scale fading happens.[7] This formal definition can run into mathematical problems when $\hat{h}(t,\tau)$ is a discrete sum of impulses, bandlimited or otherwise. One popular alternative is to define the PDP itself as a sum of impulses along the τ-axis, having the delays of the MPCs $(\tau_1,\ldots,\tau_N)$ as in (3.37), and with the power $\mathcal{P}_n$ at delay τ_n being $|a_n|^2$. This approach is especially useful in ultra-wideband channels, where the individual MPCs are resolvable in delay. For more conventional wideband channels with many MPCs per delay bin (the case considered here), the PDP can instead be a sum of impulses at

---

[6] As an example of this time duration, consider a frequency of 5 GHz ($\lambda = 6$ cm) and a vehicle moving at 100 km/hr. The time it takes for a mobile to travel a distance λ will be 2.16 ms, very large compared to the symbol duration of a wideband wireless signal.

[7] The above definition actually requires the assumption of ergodicity of the channel [76, App. 6A].

$\tau = 0, 1/W, \ldots, m/W, \ldots$, with the power at $\tau = m/W$ being the mean-square value of $c_m(t)$ in (3.40).

Another alternative for the case of many MPCs per delay bin is to treat the PDP as a smooth, continuous function of delay, implying a continuum of echoes along the delay axis, see [7]. We shall invoke that representation here, where the value of $\mathcal{P}(\tau)$ at any given τ is the sum of the powers of the MPCs arriving in a delay bin of width $1/W$ about that τ; and the normalization is such that the area under the integral over $\mathcal{P}(\tau)$ is unity. A very popular model for the PDP shape is the one-sided exponential,

$$\mathcal{P}(\tau) = (1/\tau_0)\exp(-\tau/\tau_0); \quad 0 \leq \tau < \infty, \tag{3.43}$$

where τ_0 is the decay time constant [56], and we note that the area under $\mathcal{P}(\tau)$ is 1. The popularity of this shape stems from both its simplicity and its corroboration by many measurement campaigns conducted over the years, e.g., [16, 28]. Several other measurements have found more complex PDPs that consist of several clusters, e.g., [95].

The *rms delay spread* (for convenience henceforth simply called "delay spread," DS) is the second central moment of the PDP, i.e.,

$$\tau_{\text{rms}} = \sqrt{\int_0^\infty \mathcal{P}(\tau)\tau^2 d\tau - \left(\int_0^\infty \mathcal{P}(\tau)\tau d\tau\right)^2}. \tag{3.44}$$

Many papers report the measured delay spread (and not complete PDPs) to characterize the delay dispersion. For the decaying exponential PDP shape in (3.43), τ_{rms} is found from (3.44) to be precisely τ_0.

3.8.5 The frequency correlation function, $F(\Delta f)$

The Fourier transform of $\hat{h}(t, \tau)$ in the delay variable τ, with t fixed, can be recognized as the (normalized, complex) channel frequency response at time t, which we denote as $\hat{H}(f, t)$. In Section 3.6, where we assumed no time variation and a single frequency f_c, we denoted this function as $\hat{H}(f_c; \mathbf{x})$. (As above, we suppress here the use of $\mathbf{x}$ to designate the TX-RX spatial coordinates.) In Section 3.7.3, for the mobile case, we introduced an autocorrelation function, $A(\Delta t)$, to describe the correlation between two complex time samples of $\hat{H}$ where f is fixed and the time samples are separated by Δt. Here, we consider the parallel case: t is fixed and we seek the correlation $F(\Delta f)$ between two complex f-samples of $\hat{H}$ separated by Δf. Under conditions that generally apply in wideband wireless channels, this *frequency correlation function* (FCF) can be shown to be nothing other than the Fourier transform of $\mathcal{P}(\tau)$.

If fading is highly correlated over an interval of width B Hz (i.e., $|F(\Delta f)| \approx 1$ for Δf up to B), and a deep fade from a primary TX to a cognitive radio RX is detected at some frequency f_0, then the fade can be assumed to occupy a band of approximate width B about f_0. Thus, an agile cognitive radio TX can send signals to that RX over that band without severe interference from the primary TX. The FCF is also important for the sensing phase of CR, as the detection of radiation from other possible users is often done by "scanning" the spectrum. A single-frequency (narrowband) measurement might not allow the detection of another use if the relevant transfer function is in a fading dip. The FCF tells designers how densely a spectrum has to be sampled to avoid this problem.

For the exponential PDP, (3.43), it is easy to show that $F(\Delta f) = 1/[1 + (j2\pi\Delta f \tau_0)]$. Note that, given the unit area of $\mathcal{P}(\tau)$ and the nature of the Fourier transform, the value of $F(0)$ is always 1, as is appropriate since the fading at any frequency is perfectly correlated with itself.

The coherence bandwidth B_c is defined as the smallest frequency separation such that $|F(B_c)/F(0)| < k$, where k has been given by different authors as 0.5, 0.75, and 0.9. In [32], the coherence bandwidth is related to the delay spread via

$$B_c = \frac{\arccos(k)}{2\pi} \frac{1}{\tau_{\text{rms}}}. \tag{3.45}$$

For $k = 0.5$, this yields $B_c = 1/6\tau_{\text{rms}}$. Another approximation, given in [100] for $k = 0.5$, is $B_c \approx 1/5\tau_{\text{rms}}$. For the case of the exponential PDP (and recalling that, in this case, τ_0 and τ_{rms} are the same), it is easy to show for $k = 0.5$ that $B_c = \sqrt{3}/2\pi\tau_{\text{rms}} \approx 1/3.6\tau_{\text{rms}}$. Clearly, the precise relationship between B_c and τ_{rms} depends upon the shape of the PDP, but it is always a reciprocal one.

Finally, the coherence bandwidth gives us another way to distinguish between the "narrowband" and "wideband" cases mentioned in Section 3.8.1.

- "Narrowband" is the case where the signal bandwidth is small compared to B_c; this case is commonly known as *flat fading* (transfer function is flat across the signal bandwidth).
- "Wideband" is the case where the signal bandwidth is comparable to or greater than B_c; this case is commonly known as *frequency selective fading*.

3.8.6 A model and values for the delay spread

It follows from the above derivations that the delay spread is obtained from impulse responses taken over an area in which small-scale fading occurs. In different small-scale areas, τ_{rms} can take on different values. It is useful to interpret those different values as different realizations of a random variable, and investigate its probability density function. Following [44], τ_{rms} is well-approximated as

$$\tau_{\text{rms}} = T_1 d^\epsilon y, \tag{3.46}$$

where d is in km; y is a lognormal variate whose dB value has zero mean and standard deviation σ; and T_1 is the median delay spread at $d = 1$ km. Numerical values for the parameters T_1, ϵ, and σ are tabulated in [44] for different environments.

Many measurements for rms delay spread can be found in the literature. Typical values for different environments are as follows (see also [82]).

- *Indoor residential buildings* Delay spreads of 5–10 ns are typical, but values up to 30 ns have been measured [38, 69].

- *Indoor office environments* These show typically delay spreads between 10 and 100 ns, but even values of 300 ns have been measured. The room size has a clear influence on the delay spread; also building size and shape have an impact [9, 18, 21, 22, 52, 59, 73, 103, 108].
- *Factories and airport halls* These have delay spreads that range from 50 to 200 ns [49, 91].
- *Microcells* Delay spreads range from around 5–100 ns (for LOS situations) to 100–500 ns (for NLOS) [4, 29, 63, 113, 118].
- *Tunnels and mines* Empty tunnels typically show a very small delay spread (on the order of 20 ns), while car-filled tunnels exhibit larger values (up to 100 ns) [68, 117].
- *Typical urban and suburban environments* These show delay spreads between 100 ns and 800 ns, although values up to 3 μs have also been observed [57, 61, 86, 93, 101, 118].
- *Bad urban and hilly terrain environments* These show clear examples of multiple clusters that lead to much larger delay spreads. Delay spreads up to 18 μs, with cluster delays of up to 50 μs, have been measured in various European cities, while American cities show somewhat smaller values. Cluster delays of up to 100 μs occur in mountainous terrain [8, 23, 39, 64, 70, 75, 93, 97, 110].

3.8.7 Ultra-wideband (UWB) channels

The underlay paradigm of cognitive radio requires that secondary user transmissions be so low that their power spectrum densities at primary receivers are below that of additive receiver noise, i.e., that they raise the noise floor seen by primary receivers only slightly. This approach can be effective if the secondary signals are spread over a very wide range of frequencies and then despread at the receiver as in the use of UWB signaling and detection, e.g., [114]. By formal definition, UWB signals have bandwidths of 500 MHz or more. Military applications tend to use baseband signaling, while most commercial applications reside in the frequency range from 3 to 10 GHz.

The bandwidths associated with UWB are so wide that some conditions we have assumed previously do not apply, notably, the presence of many MPCs in any delay interval of width $1/W$. With W in the ultra-wide regime, MPCs appear sparsely within a delay bin-width $1/W$, and models of the impulse response and PDP must reflect this. An obvious and important consequence is that the tap gains of the TDL can no longer be regarded as Rayleigh-fading processes.

A flurry of measurement and modeling activity for these channels occurred in the first decade of this century, primarily focused on indoor environments, both residential and commercial. Any applications of UWB for cognitive radio in outdoor environments would use similar models, though likely with different parameter values. Accordingly, we briefly describe here three published channel models for the PDP of indoor UWB channels, to highlight some similarities and differences relative to traditional wideband models.

The exponential model, reported in [12], represents $\mathcal{P}(\tau)$ as containing a term at $\tau = 0$ (fixed in local space but variable from one path to another); and a single cluster of rays spaced at $\tau = i/W$, $i = 1, 2, 3, \ldots$ The amplitudes of the rays in the cluster decay exponentially with τ, with the decay constant being variable from one path to another. For every ray in the exponential cluster, the amplitude is an average over local spatial samples, and the pdf of the spatial variability is a Gamma function.

For the exponential-lognormal model, reported in [37], the PDP shape is similar to that of the exponential model, except that the exponential cluster is multiplied by a correlated lognormal process over τ. Several parameters of this model are found to be functions of distance along the path. This model is more complicated than the exponential one, requiring more model parameters.

For the cluster model, reported in [79], the PDP is a sum of several exponential clusters, with the starting delay of each cluster being variable from one path to another. Similarly variable are the τ-positions of the rays within each cluster, rather than being uniformly spaced at intervals of $1/W$. This model is a variation on the Saleh–Valenzuela cluster model devised for wideband (as opposed to ultra-wideband) indoor channels [95].

In all three of the above PDP models, the major difference from traditional wideband models is that the ray amplitudes are not Rayleigh-fading; because each ray is a sum of just a few MPCs, these amplitudes have narrower fading distributions (e.g., a lognormal pdf with dB standard deviation 3.4 dB, as in [79]). A detailed description of, and comparisons among, the three models is given in [46].

3.9 Angle dispersion

3.9.1 Directions of arrival and departure

The MPCs are characterized not only by their delays, but also by their directions. An MPC leaves the TX in a certain direction (determined by the location of the first scatterer it will encounter), and arrives at the RX in a certain direction (similarly determined by the position of the last scatterer before arrival). Since there are many MPCs, which can depart and arrive in different directions, a multipath channel leads to angular dispersion. This dispersion is very important for CR. For one thing, it determines the ability of adaptive antennas to transmit into directions such that they will not disturb other users located in certain directions (see Chapter 2). A small angular spread offers the possibility of precisely focusing beams into particular directions.

Conversely, a *large* angular spread can lead to gain decorrelations, for antenna elements spaced by just a fraction of a wavelength. This phenomenon can aid diversity via use of multiple elements and maximal ratio combining [96, Section 10.5]; or it can increase capacity via multiple co-frequency streams, as in multiple-input/multiple-output (MIMO) links [10]. Such MIMO-based capacity gains can benefit CR systems in the interleaving mode, by offsetting capacity losses caused by remaining silent during primary transmissions. Finally, angular dispersion is strongly related to Doppler, as we have shown in Section 3.7, and therefore influences the temporal variations (correlation times) of CR channel responses.

3.9 Angle dispersion

A description of the angular dispersion can be made by generalizing the concept of impulse response. The double-directional impulse response (DDIR), which generalizes (3.37), again consists of a sum of contributions from the MPCs [105]:

$$h(t, \tau, \boldsymbol{\Omega}, \boldsymbol{\Psi}) = \sum_{n=1}^{N} a_n(t)\xi(\tau - \tau_n)\delta(\boldsymbol{\Omega} - \boldsymbol{\Omega}_n)\delta(\boldsymbol{\Psi} - \boldsymbol{\Psi}_n), \quad (3.47)$$

with $a_n(t) = |a_n|\exp(j\Phi_n(t))$, and where the $\boldsymbol{\Omega}_n$ denote the directions of departure (DODs), the $\boldsymbol{\Psi}_n$ are the directions-of-arrival (DOAs), and N is the number of MPCs.

Equation (3.47) assumes that the MPCs can be represented by ideal homogeneous plane waves, such that they are represented by delta functions in the angular domain (when taking into account wavefront curvature, the delta functions need to be replaced by more general functions). We see the strong analogy between the traditional model, (3.37), and this double-directional representation, where each MPC is additionally characterized by a DOA and DOD.

The double-directional impulse response is an abstract description of the propagation effects alone, without being impacted by the antenna geometry. For the performance analysis of communications systems, we often need the joint impulse response at the elements of an antenna array, which can be represented as a matrix of impulse responses if we have arrays at both link ends, and a vector if we have an array at one link end only.

We now show how to obtain such an impulse response matrix from the double-directional impulse response. This derivation is based on the principle that the contribution of an MPC to the total impulse response at two different antenna elements differs only by a phase factor that is determined by the direction of the MPC and the relative position of the antenna elements. We denote the transmit and receive element coordinates as $\mathbf{r}_T^{(1)}, \mathbf{r}_T^{(2)}, \ldots, \mathbf{r}_T^{(N_T)}$, and $\mathbf{r}_R^{(1)}, \mathbf{r}_R^{(2)}, \ldots, \mathbf{r}_R^{(N_R)}$, respectively. For one particular time instant (equivalently, one realization of the a_n), the impulse response from the mth transmit to the lth receive element[8] becomes (under the assumption of plane, narrowband waves)

$$h_{l,m} = h\left(\mathbf{r}_T^{(m)}, \mathbf{r}_R^{(l)}\right)$$
$$= \sum_n a_n g_T(\boldsymbol{\Omega}_n) g_R(\boldsymbol{\Psi}_n) \exp\left(j\mathbf{k}(\boldsymbol{\Omega}_n) \bullet \mathbf{r}_T^{(m)}\right) \exp\left(j\mathbf{k}(\boldsymbol{\Psi}_n) \bullet \mathbf{r}_T^{(l)}\right), \quad (3.48)$$

where g_T and g_R are the *complex* (amplitude) patterns of the transmit and receive antenna elements, respectively, $\mathbf{k}$ is the unit wave vector in the direction of the ith DOD or DOA, and $\bullet$ denotes the dot product; the summation can be replaced by an integration if the double-directional impulse response has continuous components. We thus see that it is easily possible to obtain the impulse response matrix from a double-directional impulse response (and the knowledge of antenna positions and patterns) while the converse is not true.

---

[8] If we deal with an antenna array at only one link end, the impulse responses at the antenna elements constitute a vector (instead of a matrix). However, we write the elements of this vector still as $h_{1,j}$ in order to avoid confusion with h_l, the contribution of the lth multipath component.

The DDIR is a very general description of the channel.[9] From it, we can derive the "normal" time-variant impulse response by weighting the DDIR with the antenna pattern, and then integrating over all direction angles

$$h(t,\tau) = \int\int h(t,\tau,\mathbf{\Omega},\mathbf{\Psi}) g_T(\mathbf{\Omega}) g_R(\mathbf{\Psi}) d\mathbf{\Omega} d\mathbf{\Psi}. \qquad (3.49)$$

The angular power spectrum (APS) is a characterization of how much power, on average, arrives from (or departs towards) a certain angle. For example, when considering the APS at the RX,

$$\text{APS}(\mathbf{\Psi}) = \mathbb{E}\left\{|h(t,\tau,\mathbf{\Omega},\mathbf{\Psi})|^2\right\}, \qquad (3.50)$$

where the expectation is over t, τ, and $\mathbf{\Omega}$. A similar formulation applies to the APS at the TX, APS($\mathbf{\Omega}$). One could also define a spectrum APS($\mathbf{\Psi}, \tau$) as the expectation in (3.50) over just t and $\mathbf{\Omega}$. This function reflects the physical reality that the angular power spectrum can be different for different delays. In practice, however, it is often more convenient to include the averaging over τ; this is especially appropriate for narrowband channels, where the MPCs overlap.

A major simplification of APS($\mathbf{\Psi}$) is made possible by noting that most MPC arrival directions at an RX lie in or close to the horizontal plane containing the antenna. Thus, there is no spread with respect to elevation angle, θ. In this case, APS($\mathbf{\Psi}$) can be reduced to a power spectrum over azimuth angle, ϕ, at the receiver. Referring to Section 3.7, this can be seen as equivalent to $S_\phi(\phi)$ in (3.31) or a continuous version thereof. For convenience, and with this equivalence in mind, we will use the notation APS(ϕ) in the present discussion.

Finally, a compact characterization of the angular dispersion is given by the rms angular spread, henceforth simply called "angular spread," AS [33]. We will focus on the angular spread at the RX. This quantity is the second central moment of the APS normalized to have an area of 1 over $(-\pi, +\pi]$. We have seen that, because of angular dispersion, there is space-selective fading of received amplitudes about the RX location. The correlation distance associated with this fading has an inverse relationship to the AS.

3.9.2 Models for the APS shape and angular spread

We start with the case of an elevated TX or RX, e.g, as on a base station, mostly above the clutter. Here, the most common model for the APS is a small angular spread around the nominal line-of-sight direction, usually with a Laplacian shape [89],

$$\text{APS}(\phi) = \exp\left[-\sqrt{2}\frac{|\phi - \phi_0|}{\phi_{\text{rms}}}\right], \qquad (3.51)$$

---

[9] An even more general description includes the polarization of the MPCs, see [77, 99] and Section 3.10.

where ϕ_0 is the mean azimuthal angle and $\phi_{\rm rms}$ is the AS. More complicated shapes, consisting of multiple clusters, each of which is characterized by a Laplacian APS, and having different mean angles, have also been suggested.

As for street-level TX and RX, various investigations have been made into the shape of the APS. The conventional model, namely, an APS that is uniform over all azimuth angles, stems from Jakes [56] and was widely used until the 1990s. However, more recent studies indicate that the azimuthal spread can be considerably smaller, especially in street canyons. The street-level APS for such cases often modeled as Laplacian, as in (3.51).

Similar to the case of the delay spread, many papers give only the angular spread, as obtained from measurement campaigns, though some give the per-cluster AS. While fewer papers deal with angular dispersion than with delay dispersion, the following range of values can be considered typical.

- *Indoor office environments* Cluster angular spreads between 10° and 20° have been observed for NLOS situations; for LOS situations, they are considerably smaller [14, 103].
- *Industrial environments* Angular spreads between 20° and 30° have been observed [49].
- *Microcells* Angular spreads between 5° and 20° for LOS, and 5°–40° for NLOS, have been found in [13, 87, 110].
- *Typical urban and suburban environments* In urban environments, [65, 88] measured angular spreads on the order of 3°–20° while [66] found angular spreads of up to 40°. In suburban environments, the angular spread was usually smaller than 5° due to a frequent occurrence of LOS [65].
- *Bad urban and hilly terrain environments* These show around 20° total angular spread due to the existence of multiple clusters [67, 110].
- *Rural environments* Angular spreads between 1° and 5° have been observed [87].

3.9.3 Joint dispersions

In this and previous sections, we have introduced the notions of delay spread, angular spread, and shadow fading, each of which exhibits a particular numerical value at a given location and a (spatially slow) variation of this value over the terrain. Although we have discussed these three entities separately, there are some important connections among them, e.g., shadowing and delay spread are both influenced by obstructions between a cellular base station and a given terminal location [44].

The shadowing (S in (3.7)), delay spread (DS), and angular spread (AS) each have a pdf that can be well approximated as lognormal, so that their log-values can be described by Gaussian pdfs. Furthermore, measurements have shown that these Gaussian variates are correlated. The correlation between the log-angular spread and log-delay spread has been found to be approximately 0.5 [3], while that between log-angular spread and log-shadowing is around 0.75 [5], and similarly for the correlation between log-delay spread and log-shadowing. A correlation between delay spread and angular spread has also been observed for indoor peer-to-peer environments [24].

These correlations have an impact on the planning of CR radios. For example, the correlation between shadowing and delay spread helps in wideband sensing: in the shadowing fades, where detection of a primary radio is basically more difficult, the increased delay dispersion helps to average out the small-scale fading. In turn, this helps to decrease the missed-detection probability.

We also note that values of spreads do not give a complete description of the channel dispersion characteristics. A full description of the joint second-order statistics of the dispersion is given by the double-directional delay power spectrum (DDDPS), i.e., the expected value of the squared magnitude of the DDIR [33, 60]. While many papers in the literature assume that the DDDPS can be decomposed into a product of PDP, APS at the TX, and APS at the RX, several experimental investigations indicate that such a factorization is an oversimplification. For example, the APS *does* depend on the delay [17, 51, 64, 81]. This is especially true in cases where the multipath components are arriving in clusters [35, 77, 95, 104].

3.10 Polarization

Radio wave polarization can play an important role in wireless systems, including CR. For one thing, there is the possibility of using polarization for multiplexing of two or more separate signal streams on the same frequency, using one polarization for one set of streams and an orthogonal polarization for the other. An implementation example is shown in Fig. 3.8, where each end (TX and RX) uses two cross-polarized antennas in a 2×2 version of multiple-input/multiple-output (MIMO) signaling. A big advantage of such a dual-pol approach, especially for handset design, is that the antenna structure at each end can be compact.

Alternatively, polarization can be used to achieve diversity against multipath fading. Here, the fact that there is "polarization mixing" in multipath is exploited. To elaborate, we initially assume that the transmission from a given antenna element is vertically polarized (V-pol). Due to the reflections and scatterings of MPCs, this signal will lose its polarization purity in the air and be received on both vertical and horizontal polarizations. The same applies to an antenna element that is horizontally polarized (H-pol). Thus, for example, in a 2×2 MIMO link with crossed H-V elements at each end, there

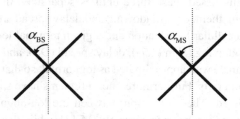

Figure 3.8 Pairs of crossed dipoles at the base and mobile, as might be used to obtain dual diversity or multiplexing. Typically, the angles ϕ_{BS} and ϕ_{MS} are both 45°. The pairs shown are in the vertical planes at the base and mobile, generally facing each other.

will be four path responses: HH, HV, VH, and HV. What has been learned over the years is that the multipath fading patterns among these four responses tend to be both of similar strength and mutually uncorrelated. Therefore, by transmitting a stream on both transmitter elements and receiving them on both receiver elements, it is possible to achieve 4-fold diversity against fading. Since the polarizations transmitted and received in Fig. 3.8 can be decomposed into H and V components, the same reasoning applies to that case.

If P_{VV} and P_{VH} are the mean incident powers launched by a vertically polarized antenna and received, respectively, by vertically and horizontally polarized antennas, then the decibel value of P_{VV}/P_{VH} is defined as the channel cross-pol discrimination (XPD). There is general consensus in the literature that the channel XPD, when expressed in dB, has a nonzero-mean Gaussian distribution. Depending upon the environment and the existence of a LOS component, the mean XPDs measured and reported in the literature vary from 0 to 12 dB, with a standard deviation of the order of 3–6 dB [98]. The XPD also depends upon the path distance, DOA, DOD, and delay of the multipath components. When dual-polarised antennas are used at both ends of a wireless link, there are four polarization channels to be considered, HH, HV, VH, and VV. This gives rise to four impulse responses to be modeled. We furthermore need to consider that antennas actually need to be characterized by two complex antenna patterns, namely, $g^{(V)}(\Omega)$ and $g^{(H)}(\Omega)$. A generalization of (3.48) thus reads

$$h_{l,m} = h\left(\mathbf{r}_T^{(m)}, \mathbf{r}_R^{(l)}\right)$$

$$= \sum_n \begin{bmatrix} g_T^{(V)}(\Omega_n) \\ g_G^{(H)}(\Omega_n) \end{bmatrix}^T \begin{bmatrix} a_n^{(VV)} & a_n^{(HV)} \\ a_n^{(VH)} & a_n^{(HH)} \end{bmatrix} \begin{bmatrix} g_R^{(V)}(\Psi_n) \\ g_R^{(H)}(\Psi_n) \end{bmatrix}$$

$$\times \exp\left(j\mathbf{k}(\Omega_n) \bullet \mathbf{r}_T^{(m)}\right) \exp\left(j\mathbf{k}(\Psi_n) \bullet \mathbf{r}_T^{(l)}\right) \qquad (3.52)$$

where the $a_n^{(VV)}$, $a_n^{(HV)}$, $a_n^{(VH)}$, and $a_n^{(HH)}$ have independent phases. Their absolute amplitudes are random variables, with $\mathbb{E}\{|a_n^{(VV)}|^2/|a_n^{(HV)}|\}$ and $\mathbb{E}\{|a_n^{(VH)}|^2/|a_n^{(HH)}|\}$ being lognormally distributed; it is furthermore often assumed that $\mathbb{E}\{|a_n^{(VV)}|^2/|a_n^{(HH)}|\} = 1$.

Knowledge of polarization effects can improve accuracies in estimating the interference from CRs into primary receivers. Polarization discrimination can also play a useful role in CR, e.g., systems might be able to transmit polarizations that primary links are not using. This of course requires minimal polarization mixing and/or CR knowledge of the relevant channel responses, as in the overlay mode.

3.11 Special environments

3.11.1 Vehicle-to-vehicle (V2V) propagation

An application that has gained particular attention in recent years is communication between vehicles. Special frequency regions in the 5 GHz band are reserved for this type of application in many countries. However, the high carrier frequency implies relatively large path loss, so that alternative approaches, including cognitive radio at lower frequencies, need to be investigated. The characteristics of V2V channels[10] differ from those of mobile cellular channels because of specific features of V2V radio propagation.

1. In V2V systems, the TX and RX antennas are at the same height, and in similar environments (peer-to-peer communications). In cellular communications, on the other hand, communication is between a base station that is high above street level, with a mobile station at street level. As a consequence, the dominant propagation mechanisms of the multipath components are different. For example, in cellular communications, propagation of waves over rooftops is important, while in V2V systems, propagation in the horizontal plane, with reflection and diffraction (e.g., at street corners), is more important. Also, for a V2V channel, scattering can occur near both link ends, while for cellular channels, the area around the base station is usually free of scatterers.
2. In a V2V channel, both TX and RX, as well as many of the important scatterers, are moving, whereas in cellular channels, only one (the terminal) is moving, and moving scatterers have less relative importance. This implies that the channel fluctuations in V2V channels are faster and that commonly used relationships between angular and Doppler spectra, as discussed in Section 3.7, are not valid anymore. Furthermore, fading distributions that show more frequent deep fades than Rayleigh fading can occur [71].

A number of measurements have found path loss coefficients $\gamma < 2$ if there is a LOS between TX and RX (see [58] and references therein). Related to this, rather high Rice factors have been found. Delay spreads typically show moderate values (on the order of a few hundred ns). A major difference between V2V and cellular communications is the short time during which the stationarity assumption is valid, i.e., the assumption that the absolute amplitudes and delays of the MPCs stay constant. More detailed discussions and extensive references to the literature can be found in [72, 83, 109].

3.11.2 Wireless sensor networks (WSNs)

Another important application is the communication between sensors that are located in homes (e.g., for improvement of energy efficiency), on the ground (for environmental monitoring), on streets (for traffic monitoring) and in all locations containing CRs (for measurement of the radio environment). Again, in this case propagation is often from peer to peer (with the exception of backhaul nodes). This leads to path loss behavior that is significantly different from traditional cellular networks. In particular, the closeness of the nodes to the ground means that ground reflections can have significant impact even for very short distances between TX and RX. So far, there is very little

---

[10] The V2V scenario (also called car-to-car (C2C)) is a subset of the broader category called mobile-to-mobile (M2M). The reader should be aware that M2M is also used to mean machine-to-machine, see Chapter 1.

literature on this important topic. Some papers [115] have investigated the impact of objects or people passing between two sensors, and thus blocking the LOS; this leads to nonstationarities of the channel. In [71] we learn of "worse than Rayleigh" fading in such environments, but, more generally, there are not enough measurement results available to establish reliable statistical models for sensor-to-sensor communications. We mention it here mainly as a topic of future research.

3.12 Summary of key model parameters

We have reviewed the models that are used to define the propagation conditions of the wireless systems operating in different locales. The list of references containing measurements is very extensive. Here we give a summary of selected references where key parameters of these models may be found and which have been standardized by national or international organizations, and/or are often used by industry practitioners and system designers.

3.12.1 Path loss models

The best-known path loss model and its empirical parameters for the cellular environment with base station height above rooftops (macrocells) is given by Hata in [53] and is based on measurements in Japan [85]. These models are based on measurements made in frequency bands below 1 GHz. An extension to the 1.8 GHz band is given in [111]; this reference also gives a model (based on both theory and measurements) for cellular environments where the base station height is comparable to, or lower than, rooftop height (microcells). Standardized values for the 2 GHz band are given in [48]. The broadcasting bands use very high towers (often located on prominent hill tops) and use large (kW) transmit powers. The field strength of these systems is calculated using the models given in [94]. Path loss models for fixed wireless access systems are given in [25, 50].

3.12.2 Ricean K-factor models

Extensive measurements of the K-factor and empirical relationships are given in [31]. These measurements were made in European cities in the 5 GHz band. Temporal K-factor measurements in the 2 GHz band and used in fixed wireless access systems are given in [25].

3.12.3 Delay dispersion models

A considerable number of standardized multipath delay profiles exist. In [30] we find profiles based on measurements in Europe for a 900 MHz carrier frequency and a 200 kHz bandwidth. A model for higher bandwidth (5 MHz) and carrier frequency (2 GHz) was standardized by the ITU in [48]; this model is mainly intended for macrocellular environments. More refined models for macro- and microcells can be found in [5, 102, 104].

For indoor environments, [19, 74] give power delay profiles suitable for systems with 20 MHz bandwidth; the models are used both for the 2 GHz and the 5 GHz bands. Ultrawideband channel models, which cover the whole bandwidth between 3 and 10 GHz, can be found in [20, 78, 79].

3.12.4 Frequency dispersion models

Most people use the classic Clarke–Jakes Doppler spectrum [15, 56]. Indeed this is the case for the ITU standard [48]. Similarly, [30] defines Doppler spectra that are either Jakes-like or Gaussian in shape for the different taps of the power delay profile in macrocellular environments. Doppler spectra peaked at the center tend to occur more frequently in fixed-wireless applications, while Doppler spectra with a "bathtub" shape correspond to a mobile station moving on a street, with scatter from all directions.

3.12.5 Comprehensive models

Several models have been developed that describe path loss, delay dispersion, and angular dispersion (and thus implicitly Doppler spectra). For both macrocellular and microcellular environments, [102] gives models for a 5 MHz bandwidth and a 2 GHz carrier frequency. An indoor model that includes delay dispersion and angular dispersion is described in [26]. More refined models that are suitable for even more environments can be found in [5, 77, 104], and [11, 81].

3.12.6 Usage of models

The models presented above, and more generally, all statistical channel models, do not completely describe the realities of propagation. "Typical" or "standardized" values of channel parameters are helpful in comparing different radio systems to each other, but they can deviate from the physical reality. Therefore, no performance guarantees can be given for CR designs based on specific channel models, nor can noninterference guarantees be provided.

3.13 Summary

Radio propagation models like those described here are critical to the prediction of both primary–CR and CR–primary interferences in cognitive radio environments, and to the design of real-time methods to control, avoid, or mitigate these interferences. In addition, the models described can be used to design CR–CR transmission strategies to cope with the secondary status of CR users. Some examples:

- models of median path loss and its distance dependence help to predict the median interference powers in primary and CR receivers;

- models of both shadow fading and multipath fading statistics (the latter augmented by models of Ricean K-factor) help to (1) determine the needed fade margins on CR–CR paths; and (2) predict the statistical variability of interference levels;
- models of the spatial variability of shadow fading help to (1) design the sensor networks used to characterize the radio environment; and (2) relate power measurements on CR–CR paths to interference levels on CR–primary paths;
- models of rms delay spread and/or coherence bandwidth help to relate power measurements at one frequency to interference levels at another;
- models of angular spread help to predict the ability of a CR system to suppress interference in the direction of the primary receiver;
- all these models make it possible for system analysts and designers to simulate propagation environments and then evaluate candidate CR algorithms for interference mitigation, leading to optimized designs and more reliable performance predictions.

We have reviewed the two main aspects of channel models, namely, path loss (large-scale effects) and impulse response (small-scale effects) and we have provided samples of the models developed and used for each. We have also cited numerous references, for more details on model parameters and for further reading in general. Most of all, we have attempted to relate the various properties of channel models to their potential use in the design of cognitive radios and CR networks. The large body of measurement and modeling work done in this field to date provides a strong foundation for the excellent future work that will surely build on it.

3.14 Further reading

The topic of propagation in the bands relevant to CR are covered quite extensively in three modern-day texts by Rappaport [92], Goldsmith [40], and Molisch [76]. All three textbooks provide detailed coverage – beyond the scope of a mere chapter – of both large-scale and small-scale effects and their impact on wireless systems. For an early treatment of these topics and their relevance to cellular engineering, the reader is referred to the pioneering paper by R. H. Clarke [15] and to the pioneering text edited by Jakes [56]. For a deeper understanding of the wireless channel functions involving time, delay, frequency, and Doppler, and the connections among these functions, the reader is referred to the seminal work by Bello [7]. For an appreciation of the emerging topic of radio channels with low-height antennas at both ends, as pertains to sensor networks and mobile-to-mobile systems, readers will find papers by Talha and Patzold [109] and Molisch et al. [83] to give useful summaries. An example study of interference in cognitive radio environments is a paper by Suraweera et al. [106], which highlights the importance of accurately estimating the path loss between primary and CR users.

Finally, readers actively engaged in the main topic of this book, or interested in pursuing it, may wonder how all the propagation channel concepts presented here can be brought together in the design and analysis of cognitive radio systems. An excellent example of how to do it is the simulation method proposed in [102]. This approach is based on the Spatial Channel Model (SCM) for 3GPP but its range of application is not limited to cellular or to any one system. The interested reader will find all of the wireless channel concepts discussed here incorporated in one way or another in that method.

References

[1]   M. Abramowitz and I. A. Stegun (eds.), *Handbook of Mathematical Functions*, Dover Publications Inc., 1972.

[2]   P. Agrawal and N. Patwari, "Correlated link shadow fading in multi-hop wireless networks," *IEEE Trans. Wireless Commun.*, **8**, no. 8, 4024–4036, August 2009.

[3]   A. Algans, K. I. Pedersen, and P. E. Mogensen, "Experimental analysis of the joint statistical properties of azimuth spread, delay spread, and shadow fading," *IEEE J. Sel. Areas Commun.*, **20**, no. 3, 523–531, April 2002.

[4]   A. A. Arowojolu, A. M. D. Turkmani, and J. D. Parsons, "Time dispersion measurements in urban microcellular environments," *Proc. IEEE Veh. Technol. Conf.*, 150–154, 1994.

[5]   H. Asplund, A. A. Glazunov, A. F. Molisch, K. I. Pedersen, and M. Steinbauer, "The COST259 directional channel model – Part II: Macrocells," *IEEE Trans. Wireless Commun.*, **5**, no. 12, 3434–3450, Dec. 2006.

[6]   D. S. Baum, D. Gore, R. Naber, *et al.*, "Measurements and characterization of broadband MIMO fixed wireless channels at 2.5 GHz," *Proc. ICPWC*, Dec. 2000.

[7]   P. Bello, "Characterization of randomly time-variant linear channels," *IEEE Trans. Commun.*, **11**, no. 4, 360–393, Dec. 1963.

[8]   J. E. Berg, J. Ruprecht, J. P. de Weck, and A. Mattsson, "Specular reflections from high-rise buildings in 900 MHz cellular systems," *Proc. IEEE Veh. Technol. Conf.*, 594–599, 1991.

[9]   C. Bergljung and P. Karlsson, "Propagation characteristics for indoor broadband radio access networks in the 5 GHz band," *Proc. IEEE International Symposium on Personal, Indoor and Mobile Radio Communications*, pp. 612–616, 1998.

[10] E. Biglieri, R. Calderbank, A. Constantinides, A. Goldsmith, A. Paulraj, and H. V. Poor, *MIMO Wireless Communications*, Cambridge University Press, 2007.

[11] G. Calcev, D. Chizhik, B. Goransson, *et al.*, "A Wideband Spatial Channel Model for System-Wide Simulations," *IEEE Trans. Veh. Technol.*, **56**, no. 2, 389–403, March 2007.

[12] D. Cassioli, M. Z. Win, and A. Molisch, "The ultra-wide bandwidth indoor channel: from statistical model to simulations," *IEEE J. Sel. Areas Commun.*, **20**, no. 6, 1247–1257, Aug. 2002.

[13] M. Chen and H. Asplund, "Measurements and models for direction of arrival of radio waves in LOS in urban microcells," *Proc. IEEE International Symposium on Personal, Indoor and Mobile Radio Communications*, pp. 100–104, 2001.

[14] C.-C. Chong, C.-M. Tan, D. I. Laurenson, S. McLaughlin, M. A. Beach, and A. R. Nix, "A new statistical wideband spatio-temporal channel model for 5-GHz band WLAN systems," *IEEE J. Sel. Areas Commun.*, **21**, no. 2, 139–150, Feb. 2003.

[15] R. H. Clarke, "A statistical theory of mobile radio reception," *Bell Syst. Tech. J.*, **47**, no. 6, 957–1000, 1968.

[16] D. C. Cox, "910 MHz urban mobile radio propagation: multipath characteristics in New York City," *IEEE Trans. Commun.*, **21**, no. 11, 1188-1194, Nov. 1973.

[17] R. J. M. Cramer, R. A. Scholtz, and M. M. Z. Win, "Evaluation of an ultra-wide-band propagation channel," *IEEE Trans. Antennas Propag.*, **50**, no. 5, 561–570, May 2002.

[18] I. Cuinas and M. G. Sanchez, "Measuring, modeling, and characterizing of indoor radio channel at 5.8 GHz," *IEEE Trans. Veh. Technol.*, **50**, no. 2, 526–535, Mar. 2001.
[19] "Delay dispersion in UWB residential/office environments," *IEEE 802.15.3a*.
[20] "Delay dispersion in UWB channels (office, residential, outdoor, industrial)," *IEEE 802.15.4a*.
[21] D. Devasirvatham, "Time delay spread and signal level measurements of 850 MHz radio waves in building environments," *IEEE Trans. Antennas Propag.*, **34**, no. 11, 1300–1305, Nov. 1986.
[22] D. Devasirvatham, "A comparison of time delay spread and signal level measurements within two dissimilar office buildings," *IEEE Trans. Antennas Propag.*, **35**, no. 3, 319–324, Mar. 1987.
[23] P. E. Driessen, "Prediction of multipath delay profiles in mountainous terrain," *IEEE J. Sel. Areas Commun.*, **18**, no. 3, 336–346, Mar. 2000.
[24] G. D. Durgin, V. Kukshya, and T. S. Rappaport, "Wideband measurements of angle and delay dispersion for outdoor and indoor peer-to-peer radio channels at 1920 MHz," *IEEE Trans. Antennas Propag.*, **51**, no. 5, 936–944, May 2003.
[25] V. Erceg, K. V. S. Hari, M. S. Smith, et al., "Channel models for fixed wireless applications," *IEEE 802.16d-03/34*.
[26] V. Erceg, L. Schamacher, P. Kyristi, et al., "TGn channel models," *IEEE 802.11-03/940r4*, May 2004.
[27] V. Erceg, L. J. Greenstein, S. Y. Tjandra, et al., "An empirically based path loss model for wireless channels in suburban environments," *IEEE J. Sel. Areas Commun.*, **17**, no. 7, 1205–1211, Jul. 1999.
[28] V. Erceg, D. G. Michelson, S. S. Ghassemzadeh, et al., "A model for the multipath delay profile of fixed wireless channels," *IEEE J. Sel. Areas Commun.*, **17**, no. 3, 399–410, March 1999.
[29] M. J. Feuerstein, K. L. Blackard, T. S. Rappaport, S. Y. Seidel, and H. H. Xia, "Path loss, delay spread, and outage models as functions of antenna height for microcellular system design," *IEEE Trans. Veh. Technol.*, **43**, no. 3, 487–498, Aug. 1994.
[30] M. Failli (ed.), Final report of COST 207, "Digital land mobile radio communications," *Commission of the European Communities*, 1989.
[31] D. S. Baum, H. El-Sallabi, T. Jämsä, et al., "Final report on link level and system level channel models," WINNER, IST-2003-507581, D5.4, 2005.
[32] B. H. Fleury, "An uncertainty relation for WSS processes and its application to WSSUS systems," *IEEE Trans. Commun.*, **44**, no. 12, 1632–1634, Dec. 1996.
[33] B. H. Fleury, "First- and second-order characterization of direction dispersion and space selectivity in the radio channel," *IEEE Trans. Information Theory*, **46**, no. 6, 2027–2044, Sep. 2000.
[34] H. T. Friis, "A note on a simple transmission formula," *Proc. IRE*, **34**, no. 5, 254, May 1946.
[35] J. Fuhl, A. F. Molisch, and E. Bonek, "Unified channel model for mobile radio systems with smart antennas," *IEE Proc. Radar, Sonar and Navigation*, **145**, no. 1, 32–41, Feb. 1998.
[36] C. Gerami, N. Mandayam, and L. Greenstein, "Backhauling in TV white spaces," *2010 IEEE Global Telecommunications Conference*, Dec. 2010, pp. 1–6.
[37] S. S. Ghassemzadeh, L. J. Greenstein, T. Sveinsson, and V. Tarokh, "UWB delay profile models for residential and commercial indoor environments," *IEEE Trans. Veh. Technol.*, **54**, no. 4, 1235–1244, July 2005.
[38] S. S. Ghassemzadeh, R. Jana, C. W. Rice, W. Turin, and V. Tarokh, "Measurement and modeling of an ultra-wide bandwidth indoor channel," *IEEE Trans. Commun.*, **52**, no. 10, 1786–1796, Oct. 2004.
[39] A. A. Glazunov, H. Asplund, and J. E. Berg, "Statistical analysis of measured short-term impulse response functions of 1.88 GHz radio channels in Stockholm with corresponding channel model," *Proc. IEEE Veh. Technol. Conf.*, 107–111, 1999.

[40] A. Goldsmith, *Wireless Communications*, Cambridge University Press, 2005.
[41] A. J. Goldsmith and L. J. Greenstein, "A measurement-based model for predicting coverage areas of urban microcells," *IEEE J. Sel. Areas Commun.*, **11**, no. 7, 1013–1023, Sept. 1993.
[42] F. Graziosi and F. Santucci, "A general correlation model for shadow fading in mobile radio systems," *IEEE Commun. Letters*, **6**, no. 3, 102–104, 2002.
[43] L. J. Greenstein and V. Erceg, "Gain reductions due to scatter on wireless paths with directional antennas," *IEEE Commun. Letters*, **3**, no. 6, 169–171, Jun. 1999.
[44] L. J. Greenstein, V. Erceg, Y. S. Yeh, and M. V. Clark, "A new path-gain/delay-spread propagation model for digital cellular channels," *IEEE Trans. Veh. Technol.*, **46**, no. 2, 477–485, May 1997.
[45] L. J. Greenstein, S. S. Ghassemzadeh, V. Erceg, and D. G. Michelson, "Ricean K-factors in narrowband fixed wireless channels: theory, experiments and statistical models," *IEEE Trans. Veh. Technol.*, **58**, no. 9, 4000–4012, Oct. 2009.
[46] L. J. Greenstein, S. S. Ghassemzadeh, S. C. Hong, and V. Tarokh, "Comparison study of UWB indoor channel models," *IEEE Trans. Wireless Commun.*, **6**, no. 1, 128–135, Jan. 2007.
[47] M. Gudmundson, "Correlation model for shadow fading in mobile radio systems," *Electronics Letters*, **27**, no. 23, 2145–2146, 7 Nov. 1991.
[48] Guidelines for Evaluation of Radio Transmission Technologies for IMT-2000, Recommendation ITU-R M.1225, 1997.
[49] D. Hampicke, A. Richter, A. Schneider, G. Sommerkorn, R. S. Thoma, and U. Trautwein, "Characterization of the directional mobile radio channel in industrial scenarios, based on wideband propagation measurements," *Proc. Veh. Technol. Conf.*, 2258–2262, 1999.
[50] K. V. S. Hari and C. Bushue, "Interim channel models for G2 MMDS fixed wireless applications," *IEEE 802.16.3c-00/49r2*.
[51] F. Harryson, J. Medbo, A. F. Molisch, *et al.*, "Efficient experimental evaluation of MIMO and set with user influence," *IEEE Trans. Wireless Commun.*, **9**, no. 2, 853–863, Feb. 2010.
[52] H. Hashemi and D. Tholl, "Statistical modeling and simulation of the RMS delay spread of indoor radio propagation channels," *IEEE Trans. Veh. Technol.*, **43**, no. 1, 110–120, Feb. 1994.
[53] M. Hata, "Empirical formula for propagation loss in land mobile radio services," *IEEE Trans. Veh. Technol.*, **29**, no. 3, 317–325, Aug. 1980.
[54] M. H. Hayes, *Statistical Digital Signal Processing and Modeling*. J. Wiley & Sons, 1996.
[55] IEEE 802.16.j standardization group, "Multi-hop relay system evaluation methodology (channel model and performance metric)," *IEEE 802.16j-06/013r3*, Feb. 2, 2007.
[56] W. C. Jakes, *Microwave Mobile Communications*, J. Wiley & Sons, 1974 (Reprinted by IEEE Press, 1994).
[57] A. Kanatas, N. Moraitis, G. Pantos, and P. Constantinou, "Wideband characterization of microcellular suburban mobile radio channels at 1.89 GHz," *Proc. IEEE Veh. Technol. Conf.*, 1060–1064, 2002.
[58] J. Karedal, N. Czink, A. Paier, F. Tufvesson, and A. F. Molisch, "Path loss modeling for vehicle-to-vehicle communications," *IEEE Trans. Veh. Technol.*, **60**, no. 1, 323–328, Jan. 2011.
[59] P. Karlsson, C. Bergljung, E. Thomsen, and H. Borjeson, "Wideband measurement and analysis of penetration loss in the 5 GHz band," *Proc. IEEE Veh. Technol. Conf.*, 2323–2328, 1999.
[60] R. Kattenbach, "Statistical modeling of small-scale fading in directional radio channels," *IEEE J. Sel. Areas Commun.*, **20**, no. 3, 584–592, Apr. 2002.
[61] J. F. Kepler, T. P. Krauss, and S. Mukthavaram, "Delay spread measurements on a wideband MIMO channel at 3.7 GHz," *Proc. IEEE Veh. Technol. Conf.*, 2498–2502, 2002.
[62] J. Kivinen, X. Zhao, and P. Vainikainen, "Empirical characterization of wideband indoor radio channel at 5.3 GHz," *IEEE Trans. Antennas Propag.*, **49**, no. 8, 1192–1203, Aug. 2001.

[63] S. Kozono and A. Taguchi, "Mobile propagation loss and delay spread characteristics with a low base station antenna on an urban road," *IEEE Trans. Veh. Technol.*, **42**, no. 1, 103–109, Feb. 1993.

[64] A. Kuchar, J. P. Rossi, and E. Bonek, "Directional macro-cell channel characterization from urban measurements," *IEEE Trans. Antennas Propag.*, **48**, no. 2, 137–146, Feb. 2000.

[65] M. Larsson, "Spatio-temporal channel measurements at 1800 MHz for adaptive antennas," *Proc. IEEE Veh. Technol. Conf.*, 376–380, 1999.

[66] P. Laspougeas, P. Pajusco, and J. C. Bic, "Radio propagation in urban small cells environment at 2 GHz: experimental spatio-temporal characterization and spatial wideband channel model," *Proc. IEEE Veh. Technol. Conf.*, 885–892, 2000.

[67] J. Laurila, K. Kalliola, M. Toeltsch, K. Hugl, P. Vainikainen, and E. Bonek, "Wideband 3D characterization of mobile radio channels in urban environment," *IEEE Trans. Antennas Propag.*, **50**, no. 2, 233–243, Feb. 2002.

[68] M. Lienard and P. Degauque, "Natural wave propagation in mine environments," *IEEE Trans. Antennas Propag.*, **48**, no. 9, 1326–1339, Sep. 2000.

[69] J. MacLellan, S. Lam, and X. Lee, "Residential indoor RF channel characterization," *Proc. IEEE Veh. Technol. Conf.*, 210–213, 1993.

[70] U. Martin, "Spatio-temporal radio channel characteristics in urban macrocells," *IEE Proc. Radar, Sonar and Navigation*, **145**, 42–49, Feb. 1998.

[71] D. Matolak and J. Frolik, "Worse-than-Rayleigh fading: experimental results and theoretical models," *IEEE Commun. Mag.*, **49**, issue 4, 140–146, 2011.

[72] C. F. Mecklenbräuker, A. F. Molisch, J. Karedal, *et al.*, "Vehicle channel characterization and its implications for wireless system design and performance," *Proc. IEEE*, **99**, no. 7, 1189–1212, July 2011.

[73] J. Medbo, H. Hallenberg, and J. E. Berg, "Propagation characteristics at 5 GHz in typical radio-LAN scenarios," *Proc. IEEE Veh. Technol. Conf.*, 185–189, 1999.

[74] J. Medbo and P. Schramm, "Channel models for HIPERLAN/2," *ETSI/BRAN document no. 3ERI085B*.

[75] W. Mohr, "Wideband propagation measurements of mobile radio channels in mountainous areas in the 1800 MHz frequency range," *Proc. IEEE Veh. Technol. Conf.*, 49–52, 1993.

[76] A. F. Molisch, *Wireless Communications* (2nd edn.). J. Wiley & Sons, 2011.

[77] A. F. Molisch, H. Asplund, R. Heddergott, M. Steinbauer, and T. Zwick, "The COST259 directional channel model part I: overview and methodology," *IEEE Trans. Wireless Commun.*, **5**, no. 12, 3421–3433, Dec. 2006.

[78] A. F. Molisch, K. Balakrishnan, C. C. Chong, *et al.*, "A comprehensive model for ultrawideband propagation channels," *IEEE Trans. Antennas Propag.*, **54**, no. 11, 3151–3166, Nov. 2006.

[79] A. F. Molisch, J. R. Foerster, and M. Pendergrass, "Channel models for ultra-wideband personal area networks," *IEEE Wireless Commun. Mag.*, **10**, no. 6, 14–21, Dec. 2003.

[80] A. F. Molisch, L. J. Greenstein, and M. Shafi, "Propagation issues for cognitive radio," *Proc. IEEE*, **97**, no. 5, 787–804, May 2009.

[81] A. F. Molisch and H. Hofstetter, "The COST 273 MIMO channel model," in L. Correia (ed.), *Mobile Broadband Multimedia Networks*, Academic Press, 2006.

[82] A. F. Molisch and F. Tufvesson, "Multipath propagation models for broadband wireless systems," in M. Ibnkahla (ed.) *Handbook of Signal Processing for Wireless Commmunications*, CRC Press, 2004.

[83] A. F. Molisch, F. Tufvesson, J. Karedal, and C. Mecklenbraueker, "A survey on vehicle-to-vehicle propagation channels," *IEEE Wireless Comm.* **16**, issue 6, 12–22, 2009.

[84] M. Nakagami, "The M-distribution: a general formula of intensity of rapid fading," in W. C. Hoffman (ed.), *Statistical Methods Radio Wave Propagation*, Pergamon Press, 1960.

[85] Y. Okumura, E. Ohmori, T. Kawano, and K. Fukuda, "Field strength and its variability in UHF and VHF land-mobile radio service," *Rev. Elec. Commun. Lab.*, **16**, no. 9, 1968.

[86] I. Oppermann, J. Talvitie, and D. Hunter, "Wide-band wireless local loop channel for urban and sub-urban environments at 2 GHz," *Proc. IEEE Int. Conf. Commun.*, 61–65, 1997.

[87] P. Pajusco, "Experimental characterization of DOA at the base station in rural and urban area," *Proc. IEEE Veh. Technol. Conf.*, 993–997, 1998.

[88] K. I. Pedersen, P. E. Mogensen, and B. H. Fleury, "Power azimuth spectrum in outdoor environments," *Electronics Letters*, **33**, 1583–1584, Aug. 1997.

[89] K. I. Pedersen, P. E. Mogensen, B. H. Fleury, F. Frederiksen, K. Olesen, and S. L. Larsen, "Analysis of time, azimuth, and doppler dispersion in outdoor radio channels," *Proc. ACTS Mobile Communications Summit*, Aalborg, Denmark, pp. 308–313, 1997.

[90] A. Rahman and P. Gburzynski, "Hidden problems with the hidden node problem," *23rd Biennial Symposium on Communications*, pp. 270-273, May 29–June 1, 2006.

[91] T. S. Rappaport, "Characterization of UHF multipath radio channels in factory buildings," *IEEE Trans. Antennas Propag.*, **37**, no. 8, 1058–1069, Aug. 1989.

[92] T. S. Rappaport, *Wireless Communications: Principles and Practice*, 2nd edition, Prentice Hall, 2001.

[93] T. S. Rappaport, S. Y. Seidel, and R. Singh, "900-MHz multipath propagation measurements for US digital cellular radiotelephone," *IEEE Trans. Veh. Technol.*, **39**, no. 2, 132–139, May 1990.

[94] Recommendation ITU P 1546-3, "Method for point-to-area predictions for terrestrial services in the frequency range 30 MHz to 3000 MHz," 2007.

[95] A. Saleh and R. A. Valenzuela, "A statistical model for indoor multipath propagation," *IEEE J. Sel. Areas Commun.*, **5**, no. 2, 128–137, Feb. 1987.

[96] M. Schwartz, W. R. Bennett, and S. Stein, *Communication Systems and Techniques*, McGraw-Hill, 1966.

[97] S. Y. Seidel, T. S. Rappaport, S. Jain, M. L. Lord, and R. Singh, "Path loss, scattering and multipath delay statistics in four European cities for digital cellular and microcellular radiotelephone," *IEEE Trans. Veh. Technol.*, **40**, no. 4, 721–730, Nov. 1991.

[98] M. Shafi, M. Zhang, A. L. Moustakas, *et al.*, "Polarized MIMO channels in 3-D: models, measurements and mutual information," *IEEE J. Sel. Areas Commun.*, **24**, no. 3, 514–527, Mar. 2006.

[99] M. Shafi, M. Zhang, A. L. Moustakas, P. J. Smith, A. F. Molisch, F. Tufvesson, and S. H. Simon, "Polarized MIMO channels in 3-D: models, measurements and mutual information," *IEEE J. Sel. Areas Commun.*, **24**, no. 3, 514–527, March 2006.

[100] B. Sklar, "Rayleigh fading channels in mobile digital communication systems. I. Characterization," *IEEE Commun. Mag.*, **35**, no. 9, 136–146, Sep. 1997.

[101] E. S. Sousa, V. M. Jovanovic, and C. Daigneault, "Delay spread measurements for the digital cellular channel in Toronto," *IEEE Trans. Veh. Technol.*, **43**, no. 4, 837–847, Nov. 1994.

[102] "Spatial channel model for multiple input multiple output (MIMO) simulations (Rel. 7)," 3GPP TR 25.996 V7.0.0, 2007.

[103] Q. H. Spencer, B. D. Jeffs, M. A. Jensen, and A. L. Swindlehurst, "Modeling the statistical time and angle of arrival characteristics of an indoor multipath channel," *IEEE J. Sel. Areas Commun.*, **18**, 347–360, Mar. 2000.

[104] M. Steinbauer and A. F. Molisch (chapter eds.), "Directional channel models," in L. Correia (ed.), *Flexible Personalized Wireless Communications*, J. Wiley & Sons, 2001.

[105] M. Steinbauer, A. F. Molisch, and E. Bonek, "The double-directional radio channel," *IEEE Antennas Propag. Mag.*, **43**, no. 4, 51–63, Aug. 2001.

[106] H. A. Suraweera, P. J. Smith and M. Shafi, "Capacity limits and performance analysis of cognitive radio with imperfect channel knowledge," *IEEE Trans. Veh. Technol.*, **59**, no. 4, 1811–1822, May 2010.

[107] S. S. Szyszkowicz, H. Yanikomeroglu, and J. S.Thompson, "On the feasibility of wireless shadowing correlation models," *IEEE Trans. Veh. Technol.*, **59**, no. 9, 4222–4236, Nov. 2010.

[108] L. Talbi and G. Y. Delisle, "Experimental characterization of EHF multipath indoor radio channels," *IEEE J. Sel. Areas Commun.*, **14**, no. 3, 431–440, Apr. 1996.

[109] B. Talha and M. Patzold, "Channel models for mobile-to-mobile cooperative communication systems," *IEEE Veh. Technol. Mag.*, **6**, no. 2, 33–43, 2011.

[110] M. Toeltsch, J. Laurila, K. Kalliola, A. F. Molisch, P. Vainikainen, and E. Bonek, "Statistical characterization of urban spatial radio channels," *IEEE J. Sel. Areas Commun.*, **20**, no. 3, 539–549, Apr. 2002.

[111] "Urban transmission loss models for mobile radio in the 900 and 1800 MHz bands," European Cooperative in the Field of Science and Technical Research EURO-COST 231, Rev. 2, The Hague, Sep. 1991.

[112] J. Weitzen and T. J. Lowe, "Measurement of angular and distance correlation properties of lognormal shadowing at 1900 MHz and its application to design of PCS systems," *IEEE Trans. Veh. Technol.*, **51**, no. 2, 265–273, Mar. 2002.

[113] J. Wiart, P. Pajusco, A. Levy, and J. C. Bic, "Analysis of microcellular wide band measurements in Paris," *Proc. IEEE International Symposium on Personal, Indoor and Mobile Radio Communications*, pp. 144–147, 1995.

[114] M. Z. Win and R. A. Scholtz, "Ultra-wide bandwidth time-hopping spread-spectrum impulse radio for wireless multiple-access communications," *IEEE Trans. Commun.*, **48**, no. 4, 679–689, Apr 2000.

[115] S. Wyne, T. Santos, A. P. Singh, F. Tufvesson, and A. F. Molisch, "Characterisation of a time-variant wireless propagation channel for outdoor short-range sensor networks," *IET Communications*, **4**, 253–264, 2010.

[116] Y. Zhang, J. Zhang, D. Don, et al., "A novel spatial autocorrelation model of shadow fading in urban macro environments," *Proc. IEEE GLOBECOM 2008*, 2008.

[117] Y. P. Zhang and Y. Hwang, "Characterization of UHF radio propagation channels in tunnel environments for microcellular and personal communications," *IEEE Trans. Veh. Technol.*, **47**, 283–296, no. 1, Feb. 1998.

[118] X. Zhao, J. Kivinen, P. Vainikainen, and K. Skog, "Propagation characteristics for wideband outdoor mobile communications at 5.3 GHz," *IEEE J. Sel. Areas Commun.*, **20**, no. 3, 507–514, Apr. 2002.

4 Spectrum sensing

Ezio Biglieri

4.1 Introduction

As discussed in Chapter 2, cognitive radio (CR) aims at maximizing the throughput of secondary users coexisting with primary users under a noninterference or a limited-interference assumption. Both assumptions require the secondary user to collect cognition about the radio environment, or *spectrum sensing*. Here the concept of *spectrum space* is extended to multiple dimensions, such as time, space, frequency, and code, and sensing may include not only detecting and classifying the regions of the spectrum space that can be used by secondary users, but also determining what type of signals are occupying the spectrum, including modulation, waveform, bandwidth, carrier frequency, etc. [50]. All these operations assume that the primary user is not aware of the presence of a secondary user.

Spectrum sensing classifies spectrum spaces as follows:

- *white space,* one which is completely empty, except for noise;
- *gray space,* one which is partially occupied by interfering signals;
- *black space,* one which is fully occupied by communication signals, interfering signals, and noise.

With reference to the CR paradigms categorized in Chapter 2, white spaces are relevant to *interweaving*, which allows secondary users to operate in spectrum regions that are unused, gray spaces to *underlaying*, which tries to keep the interference on the primary user at a tolerable level, and black spaces to *overlaying*, where the primary user transmission is overheard, and signals are processed in a way that makes the quality of this transmission unimpaired by the secondary user.

In this chapter we examine spectrum sensing for application to the first two paradigms. Most of our attention will be devoted to the detection of white spaces, which are spectrum resources left unused by the primary users. We shall also study cognitive underlaying, by examining a metric, the *interference temperature*, that measures the amount of interference that the secondary user cannot exceed when sharing spectrum with the primary user (Section 4.2). Section 4.3 provides a general introduction to cognitive interweaving, and describes a number of techniques for white-space detection. An example of application of spectrum sensing (focusing on OFDM) is given in Section 4.4. Sections 4.5 and 4.6 deal with the effects on the accuracy of spectrum sensing of an inaccurate statistical model of the channel. Only the basic aspects of spectrum sensing will be covered here, while more advanced material is contained in Chapter 5: in particular, while here we focus primarily on single-sensor detection, the next chapter will be centered on collaborative sensing.

Increasing the dimensionality of the observed-signal space
As the spectrum sensing performance must meet an assigned target, two choices are available to enhance its performance if needed. They consist of increasing either the signal-to-noise ratio (SNR) of the primary user, as measured by the secondary user, or the number of dimensions of the observed-signal space [47]. Due to the presence of shadowing and multipath fading, as discussed in Chapter 3, and to channel modeling uncertainties – see Sections 4.5 and 4.6), increasing the SNR may be a challenging task. Thus, dimensionality expansion may be a more attractive solution. This can be achieved by increasing the number of time dimensions (i.e., the number of independent observed samples), or the number of space dimensions. The latter task can be achieved by using more than one sensor, each located at a spatially independent position and working cooperatively. Cooperative sensing will be described in detail in Chapter 5.

Sensing periodicity
The sensors should periodically sense the spectrum, to detect changes in the primary signal transmission pattern. With interweaving, after a white space is detected the secondary users may start transmitting on it, but at the same time should continue to sense the spectrum and detect when the primary user resumes its transmission. Since sensing cannot occur simultaneously with transmission, it must be interleaved with data transmission. The sensing period, i.e., the time interval between two adjacent sensing sessions, should reflect the maximum time duration in which the secondary user may assume that no change has occurred in the spectrum (an appearing or disappearing primary user). This maximum duration will depend on the type of primary service (very small for public safety, larger for services where the spectrum usage varies over a larger time scale). In order to maximize the time available for transmission, secondary users should keep the time required for sensing well below the sensing period: see [15, p. 1053], [17, p. 34 ff.]. Notice that long and frequent sensing sessions improve sensing performance, but reduce the time devoted to data transmission. Conversely, short and widely scattered sensing sessions increase the time allocated to transmission, but degrade the quality of sensing. Thus, there is a tradeoff between sensing duration and secondary-user throughput [28, 34, 47].

4.2 Interference temperature for cognitive underlaying

With cognitive underlaying, the secondary users are allowed to operate on the same spectrum as primary users, provided that the amount of interference power they generate, measured at the primary receiver, does not exceed a given limit and hence can be treated as essentially harmless. Implementation of this concept hinges on the definition of a useful metric for interference tolerance, and on the accurate estimation of the gain of the interference channel between the primary receiver and the secondary transmitter. One definition of a single metric that captures the harm caused by interference and noise on the primary user is that of *interference temperature*. This was introduced by the Federal Communications Commission (FCC) for "quantifying and managing interference" [11].

The interference temperature associated with the power available at a primary-receiver antenna and generated by other emitters and noise sources is defined, in analogy with noise temperature, as

$$T_I \triangleq \frac{\mathcal{P}_I}{kW},$$

where $\mathcal{P}_I$ is the average interfering power in bandwidth W, and $k = 1.38 \times 10^{-23}$ J/K is Boltzmann's constant. If the measured T_I within a frequency band is low enough to achieve communication at a target throughput, then it is assumed that primary and secondary transmissions can coexist in that band. Specifically, if a limit interference temperature T_L is set for a particular frequency band of width B, then the secondary transmitter has to make sure that the average interference power it generates is below kWT_L [46].

As discussed in [5, 6], implementation of a CR system based on the interference temperature model will meet a number of challenges. One involves discerning primary from secondary signals: this problem can be made simpler if the primary signal structure is known, as for example with digital broadcast television. Another problem involves the measurement of T_I in the presence of a primary signal, that is, the determination of the interference floor lying underneath the primary signal. Again, this task is made simpler by the knowledge of the primary signal, as measurement can occur for example during time intervals when the latter is not present. Further, experimental results, obtained with the IEEE 802.11 protocol for wireless networking, show that if theoretical bounds are obtained by computing the temperature interference on the basis of protocol specifications, they are likely to predict a much higher tolerance to interference than that provided by the actual transmission system [31].

Yet another difficulty with the interference-temperature model described above is its dependence on *average* interference power. As observed in [7], it may occur that a narrowband secondary signal severely interferes with a wideband primary user signal, although the average interference power remains below the acceptable threshold. To counteract this shortfall, the authors of [7] advocate a regulatory allowance based on *power spectra* rather than average powers. This would allow one to prevent excessive interference by exploiting the shape of the power spectrum of the secondary or the primary user.

In [16] a regulatory constraint is advocated based on the *maximum tolerated interference power*. Within it, the *probability of harmful interference*, defined as the probability that the interference level at the primary receiver is higher than a given threshold, should not exceed a fixed value. This criterion allows one to account for the unknown number and locations of the CRs and for the randomness of the propagation medium. In fact, if the transmission channel is affected by deep fading or heavy shadowing, as described in

Chapter 3, then a low received signal level might not necessarily be caused by a primary transmitter located outside of the CR interference range.

4.3 White-space detection for cognitive interweaving

Observe from the onset the basic assumption that no infrastructure change is required for primary networks to share spectrum with secondary users. This implies that each CR detects the presence of primary users without any help from them (yet, as we shall see in the next chapter, spectrum sensing can be improved by cooperation among sensors). The following techniques for spectrum sensing will be examined in this chapter.

- *Energy sensing* This detector decides that the primary signal is present if the energy sensed exceeds a given threshold.
- *Coherent sensing* The detector, which has perfect knowledge of the primary-user signal and of the transmission channel, correlates the received signal with a noiseless copy of that signal, and compares the result against a threshold.
- *Cyclostationarity-based sensing* The detector exploits the cyclostationarity of the observed signal, caused by the periodicity of certain features of the signal, like mean and autocorrelation, possibly introduced to assist spectrum sensing. Signals can be distinguished because the observation of a wide-sense stationary process, with no periodic features, denotes absence of the primary signal.
- *Autocorrelation detection* The detector assumes that the primary signal is wide-sense stationary, and exploits its autocorrelation structure to discriminate it from the noise floor.

In the remainder of this chapter, we shall focus on the first four methods. Notice finally that a cognitive-radio receiver may investigate the presence of different types of primary-user signals, and hence needs different detectors running in parallel, each matched to a specific primary signal.

Modeling the observed signal

In its simplest form, the observed signal can be written as

$$y(t) = x(t) + z(t), \qquad (4.1)$$

where $x(t)$ denotes the primary-user signal ($x(t) = 0$ when the primary user is not transmitting), and $z(t)$ is additive disturbance. Specifically, we may assume that

$$z(t) = i(t) + w(t), \qquad (4.2)$$

where $w(t)$ is white Gaussian noise, whereas $i(t)$ denotes another additive interference, modeled as independent of $x(t)$ and $w(t)$.

A single detector must decide between the two hypotheses

$$\mathcal{H}_0 : y(t) = z(t),$$
$$\mathcal{H}_1 : y(t) = x(t) + z(t).$$

Notice that the basic model (4.1) can be enriched by introducing fading and shadowing, and explicity distinguishing between additive noise and interference, as in (4.2). In this case, we write

$$y(t) = hx(t) + i(t) + w(t), \tag{4.3}$$

where h denotes the fading affecting the path between the primary source and the sensor, one which is usually assumed to remain constant during the observation interval (the "block-fading" assumption [3], which expresses the assumption that the coherence time of the multiplicative fading process exceeds the sensing time). All random entities in (4.3) are assumed to be independent of each other. In practice, $y(t)$ will be sampled before processing, with N samples taken at a high enough rate, which allows us to deal with the discrete version of (4.1):

$$y_n = x_n + z_n, \qquad n = 0, \ldots, N-1 \tag{4.4}$$

and of (4.3):

$$y_n = hx_n + i_n + w_n, \qquad n = 0, \ldots, N-1. \tag{4.5}$$

The variable N is usually referred to as the *sensing time*, although its value may generally include not only the number of time dimensions, but also the space dimensions generated by using multiple independent sensors. Fading and shadowing usually impair the performance of a spectrum sensor: in fact, the sensor may mistake a deep fade for the absence of a primary signal.

After sampling, the detection problem is converted into a discrete-time binary hypothesis testing problem, based on the map

$$\mathcal{D}_N : \mathbb{C}^N \to \{\hat{\mathcal{H}}_0, \hat{\mathcal{H}}_1\}, \tag{4.6}$$

where $\mathcal{D}_N$ maps the N-dimensional complex vector $(y_0, \ldots, y_{N-1})$ to the set $\{\hat{\mathcal{H}}_0, \hat{\mathcal{H}}_1\}$ of possible decisions (primary signal absent or present, respectively). For a specific realization of the observation, the detector is in error when $\hat{\mathcal{H}}_i \neq \mathcal{H}_i$, $i = 0, 1$. Two errors of specific interest here are the *false-alarm* and *missed detection* probabilities, whose definitions are

$$P_{\text{FA}} \triangleq \mathbb{P}[\mathcal{D}_N = \hat{\mathcal{H}}_1 \mid \mathcal{H}_0\}, \tag{4.7}$$

$$P_{\text{MD}} \triangleq \mathbb{P}[\mathcal{D}_N = \hat{\mathcal{H}}_0 \mid \mathcal{H}_1\} \tag{4.8}$$

(the *probability of correct detection* $P_{\text{D}} \triangleq 1 - P_{\text{MD}}$ is also used). These probabilities are not independent: they are linked by a function, summarizing the detector performance and called the receiver operating characteristic (ROC), which expresses the dependency between P_{FA} and P_{D} for a given detector and given parameters on which the detection depends (for example, the signal-to-noise ratio) [22, p. 74], [35, p. 27]. In general, P_{D} increases as P_{FA} increases.

To stress the dependence of P_{FA} and P_{MD} on the distribution of the random quantities they involve, we may write them in the alternative form

$$P_{\text{FA}}(Z) \triangleq \mathbb{E}_Z\{[\mathcal{D}_N = \hat{\mathcal{H}}_1] \mid \mathcal{H}_0\}, \tag{4.9}$$

$$P_{\text{MD}}(Z, H, X) \triangleq \mathbb{E}_{H,X,Z}\{[\mathcal{D}_N = \hat{\mathcal{H}}_0] \mid \mathcal{H}_1\}, \tag{4.10}$$

where $[\pi]$ denotes the *Iverson function*, which takes the value 1 if proposition π is true, and the value 0 otherwise, and H, X, Z denote the probability distributions of h and of the sequences x_n and z_n, assumed stationary. Notice that, in our context, having a false alarm is equivalent to neglecting spectral opportunities, while missed detections lead to transmission collisions, and ultimately to throughput reductions (this was thoroughly discussed in Chapter 2). Thus, performance targets should include high throughput for secondary users (i.e., low false-alarm probability) as well as high protection of primary users from unintended interference (i.e., low probability of missed detection). One may define three classes of spectrum sensing schemes based on their keenness towards opportunistic spectrum usage: *conservative*, *aggressive*, and *hostile*, classified according to different target values of P_{FA} and P_{MD}. A conservative system operates with a high value of P_{FA}, and hence a small spectrum utilization rate, and a small probability of interfering with the primary radio. An aggressive system operates with a high spectrum utilization rate, but a small probability of interference. A hostile system has an extremely low value of P_{FA} and a large probability of interference. In [36], the following values are suggested:

- *conservative*, if $P_{\text{FA}} > 0.5$ and $P_{\text{MD}} < 0.5$,
- *aggressive*, if $P_{\text{FA}} < 0.5$ and $P_{\text{MD}} < 0.5$,
- *hostile*, if $P_{\text{FA}} < 0.5$ and $P_{\text{MD}} > 0.5$.

Optimization criteria

Several strategies are available for the optimization of the detector, based on a tradeoff between P_{MD} and P_{FA} (one of the two probabilities can always be made small at the expense of an increase of the other, but a CR system should satisfy constraints on both). We briefly describe here *Bayesian*, *minimax*, and *Neyman–Pearson* criteria. Additional details can be found for example in [35].

Bayes criterion

This is defined after assigning a cost to each of the four possible outcomes of the decision process (Table 4.1). The Bayes criterion consists of minimizing the expected cost, called *risk*. This is defined by

$$\begin{aligned}\mathcal{R} \triangleq{} & C_{00}\,\pi_0\,\mathbb{P}\{\mathcal{D}_N = \hat{\mathcal{H}}_0 \mid \mathcal{H}_0\} \\ & + C_{10}\,\pi_0\,\mathbb{P}\{\mathcal{D}_N = \hat{\mathcal{H}}_1 \mid \mathcal{H}_0\} \\ & + C_{01}\,\pi_1\,\mathbb{P}\{\mathcal{D}_N = \hat{\mathcal{H}}_0 \mid \mathcal{H}_1\} \\ & + C_{11}\,\pi_1\,\mathbb{P}\{\mathcal{D}_N = \hat{\mathcal{H}}_1 \mid \mathcal{H}_1\},\end{aligned} \tag{4.11}$$

Spectrum sensing

$\hat{\mathcal{H}}$	$\mathcal{H}$	cost
$\hat{\mathcal{H}}_0$	$\mathcal{H}_0$	C_{00}
$\hat{\mathcal{H}}_1$	$\mathcal{H}_0$	C_{10}
$\hat{\mathcal{H}}_0$	$\mathcal{H}_1$	C_{01}
$\hat{\mathcal{H}}_1$	$\mathcal{H}_1$	C_{11}

Table 4.1 Decision costs for Bayes criterion.

where π_1 and $\pi_0 = 1 - \pi_1$ are the a priori probabilities of hypotheses $\mathcal{H}_1$ and $\mathcal{H}_0$, respectively. In the special case of no cost for correct decisions ($C_{00} = C_{11} = 0$) and equal costs for false alarms and missed detections ($C_{01} = C_{10} = 1$), the risk becomes the *average error probability*.

Minimax criterion
Using definitions (4.9)–(4.10), we can rewrite (4.11) in the form

$$\mathcal{R} = C_{00}(1 - \pi_1)(1 - P_{\text{FA}}) + C_{10}(1 - \pi_1)P_{\text{FA}}$$
$$+ C_{01}\pi_1 P_{\text{MD}} + C_{11}\pi_1(1 - P_{\text{MD}}), \qquad (4.12)$$

which shows that, for a given set of costs and a given detector, the risk is an affine function of π_1. Thus, if the value of π_1 is not known, it may make sense to select a decision rule that minimizes the maximum value taken by the risk as π_1 ranges in $[0, 1]$.

Neyman–Pearson criterion
In certain circumstances, it may be unrealistic to assign costs to all four detection outcomes. If this is the case, then the Neyman–Pearson criterion can be used. This consists of minimizing P_{MD} subject to a constraint on the probability of false alarm, viz., $P_{\text{FA}} \leq \alpha$. Notice how this criterion admits a basic asymmetry in the importance of the two hypotheses.

Likelihood-ratio test
Central to the Neyman–Pearson criterion is the log-likelihood ratio

$$\Lambda \triangleq \log \frac{p(y \mid \mathcal{H}_1)}{p(y \mid \mathcal{H}_0)}, \qquad (4.13)$$

i.e., the logarithm of the ratio between the conditional pdfs of the observation under hypotheses $\mathcal{H}_1$ and $\mathcal{H}_0$. The decision in favor of $\mathcal{H}_1$ or $\mathcal{H}_0$ is made by comparing Λ against a threshold ψ:

$$\Lambda \underset{\mathcal{H}_0}{\overset{\mathcal{H}_1}{\gtrless}} \psi. \qquad (4.14)$$

Generalized likelihood-ratio test

If some of the parameters of the transmission model are unknown (for example, the fading attenuation h in (4.3)), then a useful (albeit suboptimal) approach to detection consists of using (4.14) where Λ is replaced by the *generalized* log-likelihood ratio

$$\Lambda_G \triangleq \log \frac{\sup_\theta p(y \mid \mathcal{H}_1, \boldsymbol{\theta})}{\sup_\theta p(y \mid \mathcal{H}_0, \boldsymbol{\theta})} \qquad (4.15)$$

and $\boldsymbol{\theta}$ denotes the unknown parameters. Consequently, the generalized likelihood-ratio test is

$$\Lambda_G \underset{\mathcal{H}_0}{\overset{\mathcal{H}_1}{\gtrless}} \psi_G. \qquad (4.16)$$

4.3.1 Energy sensing

This is the simplest form of spectrum sensing, due to its conceptual simplicity and reduced implementation complexity. The presence of a spectrum hole is detected by comparing the measured energy against a suitable threshold, which is highly susceptible to the noise floor, to the presence of in-band interferences, and to channel notches caused by frequency-selective fading.

The decision metric, denoted Y, is built from a sequence of N received-signal samples:

$$Y = \frac{1}{N} \sum_{n=0}^{N-1} |y_n|^2 \qquad (4.17)$$

and is obtained from the formal definition of signal energy,

$$Y = \int |y(t)|^2 \, dt$$

after sampling at a sufficiently high rate. An assumption which simplifies analysis, and one we make here, is that there is no fading and no disturbance other that Gaussian noise, whose samples are independent, circularly symmetric zero-mean complex Gaussian random variables with variance $\mathbb{E}\left[|w_n|^2\right] = \sigma_w^2$, i.e., $w_n \sim \mathcal{N}(0, \sigma_w^2)$ (the model for the signal is more complicated, as it should include information on the signal statistics as well as on the fading that may affect it). The decision strategy consists of comparing the "test statistic" Y against a suitably optimized threshold. Since the decision threshold depends on the observed-signal model, and hence on the noise variance, an error on the estimate of this variance affects the performance of the detector, which becomes vulnerable to noise power inaccuracies (more on this below). In particular, the energy detector cannot discriminate between signal and noise, which increases uncertainty especially at low signal-to-noise ratios. A possible solution to this problem consists of complementing the energy detector with a *feature detector*, which takes advantage of one or more peculiar features of the primary signal (for example, its spectral correlation, or the presence of a cyclic prefix; see below).

Performance
Known primary signal
Assume a deterministic, known primary signal. In this situation, the probability of a false alarm is

$$P_{\text{FA}}(W) = \mathbb{E}_W \mathbb{P}\left(\frac{1}{N}\sum_{n=0}^{N-1}|w_n|^2 > \theta\right)$$

$$= \mathbb{E}_W \mathbb{P}\left(\frac{2}{\sigma_w^2}\sum_{n=0}^{N-1}|w_n|^2 > \frac{2N}{\sigma_w^2}\theta\right). \quad (4.18)$$

With $W = \mathcal{N}_c(0, \sigma_w^2)$, the random variable appearing before the inequality sign in the last equation has a chi-square distribution with $2N$ degrees of freedom.

Digression on chi-square distributions
Before proceeding further with our calculations, we digress to summarize some basic facts about chi-square distributions. The chi-square distribution with $2N$ degrees of freedom, denoted χ_{2N}, is the distribution of the sum X of the squares of $2N$ independent real Gaussian RVs $\sim \mathcal{N}(0, 1)$, or, equivalently, of the squared magnitudes of N independent circularly symmetric complex Gaussian RVs $\sim \mathcal{N}_c(0, 2)$. The corresponding pdf has the form

$$f(x) = \frac{1}{2^N \Gamma(N)} x^{N-1} e^{-x/2}, \qquad x \geq 0,$$

where $\Gamma(\cdot)$ is Euler's Gamma function

$$\Gamma(a) \triangleq \int_0^\infty t^{a-1} e^{-t} \, dt.$$

The cumulative distribution function of X is

$$F(x) = 1 - Q(N, x/2),$$

where $Q(\cdot, \cdot)$ is the normalized incomplete Gamma function [33, p. 174]

$$Q(a, z) \triangleq \frac{\Gamma(a, z)}{\Gamma(a)}$$

and $\Gamma(\cdot, \cdot)$ is the incomplete Gamma function

$$\Gamma(a, z) \triangleq \int_z^\infty t^{a-1} e^{-t} \, dt.$$

X has mean value $2N$ and variance $4N$. Hence, by the central limit theorem, if X has the chi-square distribution, then as $N \to \infty$ the RV $(X - 2N)/\sqrt{4N}$ converges to the Gaussian distribution $\mathcal{N}(0, 1)$.

A generalization of the chi-square distribution occurs when the N independent complex Gaussian RVs are $\sim \mathcal{N}_c(\mu_i, 1)$, $\mu_i \neq 0$, $i = 1, \ldots, N$. In this case the distribution is called *noncentral chi-square* with noncentrality parameter $\lambda \triangleq |\mu_1|^2 + \ldots, |\mu_N|^2$, and is denoted $\chi^2_{2N}(\lambda)$. The corresponding pdf is

$$f(x) = \frac{1}{2} e^{-(x+\lambda)/2} \left(\frac{x}{\lambda}\right)^{N/2-1/2} I_{N-1}\left(\sqrt{\lambda x}\right),$$

where $I_\nu(\cdot)$ is the modified Bessel function of the first kind and order ν. The cumulative distribution function is

$$F(x) = 1 - Q_N(\sqrt{\lambda}, \sqrt{x}),$$

where $Q_N(\cdot, \cdot)$ denotes the generalized Marcum Q-function

$$Q_N(a,b) \triangleq \frac{1}{a^{N-1}} \int_b^\infty x^N \exp\left(-\frac{x^2+a^2}{2}\right) I_{N-1}(ax)\, dx. \qquad (4.19)$$

We have in particular [38, Eq. (4.71)]

$$Q_N(0,b) = \frac{\Gamma(N, b^2/2)}{\Gamma(N)} = Q(N, b^2/2). \qquad (4.20)$$

Back to known primary signal
With the notations described in the previous paragraph, we obtain

$$P_{\text{FA}} = Q(N, N\theta/\sigma_w^2), \qquad (4.21)$$

where θ is the decision threshold. The probability of missed detection is

$$P_{\text{MD}} = 1 - Q_N\left(\sqrt{\lambda}, \sqrt{2N\theta/\sigma_w^2}\right), \qquad (4.22)$$

where

$$\lambda \triangleq \frac{2}{\sigma_z^2} \sum_{n=0}^{N-1} |x_n|^2. \qquad (4.23)$$

Gaussian primary signal
A similar analysis can be based on the assumption that the primary signal is random, and in particular zero-mean Gaussian: $x_n \sim \mathcal{N}_c(0, \mathcal{P})$ [50]. The probability of false alarm is the same as in (4.21), while the probability of missed detection is given by [9, 30]

$$P_{\text{MD}} = \mathbb{P}\left(\frac{2}{\mathcal{P}+\sigma_w^2} \sum_{n=0}^{N-1} |x_n + w_n|^2 < \frac{2N}{\mathcal{P}+\sigma_w^2}\theta\right)$$

$$= 1 - Q\left(N, \frac{N}{\mathcal{P}+\sigma_w^2}\theta\right), \qquad (4.24)$$

where θ again denotes the decision threshold.

When the number N of signal samples is large, the central limit theorem can be invoked to obtain the approximations

$$\frac{2}{\sigma_w^2} \sum_{n=0}^{N-1} |w_n|^2 \sim \mathcal{N}(2N, 4N),$$

$$\frac{2}{\mathcal{P}+\sigma_w^2} \sum_{n=0}^{N-1} |x_n + w_n|^2 \sim \mathcal{N}(2N, 4N).$$

In this asymptotic regime, we obtain

$$P_{\text{FA}} \sim Q\left(\sqrt{N}\frac{\theta - \sigma_w^2}{\sigma_w^2}\right), \tag{4.25}$$

$$P_{\text{MD}} \sim 1 - Q\left(\sqrt{N}\frac{\theta - (\mathcal{P} + \sigma_w^2)}{\mathcal{P} + \sigma_w^2}\right), \tag{4.26}$$

where we have used the properties

$$\mathbb{P}(X > x) = Q\left(\frac{x - \mu}{\sigma}\right), \tag{4.27}$$

$$\mathbb{P}(X < x) = 1 - Q\left(\frac{\mu - x}{\sigma}\right), \tag{4.28}$$

valid for $X \sim \mathcal{N}(\mu, \sigma^2)$, with $Q(\cdot)$ the Gaussian tail function [33, p. 160]

$$Q(x) \triangleq \frac{1}{\sqrt{2\pi}} \int_x^\infty e^{-z^2/2}\, dz. \tag{4.29}$$

4.3.2 Coherent detection

To improve the performance of the spectrum sensor, one may consider exploiting any prior knowledge about the features of the primary signal. These may consist of modulation type and order, pulse shape, data rate, or statistical properties. Along this line, we examine first the limiting situation in which the primary signal $x(t)$ is fully known. This assumes that the primary user signal can be demodulated coherently, which requires timing and carrier synchronization (this task may be facilitated by the presence, in the primary signal, of pilots, preambles, etc.).

If there is no interference, and the noise is Gaussian and stationary, the optimal detector, i.e., the one which maximizes the received signal-to-noise ratio, consists of a filter matched to $x(t)$ followed by a threshold test (if there is fading, and the fading coefficient h is known at the receiver, then the filter should be matched to $hx(t)$).

The maximum-likelihood statistic of the signal $y_n = x_n + w_n$, $n = 0, \ldots, N-1$, obtained after sampling, is

$$Y = \frac{1}{N}\sum_{n=0}^{N-1} y_n x_n^\star, \tag{4.30}$$

(where $x^\star$ denotes the conjugate of x), so that the signal-to-noise ratio of the matched-filter output is $\gamma = \mathcal{P}/\sigma_w^2$, where

$$\mathcal{P} \triangleq \frac{1}{N}\sum_{n=0}^{N-1} |x_n|^2 \tag{4.31}$$

is the average signal power under the assumption that the primary signal is stationary and the sensing time N is long enough. The false-alarm and missed-detection probabilities are

$$P_{\text{FA}} = Q\left(\sqrt{N}\frac{\theta}{\sqrt{\mathcal{P}\sigma_w^2}}\right),$$

$$P_{\text{MD}} = Q\left(\sqrt{N}\frac{\mathcal{P}-\theta}{\sqrt{\mathcal{P}\sigma_w^2}}\right),$$

with Q the Gaussian tail function.

Unknown phase
If the carrier phase of the received signal is unknown but constant, then coherent detection can be modified by computing the test statistic

$$Y = \left|\frac{1}{N}\sum_{n=0}^{N-1} y_n x_n^\star\right|^2 = \left|\mathcal{P} + \frac{1}{N}\sum_{n=0}^{N-1} w_n x_n^\star\right|^2. \tag{4.32}$$

Imperfect time alignment
Coherent detection assumes perfect time alignment at the receiver of the transmitted data sequence, which is usually not verified in practice. If there is a maximum time offset of ν time instants, then the detector may rather form the following test statistic, based on an exhaustive search of the actual time offset:

$$Y = \max_{0\leq\ell\leq\nu} \left|\frac{1}{N}\sum_{n=0}^{N-1} y_{n+\ell} x_n^\star\right|^2. \tag{4.33}$$

Coherent detection using a pilot tone
In practice, since the sequence (x_n) is not known at the receiver if it consists of random information data, detection may be performed with the aid of *pilot tones*. With this model, the primary signal includes a known pilot tone x_n^{P}, orthogonal to the white information signal x_n and to which a fraction δ of energy is allocated. The observed-signal model becomes [37]

$$y_n = \sqrt{\delta}x_n^{\text{P}} + \sqrt{1-\delta}x_n + z_n. \tag{4.34}$$

The test statistic is

$$Y = \frac{1}{N}\sum_{n=0}^{N-1} y_n \hat{x}_n^{\text{P}}, \tag{4.35}$$

where $(\hat{x}_n^{\text{P}})$ is a sequence parallel to the pilot tone.

4.3.3 Cyclostationarity-based detection

If the signal $x(t)$ is not completely known, but some of its features are, then one may take advantage of this knowledge to form a test statistic closely matched to the signal. A feature capable of distinguishing a modulated signal from noise is obtained by observing that, generally, modulated signals are not stationary, not even in a wide sense. Many of them are *cyclostationary*, which means that some of their statistics exhibit periodicity (in particular, the mean and the autocorrelation function of a *wide-sense cyclostationary* process are periodic functions of time) [14]. *Cyclostationarity-based detection* consists of analyzing the *cyclic autocorrelation function* of the received signal, which is periodic for data signals, but aperiodic for noise and, more generally, for wide-sense stationary signals [12, 13]. The periods in the primary signals are related to carrier frequency, symbol, chip, code, hop rate, etc., and they can often be assumed as known (notice, however, that synchronization errors may affect this detection procedure). This technique works well even in very-low SNR regions or with uncertainty in the value of noise power, where energy detection might fail. In addition, it can discriminate among useful signals and interferences whenever the latter are either stationary, or cyclostationary with different periods.

For motivation, consider the linearly modulated signal

$$x(t) = \sum_n a_n g(t - nT_0), \qquad -\infty < t < \infty, \qquad (4.36)$$

where T_0 is the symbol period, $g(t)$ the modulation pulse, and the data sequence (a_n) is assumed to form an identically distributed wide-sense stationary sequence with mean 0 and autocorrelation $r_{n-m} \triangleq \mathbb{E}[a_n a_m^*]$. The mean of $x(t)$ is zero, and its autocorrelation function is given by

$$\begin{aligned} A_x(t,\tau) &\triangleq \mathbb{E}[x(t+\tau)x^*(t)] \\ &= \sum_n \sum_m r_{n-m} g(t+\tau-nT_0) g^*(t-nT_0). \end{aligned} \qquad (4.37)$$

It can be observed that, with n, m extending from $-\infty$ to ∞, $A_x(t,\tau)$ in (4.37) is a periodic function of t with period T_0. Consequently, it can be expanded in a Fourier series as

$$A_x(t,\tau) = \sum_n A_x^\alpha(\tau) e^{j2\pi\alpha t}, \qquad \alpha \triangleq n/T_0 \qquad (4.38)$$

with "Fourier coefficients" $A_x^\alpha(\tau)$ defined by

$$A_x^\alpha(\tau) \triangleq \frac{1}{T_0} \int_{-T_0/2}^{T_0/2} A_x(t,\tau) e^{-j2\pi\alpha t}\, dt, \qquad \alpha = n/T_0. \qquad (4.39)$$

Based on the above example, we define a wide-sense cyclostationary process $x(t)$ as one whose mean $\mathbb{E}[x(t)]$ and autocorrelation function $A_x(t,\tau)$ are periodic functions of t. The Fourier coefficients (4.39), interpreted as a function of τ, form the *cyclic autocorrelation function (CAF)* of $x(t)$, with α called *cyclic frequency* (the choice $\alpha = 0$ yields the standard autocorrelation function of $x(t)$). The *cyclic spectral density (CSD)* of $x(t)$ is defined as the Fourier transform of $A_x^\alpha(\tau)$:

$$S_x^\alpha(f) \triangleq \int_{-\infty}^{\infty} A_x^\alpha(\tau) e^{-j2\pi f \tau}\, d\tau. \qquad (4.40)$$

Now, if T_0 is known for the primary signal, then the CAF can be used to detect its presence by evaluating $A_x^\alpha(\tau)$. In fact, for a cyclostationary signal, a nonzero cyclic frequency (i.e., a nonzero value of α) exists such that $A_x^\alpha(\tau) \neq 0$. Equivalently, the cyclic spectral density has peaks at cyclic frequencies that are multiples of $1/T_0$, the fundamental frequency of the received signal.

Notice how this technique is critically dependent on the assumption that the period T_0 is known a priori. If this assumption is not valid, then cyclostationarity-based detection requires either the search for, or the extraction of, the cyclic frequencies, which entails a considerable increase of its complexity. This technique can also be used to discriminate among different types of transmission and of primary users. Checks for the presence of cycles in CAF and CSD can be done by performing an exhaustive search over candidate cycles, which are those yielding a statistically significant nonzero cyclic statistics. To check if $A_x^\alpha(\tau)$ is zero for a given candidate cycle α, the following estimator can be used:

$$\hat{A}_x^\alpha(\tau) = \frac{1}{N} \sum_{m=0}^{N-1} x_{m+\tau} x_m^* e^{-j2\pi\alpha\tau} \quad (4.41)$$

$$= A_x^\alpha(\tau) + \varepsilon_x^\alpha(\tau), \quad (4.42)$$

where $\varepsilon_x^\alpha(\tau)$ denotes the estimation error. If the estimator is consistent, this error vanishes asymptotically as the observation length tends to infinity. In practice, due to the finite observation length, $\hat{A}_x^\alpha(\tau) \neq 0$, and we must discriminate between the two hypotheses $\hat{A}_x^\alpha(\tau) = \varepsilon_x^\alpha(\tau)$ (no primary signal) and $\hat{A}_x^\alpha(\tau) = A_x^\alpha(\tau) + \varepsilon_x^\alpha(\tau)$. This can be done by choosing a set $\{\tau_1, \ldots, \tau_N\}$ and transforming the autocorrelation functions into N-vectors. An asymptotic analysis of the detector performance is contained in [8].

A possible (frequency-domain) test statistic for the detector at a given value of α (the "single-cycle" cyclostationarity detector for the cyclic feature of interest) is

$$Y = \int_{-f_s/2}^{f_s/2} \hat{S}_y^\alpha(f) [S_x^\alpha(f)]^* \, df, \quad (4.43)$$

where f_s is the sampling frequency, and $\hat{S}_y^\alpha(f)$ is an estimate of the spectral correlation function of the received signal.

4.3.4 Autocorrelation-based detection

This detection technique assumes that the primary signal is wide-sense stationary, and exploits its autocorrelation to discriminate it from white noise. Let the sampled observed signal have the form

$$y_n = x_n + w_n, \quad 0 \leq n \leq N - 1. \quad (4.44)$$

Then the $L \times L$ sample autocorrelation matrix of (y_n) is given by

$$\widehat{\mathbf{A}}_y = \frac{1}{N} \sum_{n=L-1}^{N-1} \mathbf{y}_n \mathbf{y}_n^\dagger, \quad (4.45)$$

where $\mathbf{y}_n$ is the column vector

$$\mathbf{y}_n = (y_n, y_{n-1}, \ldots, y_{n-L+1})^{\text{T}}. \tag{4.46}$$

As $N \to \infty$, under the assumption that the additive noise is white and the signal is uncorrelated with noise, $\widehat{\mathbf{A}}_y$ converges in probability to the autocorrelation of (y_n):

$$\mathbf{A}_y \triangleq \mathbb{E}[\mathbf{y}_n \mathbf{y}_n^\dagger] = \mathbf{A}_x + \sigma_w^2 \mathbf{I}_L, \tag{4.47}$$

where $\mathbf{A}_x$ is the autocorrelation matrix of the column vector

$$\mathbf{x}_n = (x_n, x_{n-1}, \ldots, x_{n-L+1})^{\text{T}}, \tag{4.48}$$

σ_w^2 is the noise average power, as usual, and $\mathbf{I}_L$ denotes the $L \times L$ identity matrix. If there is no primary signal, then $\mathbf{A}_y$ equals the scalar matrix $\sigma_w^2 \mathbf{I}_L$, which has zero off-diagonal elements and equal eigenvalues σ_w^2. Thus, the presence of a primary signal can be detected when one of these properties does not hold, as detailed below. Notice that the tests based on autocorrelation use only quantities associated with the observed signal, so that no information on the transmitted signal and the transmission channel are needed.

Dealing with nonwhite noise

In practice, the assumption that the noise is white is made invalid by the presence of a receiving filter. In this situation, consider the discrete filter impulse response $(h_n)_{n=0}^K$, normalized so that

$$\sum_{n=0}^{K} |h_k|^2 = 1. \tag{4.49}$$

After filtering, the received signal has the form

$$\tilde{y}_n = \sum_{k=0}^{K} h_k y_{n-k}. \tag{4.50}$$

Defining, similarly,

$$\tilde{x}_n \triangleq \sum_{k=0}^{K} h_k x_{n-k}, \tag{4.51}$$

$$\tilde{w}_n \triangleq \sum_{k=0}^{K} h_k w_{n-k}, \tag{4.52}$$

we obtain $\tilde{y}_n = \tilde{x}_n + \tilde{w}_n$. Defining further the vectors

$$\tilde{\mathbf{y}}_n \triangleq (\tilde{y}_n, \tilde{y}_{n-1}, \ldots, \tilde{y}_{n-L+1})^{\text{T}}, \tag{4.53}$$

$$\tilde{\mathbf{x}}_n \triangleq (\tilde{x}_n, \tilde{x}_{n-1}, \ldots, \tilde{x}_{n-L+1})^{\text{T}}, \tag{4.54}$$

$$\tilde{\mathbf{w}}_n \triangleq (\tilde{w}_n, \tilde{w}_{n-1}, \ldots, \tilde{w}_{n-L+1})^{\text{T}}, \tag{4.55}$$

the autocorrelation matrices

$$\mathbf{A}_{\tilde{y}} \triangleq \mathbb{E}[\tilde{\mathbf{y}}_n \tilde{\mathbf{y}}_n^\dagger], \tag{4.56}$$

$$\mathbf{A}_{\tilde{x}} \triangleq \mathbb{E}[\tilde{\mathbf{x}}_n \tilde{\mathbf{x}}_n^\dagger], \tag{4.57}$$

$$\mathbf{A}_{\tilde{w}} \triangleq \mathbb{E}[\tilde{\mathbf{w}}_n \tilde{\mathbf{w}}_n^\dagger], \tag{4.58}$$

and the $L \times (L+K)$ "filtering matrix"

$$\mathbf{H} = \begin{bmatrix} h_0 & h_1 & \cdots & h_K & 0 & \cdots & 0 \\ 0 & h_0 & \cdots & h_{K-1} & h_K & \cdots & 0 \\ & & \ddots & & & & \\ 0 & 0 & \cdots & h_0 & h_1 & \cdots & h_K \end{bmatrix}, \tag{4.59}$$

we obtain, under the usual assumption of uncorrelated signal and noise,

$$\mathbf{A}_{\tilde{y}} = \mathbf{A}_{\tilde{x}} + \mathbf{A}_{\tilde{w}}, \tag{4.60}$$

where

$$\mathbf{A}_{\tilde{w}} = \sigma_w^2 \mathbf{G}, \tag{4.61}$$

with $\mathbf{G}$ the positive definite Hermitian matrix

$$\mathbf{G} \triangleq \mathbf{H}\mathbf{H}^\dagger. \tag{4.62}$$

With the new definitions

$$\mathbf{A}_y \triangleq \mathbf{G}^{-1/2} \mathbf{A}_{\tilde{y}} \mathbf{G}^{-1/2}, \tag{4.63}$$

$$\mathbf{A}_x \triangleq \mathbf{G}^{-1/2} \mathbf{A}_{\tilde{x}} \mathbf{G}^{-1/2}, \tag{4.64}$$

we obtain

$$\mathbf{A}_y = \mathbf{A}_x + \sigma_w^2 \mathbf{I}_L \tag{4.65}$$

as in (4.47).

Eigenvalue-ratio test

This technique was introduced in [53, 54]. Denoting by $\rho_{\min}$ and $\rho_{\max}$ the minimum and maximum eigenvalues of $\mathbf{A}_x$, and by $\mu_{\min}$ and $\mu_{\max}$ the minimum and maximum eigenvalues of $\mathbf{A}_y$, respectively, we have

$$\mu_{\max} = \rho_{\max} + \sigma_w^2 \quad \text{and} \quad \mu_{\min} = \rho_{\min} + \sigma_w^2, \tag{4.66}$$

so that

$$\frac{\mu_{\max}}{\mu_{\min}} = 1 \tag{4.67}$$

when there is no primary signal, and

$$\frac{\mu_{\max}}{\mu_{\min}} > 1 \tag{4.68}$$

otherwise. Thus, a sensible decision test can be based on the ratio $\hat{\mu}_{\max}/\hat{\mu}_{\min}$, where $\hat{\mu}_{\min}$ and $\hat{\mu}_{\max}$ are the maximum and minimum eigenvalues of the sample covariance matrix $\widehat{\mathbf{A}}_y$.

Maximum-eigenvalue test
Consider again (4.47). If there is no primary signal in (y_n), then $\mathbf{A}_x = \mathbf{0}$, otherwise $\mathbf{A}_x \neq \mathbf{0}$. If $\mu_{\max}$ and $\rho_{\max}$ denote, as usual, the maximum eigenvalues of $\mathbf{A}_y$ and $\mathbf{A}_x$, respectively, then $\mu_{\max} = \rho_{\max} + \sigma_w^2$, with $\rho_{\max} = 0$ if and only if $\mathbf{A}_x = \mathbf{0}$. This can also be expressed by saying that, if the primary signal is present, then $\mu_{\max} > \sigma_w^2$, while without a signal $\mu_{\max} = \sigma_w^2$. Based on this observation, the maximum eigenvalue can be used as a test of the presence of the primary signal [51].

Energy/minimum-eigenvalue ratio test
This test [54] compares the energy of the observed signal (y_n) with the minimum eigenvalue of $\mathbf{A}_y$. If their ratio exceeds a given threshold, then the primary signal is assumed to be present.

Diagonal/off-diagonal ratio test
To quantify the terms lying outside of the main diagonal in the correlation matrix $\mathbf{A}_y = (r_{nm})_{n,m=1}^L$, we may define the quantities

$$U_{\mathrm{d}}^\alpha \triangleq \frac{1}{L} \sum_{n=1}^{L} |r_{n,n}|^\alpha, \tag{4.69}$$

$$U_{\mathrm{f}}^\alpha \triangleq \frac{1}{L} \sum_{n=1}^{L} \sum_{m=1}^{L} |r_{n,m}|^\alpha, \tag{4.70}$$

with $\alpha = 1, 2$. With no signal, we obtain $U_{\mathrm{f}}^\alpha = U_{\mathrm{d}}^\alpha$, while with signal $U_{\mathrm{f}}^\alpha > U_{\mathrm{d}}^\alpha$. The detection tests consists of comparing the ratio $U_{\mathrm{f}}^\alpha/U_{\mathrm{d}}^\alpha$ with a suitable threshold [52].

4.4 An application: spectrum sensing with OFDM

As an application, we examine here the special case of spectrum sensing with orthogonal frequency-division multiplexing (OFDM) [18, Chapter 12], a scheme used in a number of current and future commercial communication systems (WiFi, WiMAX, LTE, and DVB-T). OFDM signals are cyclostationary, and hence spectrum sensors exploiting cyclostationarity can be used. However, better sensors can be designed that are tailored to the specific signal structure of OFDM, and, in particular, the presence in it of a cyclic prefix (CP) of known length. This duplicates the last information data at the beginning of an OFDM symbol, and hence generates a correlation in the symbol stream. We study, following [1], coherent and noncoherent detection.

We first derive a vector model of the observed signal. Figure 4.1 shows a typical observation frame. The noisy observed sequence $\mathbf{y} = \mathbf{x} + \mathbf{z}$ includes N samples taken from K OFDM symbols. Each symbol consists of a data sequence D with length N_D

4.4 An application: spectrum sensing with OFDM

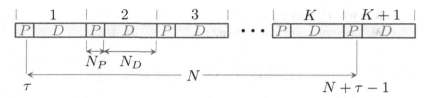

Figure 4.1 Received OFDM signal. P denotes the cyclic prefix with length N_P, and D the information data with length N_D.

preceded by a cyclic prefix P with length $N_P \leq N_D$. Assuming that the receiver is not synchronized with the transmitter, we denote by τ the synchronization mismatch, i.e., the epoch of the first sample (in general, $0 \leq \tau \leq N_P + N_D$, with $\tau = 0$ if synchronization is perfect). The total number of samples is $N = K(N_P + N_D)$, and $\mathbf{y}$ includes samples from $K+1$ OFDM symbols (or K only when $\tau = 0$). We assume that there is no clock drift along the sequence of K symbols. Let $\mathbf{d}_i$ denote the N_D-vector of data associated with the ith OFDM symbol $\mathbf{x}_i$. The latter is obtained by concatenating to $\mathbf{d}_i$ the last N_P elements of $\mathbf{d}_i$. In vector form,

$$\mathbf{x}_i = \mathbf{U}\mathbf{d}_i, \qquad (4.71)$$

where $\mathbf{U}$ is the $(N_P + N_D) \times N_D$ matrix

$$\mathbf{U} \triangleq \begin{bmatrix} \mathbf{0}_{N_P \times (N_D - N_P)} & \mathbf{I}_{N_P} \\ \mathbf{I}_{N_D} & \end{bmatrix}. \qquad (4.72)$$

Here, $\mathbf{0}_{m \times n}$ denotes the $m \times n$ matrix all of whose entries are zero, and $\mathbf{I}_n$ the $n \times n$ identity matrix. Define further the block-diagonal $(K+1)(N_P + N_D) \times (K+1)N_D$ matrix consisting of $K+1$ copies of $\mathbf{U}$:

$$\mathbf{T} \triangleq \mathrm{diag}[\underbrace{\mathbf{U}\,\mathbf{U}\,\cdots\,\mathbf{U}}_{K+1}]. \qquad (4.73)$$

The vector $\mathbf{x}$ consisting of $K+1$ OFDM symbols is obtained by removing the first τ and the last $(N_P + N_D) - \tau$ samples from $\mathbf{T}\mathbf{d}$. Thus, we can write

$$\mathbf{x} = \mathbf{T}_\tau \mathbf{d}, \qquad (4.74)$$

where $\mathbf{d} \triangleq [\mathbf{d}_1^T\ \mathbf{d}_2^T\ \cdots\ \mathbf{d}_{K+1}^T]^T$, and $\mathbf{T}_\tau$ is the $K(N_P + N_D) \times (K+1)N_D$ matrix obtained by deleting the first τ and the last $(N_P + N_D) - \tau$ rows of $\mathbf{T}$.

4.4.1 Neyman–Pearson detection

Invoking the central limit theorem, we assume the data vector $\mathbf{d}$ to be white Gaussian, i.e., $\mathbf{d} \sim \mathcal{N}_c(\mathbf{0}, \sigma_d^2 \mathbf{I})$, where σ_d^2 is the variance of the complex signal samples. The conditional autocorrelation of $\mathbf{x}$ is

$$\mathbb{E}[\mathbf{x}\mathbf{x}^\dagger \mid \tau] = \mathbb{E}[\mathbf{T}_\tau \mathbf{d}\mathbf{d}^\dagger \mathbf{T}_\tau^T] = \sigma_d^2 \mathbf{T}_\tau \mathbf{T}_\tau^T, \qquad (4.75)$$

so that $\mathbf{x}|\tau \sim \mathcal{N}_c(\mathbf{0}, \sigma_d^2 \mathbf{T}\mathbf{T}_\tau^T)$. We consider separately the cases of known and unknown values of σ_w^2 and σ_d^2.

Known σ_w^2 and σ_d^2

We have, assuming τ unknown,[1]

$$p(\mathbf{y} \mid \mathcal{H}_0) = \frac{1}{\pi^N \sigma_w^{2N}} \exp\left(-\frac{\|\mathbf{y}\|^2}{\sigma_w^2}\right), \qquad (4.76)$$

$$p(\mathbf{y} \mid \mathcal{H}_1) = \frac{1}{\pi^N \det \mathbf{Q}_\tau^{-1}} \exp\left(-\mathbf{y}^\dagger \mathbf{Q}_\tau \mathbf{y}\right), \qquad (4.77)$$

where we have defined

$$\mathbf{Q}_\tau \triangleq \sigma_d^2 \mathbf{T}_\tau \mathbf{T}_\tau^T + \sigma_w^2 \mathbf{I}. \qquad (4.78)$$

Thus, the log-likelihood ratio for the decision between $\mathcal{H}_0$ and $\mathcal{H}_1$ has the form

$$\begin{aligned} Y &\triangleq \log \frac{p(\mathbf{y} \mid \mathcal{H}_1)}{p(\mathbf{y} \mid \mathcal{H}_0)} \\ &= \log \frac{\sigma_w^{2N}}{\det \mathbf{Q}_\tau} - \mathbf{y}^\dagger \left(\mathbf{Q}_\tau^{-1} - \sigma_w^{-2}\mathbf{I}\right) \mathbf{y}. \end{aligned} \qquad (4.79)$$

Direct computation of $\det \mathbf{Q}_\tau$ and of $\mathbf{y}^\dagger \left(\mathbf{Q}_\tau^{-1} - \sigma_w^{-2}\mathbf{I}\right) \mathbf{y}$, required to evaluate Y, may entail a considerable complexity when N is large (see [1] for a simplification exploiting the sparseness of the matrix $\mathbf{Q}_\tau$).

As a special case, observe that with no cyclic prefix ($N_P = 0$, which corresponds to no usable structure of the observed signal), we have $\mathbf{T}_\tau \mathbf{T}_\tau^T = \mathbf{I}$ and $\mathbf{y}|\mathcal{H}_1 \sim \mathcal{N}_c(\mathbf{0}, (\sigma_d^2 + \sigma_z^2)\mathbf{I})$, which results into the test statistic

$$Y = \|\mathbf{y}\|^2$$

and shows that the energy detector is optimal in these conditions.

Unknown σ_w^2 and σ_d^2

If these two parameters are not known, but their a priori distribution is available, one may still use the likelihood-ratio test after eliminating them by standard marginalization:

$$p(\mathbf{y} \mid \mathcal{H}_0) = \mathbb{E}_{\sigma_w^2}\, p(\mathbf{y} \mid \mathcal{H}_0, \sigma_w^2), \qquad (4.80)$$

$$p(\mathbf{y} \mid \mathcal{H}_1) = \mathbb{E}_{\sigma_w^2, \sigma_d^2}\, p(\mathbf{y} \mid \mathcal{H}_1, \sigma_w^2, \sigma_d^2), \qquad (4.81)$$

which assumes the knowledge of the a priori distributions of σ_w^2 and σ_d^2. Another option is the use of the generalized likelihood ratio test (4.16), which consists here of replacing σ_w^2 and σ_d^2 with their maximum-likelihood estimates in (4.76)–(4.77).

---

[1] But not random: the extension to a random τ requires the marginalization of $p(\mathbf{y}|\mathcal{H}_1, \tau)$ with respect to τ. The case of uniformly distributed τ is treated in [1].

4.4.2 Detection based on second-order statistics

The notation will be made simpler here if we redefine N as $N = K(N_P + N_D) + N_D$. Now, consider the sample products

$$r_i \triangleq y_i^\star y_{i+N_D}, \qquad i = 0, \ldots, K(N_P + N_D) - 1. \tag{4.82}$$

A key observation here is that, because of the repetition of data in the cyclic prefix, y_i and y_{i+N_D} may not be independent under $\mathcal{H}_1$, and hence the expected value of r_i is nonzero even if $\mathbb{E}\, y_i = 0$. This property can be exploited for detection. Define the average sample value products

$$A_i \triangleq \frac{1}{K} \sum_{k=0}^{K-1} r_{i+k(N_P+N_D)}, \qquad i = 0, \ldots, N_P + N_D - 1. \tag{4.83}$$

Under hypothesis $\mathcal{H}_0$, all A_i are identically distributed, while, under $\mathcal{H}_1$, N_P samples $|A_i|$ will exhibit a different (and significantly larger for small noise) value than the remaining N_D. In fact, for N_P consecutive indices i we have $x_i = x_{i+N_D}$, while x_i and x_{i+N_D} are independent for the other N_D indices. Thus, spectrum sensors can be designed which detect whether or not the A_i are independent and identically distributed, or their statistics depend on i.

Defining the vector $\mathbf{A} \triangleq [A_0, \ldots, A_{N_P+N_D-1}]^\mathrm{T}$, and using the generalized likelihood-ratio test (4.16), we obtain the test statistic

$$Y \triangleq \log \frac{\max_{\sigma_w^2, \sigma_d^2} p(\mathbf{A} \mid \mathcal{H}_1, \sigma_w^2, \sigma_d^2)}{\max_{\sigma_w^2, \sigma_d^2} p(\mathbf{A} \mid \mathcal{H}_0, \sigma_d^2)}. \tag{4.84}$$

Details of the structure and the performance of these sensors are described in [1].

Other spectrum sensors, based on second-order statistics of OFDM signals, have also been advocated. Hereafter, we describe some of them (a quantitative comparison of their performance is contained in [1]).

Autocorrelation detector

The test statistic proposed in [4] consists of the empirical mean of the products r_i, normalized by the received power:

$$Y \triangleq \frac{\sum_{i=0}^{(N_P+N_D)-1} \overline{A}_i}{N^{-1}(N_P + N_D) \sum_{i=0}^{N-1} |y_i|^2}, \tag{4.85}$$

where $\overline{A}_i \triangleq \Re\{A_i\}$. This detector does not need the value of the noise variance σ_z^2. On the other hand, it does not exploit the fact that, under $\mathcal{H}_1$, the observed signal is nonstationary.

Cyclic-prefix detector

The test statistic of this detector is [27]

$$Y \triangleq \left| \frac{1}{N} \sum_{i=0}^{(N_P+N_D)-1} A_i + c \right|^2, \qquad (4.86)$$

where

$$c \triangleq \frac{\frac{N_P}{N_P + N_D} \sigma_z^2}{\left(1 + 2\left(\frac{N_P}{N_P + N_D}\right)^2\right) \frac{\sigma_d^2}{\sigma_z^2} + 2}. \qquad (4.87)$$

Even this detector does not exploit the nonstationarity of the observed signal. Notice how the calculation of this test statistic requires the knowledge of σ_d^2 and σ_z^2. To remove this requirement, one may choose to set $c = 0$. However, the decision threshold would still depend on the noise variance σ_z^2.

Sliding-window detector

This detector [21] uses a sliding window summing r_i over N_P consecutive samples and taking the maximum. The test statistic is

$$Y \triangleq \left| \sum_{i=\tau}^{\tau+N_P-1} r_i \right|. \qquad (4.88)$$

A simple extension of this detector [1] is obtained by taking K OFDM symbols instead of one:

$$Y \triangleq \left| \sum_{i=\tau}^{\tau+N_P-1} A_i \right|. \qquad (4.89)$$

Setting the decision threshold of this detector requires knowledge of the value of σ_z^2.

4.5 Effects of imperfect knowledge of noise power

The calculations presented before in this chapter can be used to compare the performance of various detectors based on their receiver operating characteristics (ROCs), which in turn are highly dependent on the SNR and on the sensing time N. Increasing the latter yields a performance improvement, but only under the optimistic assumption that the model used in the analysis is infinitely accurate. Since this assumption cannot hold in the real world (for example, noise is never perfectly Gaussian, or white, or wide-sense stationary), it happens, as shown in this section, that model uncertainties may impose fundamental limitations on the performance of the detectors, caused, as we shall discuss soon, by a lack of robustness in the detector. In this section and the next we shall examine how imperfect channel modeling can affect spectrum sensing.

Consider again the sampled-signal observation

$$y_n = h x_n + z_n, \qquad (4.90)$$

where x_n is the primary signal, affected by block fading h, and z_n is additive disturbance. As usual, the primary signal is assumed to be independent of both noise and fading. Now, derivations based on (4.90) must assume that (y_n), h, (x_n), and (z_n) are modeled with infinite precision, which requires, for example, that receiver nonlinearities, quantization noise, and interferences from other signal sources can be disregarded as irrelevant. A more realistic view of the problem would require, for example, that the noise process (z_n) be modeled as having a distribution Z taken from a set of possible distributions Z, which we may call the *noise uncertainty set*. A similar approach would be followed for the other entities in (4.90), modeled as having distributions H and X taken from suitable uncertainty sets H and X.

A detector whose decision strategy is $\mathcal{D}_N$ is said to be *robust* for target probabilities P_{FA} and P_{MD} if

$$\sup_{Z \in Z} P_{\text{FA}}(Z) \le P_{\text{FA}}, \qquad (4.91)$$

$$\sup_{H \in H, X \in X, Z \in Z} P_{\text{FA}}(H, X, Z) \le P_{\text{MD}}. \qquad (4.92)$$

Conversely, the detector is said to be *nonrobust* at a certain SNR if, even when N is made arbitrarily large, there are pairs $(P_{\text{FA}}, P_{\text{MD}})$, $0 < P_{\text{FA}} < 1/2$ and $0 < P_{\text{MD}} < 1/2$, that cannot be achieved. Mathematically, a detector is robust if and only if the sets of means of Y under both hypotheses do not overlap, i.e., $A_N \cap B_N = \varnothing$, where

$$A_N \triangleq \{\mathbb{E}_Z[Y \mid \mathcal{H}_0] : Z \in Z\}, \qquad (4.93)$$

$$B_N \triangleq \{\mathbb{E}_Z[Y \mid \mathcal{H}_1] : H \in H, X \in X, Z \in Z\}, \qquad (4.94)$$

and X includes signal distributions with the same average power $\mathcal{P}$. It turns out that robustness is easier to achieve if the SNR is high. In fact, in this situation the distributions of the test statistics under both hypotheses are sufficiently separated, and modeling uncertainties may not mix them. At low SNR, the sets of distributions of Y under both hypotheses may completely overlap. If an SNR threshold γ_t exists such that the detector becomes nonrobust for all SNRs lower than γ_t, then such a γ_t is called the *SNR wall* for the detector. The SNR wall quantifies the degree of robustness of the detector. In the balance of this section we examine three examples of SNR walls. For a more comprehensive discussion, the reader is referred to [39–42].

4.5.1 Energy sensing

Consider the asymptotic performance of the energy detector, described by (4.25)–(4.26). Assume a Gaussian primary signal, $z(t) = w(t)$, and $w(t)$ Gaussian with zero mean and power known only in a certain interval. Eliminating θ from (4.25)–(4.26), we obtain

$$N = \left[Q^{-1}(P_{\text{FA}}) - Q^{-1}(1 - P_{\text{MD}})(1+\gamma)\right]^2 \gamma^{-2}, \qquad (4.95)$$

where $\gamma \triangleq \mathcal{P}/\sigma_w^2$ denotes the SNR. The last equation shows that, under the assumption of perfectly known σ_w^2, signals can be detected at any SNR by choosing N appropriately.

Now, suppose that there is uncertainty concerning the value taken by the noise power σ_w^2. This is summarized by assuming an uncertainty interval such that $\sigma^2 \in [(1/\varepsilon)\sigma_w^2, \varepsilon\sigma_w^2]$, with $\varepsilon > 1$ a parameter quantifying the uncertainty. We obtain

$$P_{\text{FA}} = \max_{\sigma^2 \in [(1/\varepsilon)\sigma_w^2, \sigma_w^2]} Q\left(\sqrt{N}\frac{\theta - \varepsilon\sigma^2}{\sigma^2}\right) \tag{4.96}$$

$$= Q\left(\sqrt{N}\frac{\theta - \varepsilon\sigma_w^2}{\varepsilon\sigma_w^2}\right) \tag{4.97}$$

and

$$P_{\text{MD}} = 1 - \min_{\sigma^2 \in [(1/\varepsilon)\sigma_w^2, \sigma_w^2]} Q\left(\sqrt{N}\frac{\theta - (\mathcal{P} + \sigma^2)}{(\mathcal{P} + \sigma^2)}\right) \tag{4.98}$$

$$= 1 - Q\left(\sqrt{N}\frac{\theta - (\mathcal{P} + \varepsilon^{-1}\sigma^2)}{(\mathcal{P} + \varepsilon^{-1}\sigma^2)}\right). \tag{4.99}$$

Eliminating θ under the low-SNR approximation $1 + \gamma \approx 1$, we obtain

$$N \sim \frac{[Q^{-1}(P_{\text{FA}}) - Q^{-1}(1 - P_{\text{MD}})]^2}{[\gamma - (\varepsilon - \varepsilon^{-1})]^2}, \tag{4.100}$$

which shows that, as $N \to \infty$, $\gamma \downarrow (\varepsilon - \varepsilon^{-1})$, and hence the presence of an SNR wall is experienced: the energy detector cannot work reliably if $\mathcal{P} \leq (\varepsilon - \varepsilon^{-1})\sigma_w^2$, i.e., the signal power does not exceed the uncertainty in the noise power.

4.5.2 Pilot-tone-aided coherent sensing

In this scenario, the primary signal has a fraction of its power allocated to a known pilot tone or to a known preamble. The model is

$$y_n = \sqrt{\delta}x_n^{(\text{P})} + \sqrt{1-\delta}x_n^{(\text{D})} + w_n, \tag{4.101}$$

where $x_n^{(\text{P})}$ denotes the pilot tone or the preamble, $x_n^{(\text{D})}$ the information data, and w_n the additive white noise. Without fading, the matched-filter detector operates by forming the test statistic

$$Y = \frac{1}{N}\sum_{n=0}^{N-1} y_n \tilde{x}_n^{(\text{P})}, \tag{4.102}$$

where $\tilde{x}_n^{(\text{P})}$ is a vector in the direction of the pilot tone. Duplicating the calculations leading to (4.100), we obtain in this case

$$N \approx \left[Q^{-1}(P_{\text{D}}) - Q^{-1}(P_{\text{FA}})\right]^2 \delta^{-1}\gamma^{-1}, \tag{4.103}$$

which shows the robustness of the coherent detector to uncertainties in the noise distribution. The SNR wall in this case is caused by the time selectivity of the fading process. Consider the simple situation in which the multiplicative block-fading remains constant during a coherence interval of N_c samples, while the detector observes the signal within M coherent blocks. In this case the test statistic may be

$$Y = \frac{1}{M} \sum_{m=0}^{M-1} \left[\frac{1}{\sqrt{N_c}} \sum_{n=1}^{N_c} y_{mN_c+n} \, \tilde{x}_{mN_c+n}^{(P)} \right]^2 \quad (4.104)$$

obtained from a coherent combination within each coherence block – which increases the signal power by a factor N_c – and a standard energy detector. As a result, the matched filter will be nonrobust if

$$N_c \delta \gamma \leq \varepsilon - \varepsilon^{-1}. \quad (4.105)$$

4.5.3 Cyclostationarity-based detection

Due to the independence assumption, we have

$$\hat{S}_y^\alpha(f) = \hat{S}_x^\alpha(f) + \hat{S}_w^\alpha(f). \quad (4.106)$$

Furthermore, as $N \to \infty$, then $\hat{S}_x^\alpha(f) \to S_x^\alpha(f)$. Therefore, as $N \to \infty$, the test statistic (4.43) under hypothesis $\mathcal{H}_1$ converges as follows:

$$Y \to \int_{-f_s/2}^{f_s/2} |S_x^\alpha(f)|^2 \, df, \quad (4.107)$$

while under $\mathcal{H}_0$, since noise is wide-sense stationary and hence $S_w^\alpha(f) = 0$ for $\alpha \neq 0$,

$$Y \to \int_{-f_s/2}^{f_s/2} S_x^\alpha(f) [S_w^\alpha(f)]^* \, df = 0. \quad (4.108)$$

The above proves that this detection technique can be made more reliable by increasing the observation length, and hence that the single-cycle cyclostationarity-based detector is robust to noise-power level uncertainties. On the other hand, it turns out that this detector is *not* robust to fading statistics uncertainties. To prove this, assume a selective-fading model based on a linear time-varying filter whose parameters change on the time scale dictated by the channel coherence time [3]. Now, a fading filter exists such that for any $\alpha \neq 0$ the test statistic is zero. This being the case, the power spectral density becomes the only useful test statistic, and hence the energy-detector SNR wall limits the robustness of the cyclostationarity-based detector. The estimate of the spectral correlation function of a signal affected by a fading whose filter has frequency response $H(f)$ is

$$\hat{S}_y^\alpha(f) = H(f + \alpha/2) H^*(f - \alpha/2) \hat{S}_x^\alpha(f) + \hat{S}_w^\alpha(f). \quad (4.109)$$

Assuming a large enough N, we have $\hat{S}_x^\alpha(f) = S_x^\alpha(f)$ and $\hat{S}_w^\alpha(f) = 0$, which yields the test statistic under $\mathcal{H}_1$

$$Y = \int_{-f_s/2}^{f_s/2} |H(f + \alpha/2) H^*(f - \alpha/2) S_x^\alpha(f)|^2 \, df.$$

This equation shows that a fading filter $H(f)$, and hence a fading process, exists that kills or changes the shape of the spectral correlation function for any $\alpha \neq 0$ [40].

4.6 Effects of an inaccurate model of interference

The basic model of the observed signal, enriched by the introduction of the additive interference i_n, is

$$y_n = x_n + i_n + w_n, \qquad n = 0, \ldots, N-1. \tag{4.110}$$

All random quantities appearing in (4.110) are assumed to be independent of each other. The two probabilities of false-alarm and missed-detection become

$$P_{\text{FA}}(\text{W}, \text{I}) \triangleq \mathbb{E}_{\text{W},\text{I}}\{[\mathcal{D}_N = \hat{\mathcal{H}}_1] \mid \mathcal{H}_0\}, \tag{4.111}$$

$$P_{\text{MD}}(\text{W}, \text{I}, \text{X}) \triangleq \mathbb{E}_{\text{W},\text{I},\text{X}}\{[\mathcal{D}_N = \hat{\mathcal{H}}_0] \mid \mathcal{H}_1\}, \tag{4.112}$$

where W, I, and X denote the probability density functions of w_n, i_n, and x_n, respectively. Calculation of (4.111) and (4.112) assume that W, I, and X are perfectly modeled, which is a situation occurring very seldom in practice. A realistic assumption is that little (or even very little) is known about the interference statistics. Here we examine the situation in which $i(t)$ is incompletely known: we assume in particular that only its amplitude range and maximum power are known, and derive the conditions under which coherent and energy sensors fail to achieve the desired performance. Analysis can be carried out using moment-bound theory [10, 24]. This allows us to characterize all pdfs satisfying the known constraints on the interference, and derive those yielding maximum and minimum values for P_{FA} and P_{MD}. We first summarize the basics of moment-bound theory in a form suitable to solve our problem; next we show how the performance of coherent and energy sensors can be bounded above and below, and how these sensors behave asymptotically.

4.6.1 Basics of moment-bound theory

Let I denote a random variable (RV) with cumulative distribution function (CDF) defined in the finite interval $\mathcal{J} \triangleq [a, b]$. Let $k_1(x)$ and $k_2(x)$ be two continuous functions defined over $\mathcal{J}$. The *moment space* $\mathcal{M}$ is defined as the (closed, bounded, and convex) set of the pairs

$$\left(\int_{\mathcal{J}} k_1(x)\, dG(x), \int_{\mathcal{J}} k_2(x)\, dG(x) \right) \tag{4.113}$$

as $G(\cdot)$ runs over all CDFs defined over $\mathcal{J}$. The key result we need is the following [10]: $\mathcal{M}$ is the convex hull of the curve in $\mathbb{R}^2$

$$\mathcal{C} \triangleq \{(k_1(x), k_2(x)) \mid x \in \mathcal{J}\}. \tag{4.114}$$

A heuristic justification of this result goes as follows [49]. Since $\mathcal{J}$ is compact, and the set of all CDFs defined on $\mathcal{J}$ is convex, the set of the two linear functionals

$\int_{\mathcal{I}} k_i(x)\, dG(x)$, $i = 1, 2$, is also compact and convex. By restricting these CDFs to be unit step functions $u(x - x_0)$, with $x_0 \in \mathcal{I}$, the observation that

$$\int_{\mathcal{I}} k_i(x)\, du(x - x_0) = k_i(x_0), \qquad i = 1, 2$$

shows that the curve $\mathcal{C}$ is in the moment space $\mathcal{M}$. By considering convex combinations of these unit step functions, we see that the moment space contains the convex hull of $\mathcal{C}$. Conversely, since any point of the moment space can be expressed as a convex combination of points of the curve, the convex hull of $\mathcal{C}$ contains the moment space, which proves our result.

As an example, suppose that we define $k_1(x) = x^2$ and $k_2(x) = h(x)$, where $h(\cdot)$ is a known continuous function defined in $[-1, 1]$, and that

$$m_1 \triangleq \mathbb{E}[I^2] \tag{4.115}$$

is known, while we want to compute "sharp" upper and lower bounds (i.e., bounds that cannot be further tightened) to

$$m_2 \triangleq \mathbb{E}[h(I)]. \tag{4.116}$$

This problem can be solved as follows. Build the curve $\mathcal{C} \triangleq (x^2, h(x))$ as $x \in \mathcal{I}$. The convex hull of $\mathcal{C}$ contains the pairs m_1, m_2 corresponding to all CDFs defined over $\mathcal{I}$. In particular, for any given value of m_1, the values of the lower and upper envelope at m_1 yield sharp lower and upper bounds to m_2 (see Fig. 4.2 for an illustration). The dashed curve has equation $(x^2, h(x))$, while the continuous lines represent the upper and lower envelope delimiting the convex hull for any value of m_1, and hence describing the bounds on m_2 achieved by the limiting CDFs.

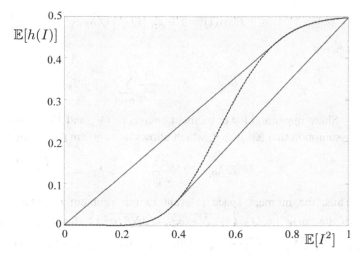

Figure 4.2 Computing moment bounds: range of $\mathbb{E}[h(I)]$ vs. $\mathbb{E}[I^2]$. The dashed line is the curve $\mathcal{C} = (x^2, h(x))$. The continuous lines delimit the convex hull of $\mathcal{C}$.

Several techniques are available for the computation of these bounds. Among these, we can mention the quadrature-rule approach described in [2, 48], the semianalytic approach of [32], and the graphical approach of [49].

4.6.2 Energy sensing

We assume a deterministic primary signal with average power $\mathcal{P}$. In this situation, the probability of a false alarm has the form

$$P_{\text{FA}}(W, I) = \mathbb{E}_{W,I} \mathbb{P}\left(\frac{2}{\sigma_w^2} \sum_{n=0}^{N-1} |i_n + w_n|^2 > \frac{2N}{\sigma_w^2} \theta\right). \qquad (4.117)$$

With $W = \mathcal{N}(0, \sigma_w^2)$, the random variable appearing before the inequality sign in the last equation has a conditional noncentral chi-square distribution with $2N$ degrees of freedom and noncentrality parameter

$$\lambda_0 \triangleq \frac{2}{\sigma_w^2} \sum_{n=0}^{N-1} |i_n|^2. \qquad (4.118)$$

Thus,

$$P_{\text{FA}}(I) = \mathbb{E}_I \left[Q_N\left(\sqrt{\lambda_0}, \sqrt{2N\theta/\sigma_w^2}\right)\right], \qquad (4.119)$$

where Q_N denotes the generalized Marcum Q-function, and $\mathbb{E}_I$ the expectation taken with respect to the distribution of the random variable I:

$$\mathbb{E}_I[f] \triangleq \int_0^\infty f(x) \, dF_I(x).$$

The conditional probability of missed detection can be derived in a similar way, to yield

$$P_{\text{MD}}(I) = 1 - \mathbb{E}_I \left[Q_N\left(\sqrt{\lambda_1}, \sqrt{2N\theta/\sigma_w^2}\right)\right], \qquad (4.120)$$

where

$$\lambda_1 \triangleq \frac{2}{\sigma_w^2} \sum_{n=0}^{N-1} |x_n + i_n|^2. \qquad (4.121)$$

Sharp upper and lower moment bounds to P_{FA} and P_{MD} can be computed under the assumption that $\mathbb{E}[i_n] = 0$, which allows us to obtain the moments

$$\mathbb{E}[\lambda_0] = 2N \frac{\sigma_I^2}{\sigma_w^2} \qquad \mathbb{E}[\lambda_1] = 2N \frac{\mathcal{P} + \sigma_I^2}{\sigma_w^2}. \qquad (4.122)$$

Thus, the moment space relevant to our problem is obtained as the convex hull of the curve $(x, Q_N\left(\sqrt{2Nx/\sigma_w^2}, \sqrt{2N\theta/\sigma_w^2}\right))$ (for P_{FA}: see Fig. 4.3) and $(x, 1 - Q_N\left(\sqrt{2N(\mathcal{P}+x)/\sigma_w^2}, \sqrt{2N\theta/\sigma_w^2}\right))$ (for P_{MD}: see Fig. 4.4). These plots allow the derivation of upper and lower bounds to P_{FA} and P_{MD} that are actually achievable through some probability distribution of the interference I.

4.6 Effects of an inaccurate model of interference

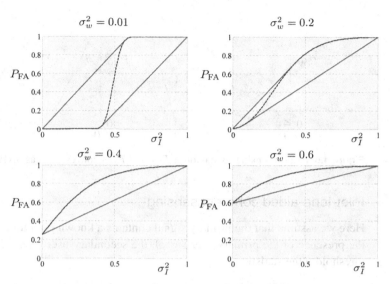

Figure 4.3 Energy sensing. Evolution of P_{FA} convex hull with noise power for $\theta = 0.5$.

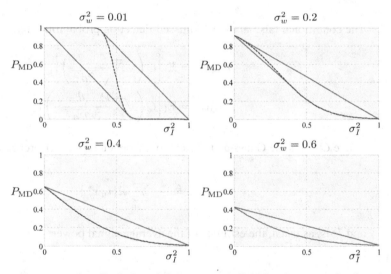

Figure 4.4 Energy sensing. Evolution of P_{MD} convex hull with noise power for $\mathcal{P} = 1$ and $\theta = 1.5$.

Asymptotic analysis, for vanishingly small noise power, can be carried out after observing that, as $\sqrt{xy} \to \infty$,

$$Q_M(x, y) \sim u\left(1 - \sqrt{y/x}\right), \tag{4.123}$$

where $u(\cdot)$ denotes the unit-step function (see [43]). Thus, for infinite SNR, the moment bounds of P_{FA} and P_{MD} behave as shown in Fig. 4.5.

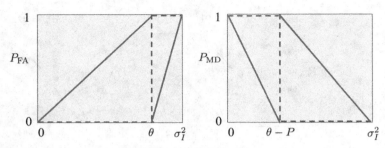

Figure 4.5 Energy sensing. Behavior of P_{FA} and P_{MD} convex hulls as SNR$\to \infty$.

4.6.3 Pilot-tone-aided coherent sensing

Here we assume that the primary signal contains a known pilot tone, and that, to detect the presence of the primary user signal, the secondary-user sensor compares against a threshold θ the statistic

$$Y = \frac{1}{N}\sum_{n=0}^{N-1}\Re\{y_n x_n^*\} = \frac{1}{N}\sum_{n=0}^{N-1}\Re\{|x_n|^2 + i_n x_n^* + w_n x_n^*\}. \tag{4.124}$$

The conditional false-alarm and missed-detection probabilities have the form

$$P_{\text{FA}}(\mathrm{I}) = \mathbb{E}_{\mathrm{I}}\left[Q\left(\sqrt{N}\frac{\theta - I}{\sqrt{\mathcal{P}\sigma_w^2}}\right)\right], \tag{4.125}$$

$$P_{\text{MD}}(\mathrm{I}) = \mathbb{E}_{\mathrm{I}}\left[Q\left(\sqrt{N}\frac{\mathcal{P} - \theta + I}{\sqrt{\mathcal{P}\sigma_w^2}}\right)\right],$$

where $Q(\,\cdot\,)$ is the Gaussian tail function, I denotes the interference term

$$I \triangleq \frac{1}{N}\sum_{n=0}^{N-1}\Re\{i_n x_n^*\}, \tag{4.126}$$

and $\mathcal{P}$ is, as usual, the estimate of the average signal power:

$$\mathcal{P} \triangleq \frac{1}{N}\sum_{n=0}^{N-1}|x_n|^2. \tag{4.127}$$

We finally assume for simplicity that $I \in [-1,1]$, and that I has an even probability density function (this implies in particular that its mean value is zero). The assumption of a symmetric pdf $p_I(\,\cdot\,)$ for the interference term I allows us to write P_{FA} and P_{MD} in a symmetric form, where now $I \in [0,1]$:[2]

---

[2] If the symmetry assumption for the interference is invalid, all the developments below can be rephrased, mutatis mutandis, to account for this situation.

4.6 Effects of an inaccurate model of interference

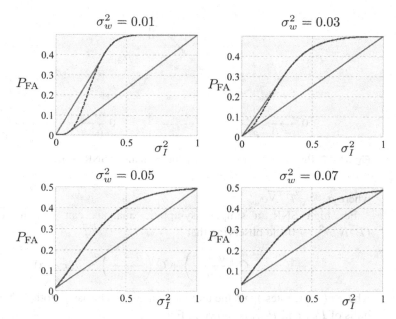

Figure 4.6 Coherent sensing. Evolution of P_{FA} convex hull with noise power for $\theta = 0.5$.

$$P_{\text{FA}}(I) = \frac{1}{2} \mathbb{E}_I \left[Q\left(\sqrt{N} \frac{\theta - I}{\sqrt{\mathcal{P}\sigma_w^2}}\right) + Q\left(\sqrt{N} \frac{\theta + I}{\sqrt{\mathcal{P}\sigma_w^2}}\right) \right], \quad (4.128)$$

$$P_{\text{MD}}(I) = \frac{1}{2} \mathbb{E}_I \left[Q\left(\sqrt{N} \frac{\mathcal{P} - \theta - I}{\sqrt{\mathcal{P}\sigma_w^2}}\right) + Q\left(\sqrt{N} \frac{\mathcal{P} - \theta + I}{\sqrt{\mathcal{P}\sigma_w^2}}\right) \right]. \quad (4.129)$$

Figure 4.6 shows the evolution of the convex hull of P_{FA} with the noise power (here $\theta = 0.5$). The convex hull contains all possible values of P_{FA} for a given value of $\sigma_I^2 = \mathbb{E}[I^2]$. Similar results hold for P_{MD}.

It may be observed that upper and lower envelopes of both P_{FA} and P_{MD} converge at a single point for $\sigma_I^2 = 0$ (corresponding to $f_I(x) = \delta(x)$) and for $\sigma_I^2 = 1$ (corresponding to the unique symmetric density with that variance, i.e., $f_I(x) = 0.5\, \delta(x+1) + 0.5\, \delta(x-1)$. A simple calculation shows that the starting points of the envelopes are

$$P_{\text{FA}}\big|_{\sigma_I^2=0} = Q\left(\frac{\theta}{\sigma}\right), \quad (4.130)$$

$$P_{\text{MD}}\big|_{\sigma_I^2=0} = Q\left(\frac{1-\theta}{\sigma}\right), \quad (4.131)$$

and the endpoints

$$P_{\text{FA}}\big|_{\sigma_I^2=1} = \frac{1}{4}\left[Q\left(\frac{\theta-1}{\sigma}\right) + Q\left(\frac{\theta+1}{\sigma}\right)\right], \quad (4.132)$$

$$P_{\text{MD}}\big|_{\sigma_I^2=1} = \frac{1}{4}\left[Q\left(\frac{-\theta}{\sigma}\right) + Q\left(\frac{2-\theta}{\sigma}\right)\right], \quad (4.133)$$

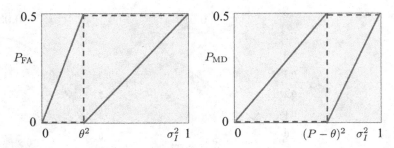

Figure 4.7 Behavior of P_{FA} and P_{MD} convex hulls as SNR$\to \infty$.

where $\sigma \triangleq \sqrt{\mathcal{P}/N}\sigma_w$.

For high SNR, a simple asymptotic analysis can be carried out by letting $(\mathcal{P}/N)\sigma_w^2 \to 0$ and observing that

$$Q\left(\frac{a+x}{\sigma}\right) + Q\left(\frac{a-x}{\sigma}\right) \sim u(x-a),$$

where $u(\cdot)$ denotes again the unit-step function. The asymptotic behavior of the convex hulls of P_{FA} and P_{MD} is shown in Fig. 4.7.

4.7 Summary

This chapter was primarily devoted to spectrum sensing under the interweaving CR paradigm (a brief mention of underlaying was based on the concept of interference temperature and its application). Spectrum sensing, which consists of detecting the presence of active primary users, was described as a binary detection problem. A number of techniques to solve this problem, and a discussion of their complexity and performance, were included (energy sensing, coherent detection, cyclostationarity-based detection, autocorrelation-based detection). Application of spectrum sensing to OFDM was given special attention. The effects on spectrum-sensing performance of an imperfect knowledge of noise power and of other sources of disturbance were finally examined.

4.8 Further reading

The basics of detection theory, as well as further details, can be found for example in [35] or in [22]. The survey papers [19, 30, 46] and the doctoral dissertation [29] describe a number of spectrum sensing techniques (wavelet-based detection, multitaper spectral estimation, etc.) that were not covered in this chapter. Chapter 5 will be devoted to more advanced techniques.

Due to its large number of potential applications, the problem of detecting a deterministic signal in noise by means of an energy-measuring device has attracted much attention. Ref. [45] is among the first papers containing its solution for a signal in additive white Gaussian noise. If the signal is observed at the output of an ergodic fading channel, then the performance metrics P_{FA} and P_{MD} must be averaged over the fading statistics. Kostylev [23] and Digham, Alouini, and Simon [9] have derived P_{FA} and P_{MD} for Rayleigh, Rice, and Nakagami-m fading. Herath, Rajatheva, and Tellambura [20] have examined the performance of the energy detector with diversity reception.

A topic not covered here concerns the channel model generated by collisions taking place when a secondary radio accesses a spectrum region occupied by the primary user. As discussed in [26], if a packet transmitted over a secondary-user subchannel is received erroneously (as revealed by the internal checksum of the packet which does not match), then it may not be passed to the higher layers in the protocol stack. Under the assumption that no ARQ is used, the application layer sees this as an *erasure*. Thus, an appropriate channel model for the choice of an error-control scheme should include, in addition to additive noise and fading, also packet erasures (see [25, 44]).

References

[1] E. Axell and E. G. Larsson, "Optimal and sub-optimal spectrum sensing of OFDM signals in known and unknown noise variance," *IEEE J. Sel. Areas Commun.*, **29**, no. 2, 290–304, Feb. 2011.

[2] S. Benedetto and E. Biglieri, *Principles of Digital Transmission With Wireless Applications*, Kluwer Academic /Plenum Publishers, 1999.

[3] E. Biglieri, J. Proakis, and S. Shamai (Shitz), "Fading channels: information-theoretic and communications aspects," *IEEE Trans. Inf. Theory*, **44**, no. 6, 2619–2692, Oct. 1998.

[4] S. Chaudhari, V. Koivunen, and H. V. Poor, "Autocorrelation-based decentralized sequential detection of OFDM signals in cognitive radios," *IEEE Trans. Signal Process.*, **57**, no. 7, 2690–2700, July 2009.

[5] T. C. Clancy, "Formalizing the interference temperature model," *J. Wireless Commun. Mobile Comput.*, **7**, no. 9, 1077–1086, Nov. 2007.

[6] T. C. Clancy, "On the use of interference temperature for dynamic spectrum access," *Ann. Telecomm.*, **64**, no. 7-8, 573–592, Aug. 2009.

[7] T. C. Clancy and D. Walker, "Spectrum shaping for interference management in cognitive radio networks," *SDR Forum Technical Conference*, Orlando, FL, Nov. 13–17, 2006.

[8] A. V. Dandawaté and G. B. Giannakis, "Statistical tests for presence of cyclostationarity," *IEEE Trans. Signal Process.*, **42**, no. 9, 2355–2369, Sep. 1994.

[9] F. F. Digham, M.-S. Alouini, and M. K. Simon, "On the energy detection of unknown signals over fading channels," *IEEE Trans. Commun.*, **55**, no. 1, 21–24, Jan. 2007.

[10] M. Dresher, "Moment spaces and inequalities," *Duke Math. J.*, **20**, 261–271, June 1953.

[11] Federal Communications Commission, *Establishment of an Interference Temperature Metric to Quantify and Manage Interference and to Expand Available Unlicensed Operation in Certain Fixed, Mobile and Satellite Frequency Bands*. ET Docket No. 03-237, Nov. 2003.

[12] W. A. Gardner, "Signal interception: a unifying theoretical framework for feature detection," *IEEE Trans. Commun.*, **36**, no. 8, 897–906, Aug. 1988.

[13] W. A. Gardner, "Exploitation of spectral redundancy in cyclostationary signals," *IEEE Signal Process. Mag.*, **8**, no. 2, 14–36, Apr. 1991.

[14] W. A. Gardner (ed.), *Cyclostationarity in Communications and Signal Processing*, IEEE Press, 1994.
[15] A. Ghasemi and E. S. Sousa, "Spectrum sensing in cognitive radio networks: the cooperation-processing tradeoff," *Wirel. Commun. Mob. Comput.*, **7**, no. 9, 1049–1060, Sept. 2007.
[16] A. Ghasemi and E. S. Sousa, "Interference aggregation in spectrum-sensing cognitive wireless networks," *IEEE J. Sel. Topics Signal Process.*, **2**, no. 1, 41–56, Feb. 2008.
[17] A. Ghasemi and E. S. Sousa, "Spectrum sensing in cognitive radio networks: requirements, challenges and design trade-offs," *IEEE Commun. Mag.*, **46**, no. 4, 32–39, Apr. 2008.
[18] A. Goldsmith, *Wireless Communications*. Cambridge University Press, 2005.
[19] S. Haykin, D. J. Thomson, and J. H. Reed, "Spectrum sensing for cognitive radio," *Proc. IEEE*, **97**, no. 5, 849–877, May 2009.
[20] S. P. Herath, N. Rajatheva, and C. Tellambura, "Energy detection of unknown signals in fading and diversity reception," *IEEE Trans. Commun.*, **59**, no. 9, 2443–2453, Sep. 2011.
[21] Huawei Technolgies and UESTC, "Sensing schemes for DVB-T," *IEEE Std. 802.22-06/0127r1*, July 2006.
[22] S. M. Kay, *Fundamentals of Statistical Signal Processing: Detection Theory*, Prentice-Hall PTR, 1998.
[23] V. I. Kostylev, "Energy detection of a signal with random amplitude," *Proc. IEEE Int. Conf. Commun. (ICC 2002)*, **3**, 1606–1610, Apr. 28–May 2, 2002.
[24] M. G. Kreĭn and A. A. Nudel'man, *The Markov Moment Problem and Extremal Problems*. American Mathematical Society, 1977.
[25] H. Kushwaha, Y. Xing, R. Chandramouli, and H. Heffes, "Reliable multimedia transmission over cognitive radio networks using fountain codes," *Proc. IEEE*, **96**, no. 1, 155–165, Jan. 2008.
[26] H. Kushwaha, Y. Xing, R. Chandramouli, and K. P. Subbalakshmi, "Erasure tolerant coding for cognitive radios," in Q. H. Mahmoud (ed.), *Cognitive Networks – Towards Self-Aware Networks*. J. Wiley & Sons, 2007.
[27] Z. Lei and F. Chin, "OFDM signal sensing for cognitive radios," *19th International Symp. on Personal, Indoor, and Mobile Radio Communications (PIMRC'08)*, Cannes, France, Sep. 15–18, 2008.
[28] Y.-C. Liang, Y. Zeng, E. C. Y. Peh, and A. T. Hoang, "Sensing-throughput tradeoff for cognitive radio networks," *IEEE Trans. Wireless Commun.*, **7**, no. 4, 1326–1337, Apr. 2008.
[29] J. Lundén, *Spectrum Sensing for Cognitive Radio and Radar Systems*, D.Sc. Dissertation, Helsinki University of Technology, Espoo, Finland, 2009.
[30] J. Ma, G. Y. Li, and B. H. Juang, "Signal processing in cognitive radio," *Proc. IEEE*, **97**, no. 5, 805–823, May 2009.
[31] J. T. MacDonald and D. R. Ucci, "Interference temperature limits of IEEE 802.11 protocol radio channels," *IEEE Electro/Information Technology Conference (EIT'07)*, Chicago, IL, May 17–20, 2007.
[32] J. W. Matthews, "Sharp error bounds for intersymbol interference," *IEEE Trans. Inf. Theory*, **IT-19**, no. 4, 440–447, July 1973.
[33] F. W. J. Olvert, D. W. Lozier, R. F. Boisvert, and C. W. Clark, *NIST Handbook of Mathematical Functions*, NIST and Cambridge University Press, 2010.
[34] Y. Pei, A. T. Hoang, and Y.-C. Liang, "Sensing-throughput tradeoff in cognitive radio networks: how frequently should spectrum sensing be carried out?," *IEEE 18th International Symposium on Personal, Indoor and Mobile Radio Communications (PIMRC 2007)*, Athens, Greece, Sep. 3–7, 2007.

[35] H. V. Poor, *An Introduction to Signal Detection and Estimation* (2nd edn.). Springer-Verlag, 1994.
[36] Z. Quan, S. Cui, and A. H. Sayed, "Optimal linear cooperation for spectrum sensing in cognitive radio networks," *IEEE J. Sel. Topics Signal Process.*, **2**, no. 4, 2431–2436, Apr. 2010.
[37] A. Sahai, R. Tandra, S. M. Mishra, and N. Hoven, "Fundamental design tradeoffs in cognitive radio systems," *Proc. 1st Int. Workshop Technol. Policy for Accessing Spectrum (TAPAS 2006)*, Boston, MA, Aug. 2–5, 2006.
[38] M. K. Simon and M.-S. Alouini, *Digital Communication over Fading Channels*, (2nd edn.). J. Wiley & Sons, 2005.
[39] R. Tandra, *Fundamental Limits on Detection in Low SNR*. MS Thesis, University of California Berkeley, Spring 2005.
[40] R. Tandra and A. Sahai, "SNR walls for feature detectors," *Proc. 2nd IEEE Int. Symp. New Frontiers in Dynamic Spectrum Access Networks (DySpan 2007)*, Dublin, Ireland, pp. 559–570, Apr. 2007.
[41] R. Tandra and A. Sahai, "SNR walls for signal detection," *IEEE J. Sel. Topics Signal Process.*, **2**, no. 1, 4–17, Feb. 2008.
[42] R. Tandra and A. Sahai, "Overcoming SNR walls through macroscale features," *Forty-Sixth Annual Allerton Conference*, Allerton House, UIUC, IL, pp. 583–590, Sep. 23–26, 2008.
[43] N. M. Temme, "Asymptotic and numerical aspects of the noncentral chi-square distribution," *Computers Math. Applic.*, **25**, no. 5, 55–63, 1993.
[44] P. Tortelier and M. M. Azeem, "Improving the erasure recovery performance of short codes for opportunistic spectrum access," *14th International Symposium on Wireless Personal Multimedia Communications (WPMC'11)*, Brest, France, Oct. 3–7, 2011.
[45] H. Urkowitz, "Energy detection of unknown deterministic signals," *Proc. IEEE*, **55**, no. 4, 523–531, Apr. 1967.
[46] B. Wang and K. J. R. Liu, "Advances in cognitive radio networks: a survey," *IEEE J. Sel. Topics Signal Process.*, **5**, no. 1, 5–23, Feb. 2011.
[47] H. Wang, G. Noh, D. Kim, S. Kim, and D. Hong, "Advanced sensing techniques of energy detection in cognitive radios," *J. Commun. Networks*, **12**, no. 1, 19–29, Feb. 2010.
[48] K. Yao and E. M. Biglieri, "Multidimensional error bounds for digital communication systems," *IEEE Trans. Inf. Theory*, **IT-26**, no. 4, 454–464, July 1980.
[49] K. Yao and R. M. Tobin, "Moment space upper and lower error bounds for digital systems with intersymbol interference," *IEEE Trans. Inf. Theory*, **IT-22**, no. 1, 65–74, Jan. 1976.
[50] T. Yücek and H. Arslan, "A survey of spectrum sensing algorithms for cognitive radio applications," *IEEE Comm. Surveys & Tutorials*, **11**, no. 1, 116–130, first quarter 2009.
[51] Y. Zeng, C. L. Koh, and Y. C. Liang, "Maximum eigenvalue detection: theory and applications," *Proc. IEEE International Conference on Communications (ICC)*, Beijing, China, pp. 4160–4164, May 2008.
[52] Y. Zeng and Y. C. Liang, "Covariance based signal detections for cognitive radios," *Proc. 2nd IEEE Int. Symp. New Frontiers in Dynamic Spectrum Access Networks (DySpan 2007)*, Dublin, Ireland, pp. 202–207, Apr. 2007.
[53] Y. Zeng and Y. C. Liang, "Maximum–minimum eigenvalue detection for cognitive radio," *IEEE PIMRC*, Athens, Greece, Sep. 2007.
[54] Y. Zeng and Y. C. Liang, "Eigenvalue-based spectrum sensing algorithms for cognitive radio," *IEEE Trans. Commun.*, **57**, no. 6, 1784–1793, June 2009.

5 Spectrum exploration and exploitation

Jarmo Lundén, Visa Koivunen, and H. Vincent Poor

5.1 Introduction

5.1.1 Chapter motivation

Chapter 4 introduced the basic spectrum sensing framework for detecting the presence of primary users in a cognitive radio system. In this chapter, we develop this framework further by taking a look at how a cognitive radio system or a cognitive radio network can optimize its spectrum sensing and access functions in order to explore and exploit the spectrum in the most efficient and effective manner.

Chapters 1 and 2 described three cognitive radio network paradigms, i.e., the underlay, overlay, and interweave paradigms. In this chapter, we focus almost exclusively on the interweave paradigm in which the secondary users try to exploit the spatio-temporal spectral opportunities (i.e., spectrum holes) resulting from intermittent and uneven spatial and temporal use of the spectrum by the primary users. As described in Chapter 4, these spectrum holes are typically considered to be white spaces, i.e., completely empty of any signals, except for noise. In addition, we will consider a few approaches suitable also for the underlay paradigm. In particular, we will consider game-theoretic techniques and algorithms for spectrum access and sharing in cognitive radio networks that operate under interference temperature constraints, as described in Chapter 4. The remaining third paradigm, the overlay paradigm, is not explicitly considered in this chapter. In the overlay paradigm, not only do the secondary users have to be aware of the primary user's transmissions, they must also possess the knowledge of the primary user's codebook and the complete channel state information between the primary transmitter and the primary receiver as well as between the secondary transmitter and the primary receiver. This makes operating under the overlay paradigm particularly challenging.

As noted above, cognitive radios operating under the interweave paradigm aim at identifying and exploiting underutilized spectrum. Idle spectrum is a local concept that, in addition to depending on time and frequency, depends on the relative locations of the primary and secondary receivers and transmitters as well as on the propagation conditions among them. The more information the cognitive radio system has about the current radio environment and the other spectrum users, the more effectively and efficiently it can exploit the spectrum and the less interference it causes to the other users. Such information is called *situational awareness*. Situational awareness at the physical radio communication level refers, for example, to knowing the current and past locations and actions of the other users of the spectrum as well as possibly the future locations and actions of these users. Moreover, situational awareness refers to knowing the local physical and radio environments and propagation conditions. Knowledge of primary and secondary waveforms is also beneficial. Knowledge of waveforms (modulation, coding, power levels, access schemes, synchronization, etc.) facilitates distinguishing between primary and secondary users, characterizing the interference in a given location, as well as identifying areas where harmful interference occurs, among other things. In essence, situational awareness includes any information that allows making more intelligent decisions on which actions to choose as well as predicting what the effects of these actions will be. Situational awareness allows the cognitive user to choose the optimum transmission parameters, such as frequency band, power, modulation, coding, beampatterns, etc., that maximize its own throughput and minimize the interference to the other users of the spectrum. Consequently, situational awareness is the cornerstone for building more intelligent and efficient systems that ultimately lead to more efficient and effective spectrum exploitation and to less interference for all spectrum users. In this chapter, we will review various techniques that aim at acquiring and exploiting situational awareness in a cognitive radio system; i.e., techniques that aim at estimating, learning, or building a (dynamic) model for primary user occupancy and for the behavior of the other users of the spectrum.

The most important form of situational awareness for cognitive radio systems is the state of the radio frequency spectrum. In interweave cognitive radio systems, spectrum sensing is a key component in spectrum exploration where awareness of the state of the spectrum is acquired and the statistical behavior of the primary user traffic as a function of time, location, and frequency is learned. Sensing policies or strategies are used to allocate the sensing and learning tasks among the sensors and over the subbands of interest, taking into account energy consumption constraints. The more effectively the spectrum can be explored through spectrum sensing, the more effectively available spectral opportunities are found and the more effectively they can be exploited, thus resulting in a more efficient use of the scarce spectral resources. Hence, exploitation of underutilized spectrum is tightly coupled with sensing and exploration. Spectrum exploitation takes advantage of the obtained knowledge on the dynamic behavior of spectral opportunities in accessing the spectrum. It determines whether and how to access and who gets access to a particular part of the spectrum based on the exploration outcomes. The coupling between these tasks can be seen by considering the sensing errors and their implications for medium access. The probabilities of false alarm and missed detection in spectrum sensing have an impact on throughput and collision probabilities in accessing the cognitive network. False alarms refer to situations in which the spectrum is considered to be occupied when it is not. Consequently, the secondary network data rate is reduced by false alarms and the exploitation of idle spectrum becomes less efficient. Missed detection, on the other hand, means that the spectrum is considered to be idle even when the primary user is in fact active. Hence, collisions may take place and the quality of service for the primary user may be reduced, interference constraints may be violated, and retransmissions are required. Moreover, due to interference constraints and possible changes in primary user activity, the longer it takes to sense and explore the spectrum, the shorter is the time the idle spectrum can be exploited until it needs to be sensed again. From this perspective, it is obvious that optimizing sensing policies and medium access should be performed jointly, using an objective function related to the throughput of the cognitive network. An example optimization approach would involve maximization of the sum throughput of the cognitive radio network under constraints on interference caused to the primary users as well as fairness and energy consumption constraints for secondary users.

At the core of multiuser cognitive radio systems is the interaction among the secondary users. The secondary users may have different capabilities, interests, and goals, which complicates the design of spectrum sensing and access policies. In general, the secondary users in a cognitive radio network are competing with each other for the available spectral opportunities. On the other hand, cooperation among the secondary users opens up avenues for making the sensing more reliable and improving the network performance and spectrum exploitation. For example, cooperation among spatially dispersed secondary users allows shortening the sensing time through distributed detection. Problems caused by fading and shadowing as well as hidden node problems can be avoided through the spatial diversity gains in distributed detection. In addition, cooperation allows distributing the sensing and network management tasks among the network members as well as minimizing the mutual interference. However, in general, cooperation also increases the transmission overhead through increased control messaging.

Cognitive radio system models may be classified using various criteria. We can have single-user and multiuser models depending on whether the other secondary users are explicitly taken into account in the model. Moreover, the spectrum model may be either a single-band or a multiband model. Multiuser models can be further divided into infrastructure (base station)-based and ad hoc models. The network may operate and perform tasks, such as spectrum sensing, in a centralized or a decentralized manner. Moreover, the interaction between the network users may be cooperative, competitive, or a combination of these two (mixed models). This is further affected by the level of synchronization among the network users. That is, the sensing and exploitation tasks may be performed synchronously or asynchronously by different users depending on the system model. A typical synchronized system may employ a periodic slot structure with alternating transmission periods and quiet periods for sensing, whereas in an asynchronized system there is more freedom in sensing and access tasks and their relative positioning in the time domain. In this chapter, our main focus is in the optimization of spectrum exploration and exploitation in multiuser, multiband models in which the users are in general both competing and cooperating with each other. We will consider both infrastructure-based and ad hoc models. Moreover, we will review literature related to other models as well, such as single-user single-band and multiband models, since it provides insight into how to design the algorithms in the more complicated models.

5.1.2 Preview of the chapter

This chapter consists of two major sections. In the first part, in Section 5.2, we will take a look at advanced spectrum sensing techniques including distributed detection as well as sequential and quickest detection. These techniques are needed in exploring the spectrum and identifying available spectral opportunities. Moreover, they are important for optimizing the spectrum sensing process so that it provides both more reliable sensing results and shorter average sensing time. In the second part of the chapter, we will turn our attention to the design of sensing and access policies for optimized spectrum exploration and exploitation. In Section 5.3, we will introduce dynamic programming, bandit problems, reinforcement learning, and game theory, and discuss their application to sensing and access policy design in cognitive radio systems. In addition, we will look at some other interesting approaches for optimizing the sensing and transmission parameters in a cognitive radio system and briefly discuss location awareness and its importance to cognitive radio systems. Finally, the chapter is concluded with a brief summary in Section 5.4 and a discussion of some resources for further reading in Section 5.5.

5.2 Advanced spectrum sensing techniques

5.2.1 Distributed detection in spectrum sensing

Distributed detection involving multiple geographically displaced sensors and a fusion center has been a topic of great interest in the sensor network and radar research communities [166, 206]. Distributed sensor systems were originally motivated by their applications in military surveillance and intelligence systems. Recently, they have attracted considerable attention for their application to cooperative spectrum sensing in cognitive radio systems, see e.g., [36, 86, 103, 123, 238]. Distributed spectrum sensing is a very attractive technology for identifying underutilized spectrum and creating awareness of the state of the radio environment since available spectrum is a local resource that varies as a function of time, location, and frequency. Understanding the state of the spectrum is crucial for interference management, too. In this section, we focus on distributed, cooperative sensing for identifying idle spectrum.

Spatial diversity in spectrum sensing
Cooperative spectrum sensing has several advantages. It provides diversity gains in the face of multipath channel effects and demanding propagation environments such as fading, shadowing, and the hidden node problem. Propagation modeling and effects in the cognitive radio context were considered in detail in Chapter 3. Wireless signals are exposed to slow fading caused by large objects in the signal path and to fast fading caused by motion and multipath propagation of the signal. This makes it harder to detect the signals as the signal attenuation caused by channel fading may be of the order of tens of dB.

The gains obtained by cooperative sensing follow from spatial diversity that is similar to that exploited in multiantenna (MIMO) wireless communications to improve the rate or reliability of a radio link [21]. It is very unlikely that all the sensing channels between the primary transmitter and spectrum sensing nodes would be in deep fades simultaneously. In fact, if the sensor displacement is sufficient, i.e., sensors are further than the channel coherence distance apart, fading can be assumed to be independent from sensor to sensor. Diversity gains are best achieved in rich scattering environments and in channels where there are significant non-line-of-sight (NLOS) components. Moreover, distributed sensing leads to improvement in the detector performance which is typically measured in terms of false alarm and detection probabilities or detection time at a specified performance level. Multiple sensors also provide inherent redundancy which improves the reliability and robustness of the system. However, we should observe here that, while increasing the number of cooperating sensing nodes improves the final decision performance, the amount of system resources needed in the fusion process (total power needed to transmit measurements, amount of overhead traffic in the "secondary" network linking sensors and fusion center, and memory requirements at the fusion center) also increases. Thus, there might exist an optimum number of cooperating sensors, beyond which the detection performance improvement does not compensate for the incurred penalty in the use of resources [43, 212]. Other relevant issues here are how to deliver information to the fusion center, and how low-SNR sensing nodes and errors in the transmission to the fusion center degrade the overall performance.

Spectrum sensors can be included in mobile terminals that want to exploit under-utilized spectrum. Consequently, multiple mobile terminals can form a distributed sensing or detection system. Cooperative sensing facilitates using simpler and more energy-efficient detectors in each sensing node and consequently extend battery life in battery-operated terminals. On the other hand, distributed sensing typically requires transmitting control information and sensing results to the fusion center or to the other nodes. Obviously, this adds to the energy consumption in the secondary network. Furthermore, policies used for accessing the spectrum also play a crucial role in determining the power consumption. Therefore, evaluating overall energy consumption requires more holistic system level considerations.

There are many ways to characterize the gains obtained through spatial diversity. In order to illustrate the gains we present a quantitative measure that is particularly suitable for cooperative detection of primary user activity in cognitive radio networks. In particular, we study the probability of detection P_D (defined in Chapter 4) as a function of the logarithm of the SNR γ, i.e., a special case of the receiver operating characteristic (ROC) curve. We define the spatial diversity order as follows [146]:

$$SD(N_{SU}) = \max \frac{\partial}{\partial \gamma_{dB}} P_D(\gamma_{dB}, N_{SU}), \qquad (5.1)$$

where $\gamma_{dB} \triangleq 10 \log_{10} \gamma$, $\frac{\partial}{\partial \gamma_{dB}}$ denotes the partial derivative with respect to γ_{dB}, and N_{SU} is the number of sensors. The ROC curves in Fig. 5.1 illustrate the detector performance for equal gain combining (EGC) cooperative energy detection (radiometry) (see (5.9)–(5.11) below) for iid Rayleigh fading channels. It can be clearly seen from the figure how the slope of the probability of detection curve gets steeper as the number of radios sensing the same band grows. Using more radios to sense each subband yields diminishing returns, however.

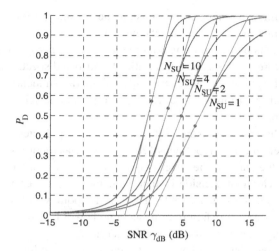

Figure 5.1 The spatial diversity order is defined as the maximum slope of the probability of detection curve as a function of the logarithmic SNR γ_{dB} [146]. In this example there are N_{SU} iid Rayleigh channels and the employed detection scheme is energy detection with $P_{FA} = 0.01$. It can be seen that the maximum slope of the probability of detection curve increases as the number of sensors increases.

Centralized vs. decentralized sensing

Distributed systems can be either centralized or decentralized. In a centralized system all the local sensors transmit all the observed local data to a central processor that combines the data to perform decision making or parameter estimation tasks using optimal signal processing techniques. Decentralized systems, on the other hand, employ intelligent sensors that process the local data before transmitting it to a central processor called the fusion center (FC). Local binary decisions or some other compact representation summarizing the state of the spectrum may be transmitted to the fusion center. Consequently, the fusion center of a decentralized system has only partial information received from the sensors, whereas the central processor has all the information in a centralized system. Typically, any quantization of the sensed data leads to performance losses.

Each spectrum sensor may also convey some quality information about the local sensing statistics such as SNR and a local sensor performance index. Such side information may be exploited in combining the results from multiple sensors and weighting the reliability of the local decision statistics from each sensor. It may also help in interference management; if the power levels transmitted by the primary user are known and the channels can be assumed to be reciprocal, then better power control and medium access may be achieved.

The decentralized distributed detection approach is appealing in the cognitive radio context where the local sensors typically have the capability to process the acquired data and extract relevant information from it. As noted above, in cognitive radio applications the local spectrum sensors might typically be installed on battery-operated devices. Hence, in order to conserve energy, the amount of data transmitted should be kept to a minimum without compromising the sensing performance. In addition, the capacity requirements a central scheme imposes on the channel used for sharing the sensing results by transmitting data to the central processor may be prohibitive in practice. Therefore, we focus on decentralized distributed spectrum sensing for cognitive radios. In such systems the nodes typically collaborate in identifying and exploiting the underutilized spectrum. Alternatively, the nodes can be selfish and compete for the unoccupied spectrum.

In distributed sensing systems, the main topologies with a fusion center are serial, parallel, and tree topologies [206]. In a serial topology, the intermediate information about the state of the spectrum is passed from user to user and it is updated at each stage. The final decision is then made by the last user which has the information from all the other users available. The serial topology is less robust than other topologies since a single link failure or missing node will lead to severe performance degradation. Hence, it is not considered any further here.

In a parallel topology local detectors may send hard or soft decisions to the fusion center. Typically the sensors do not communicate with each other and there is no feedback from the fusion center to sensors either. The sensors use a mapping rule

$$u_i = \nu_i(y_i), \quad i = 1, \ldots, N_{\text{SU}}, \tag{5.2}$$

where y_i is the observation made by sensor i and N_{SU} is the number of sensors, and then pass the local mappings u_i to the fusion center. The mapping may also perform compression of the data by extracting relevant information from multiple observations. For example, each sensor could compute a minimal sufficient statistic or local likelihood ratio from the data or make a local binary decision using an appropriate detector.

Sensing nodes that are close to each other may also form a cluster that sends summary information about the state of the spectrum to the fusion center via a cluster head, thereby forming a tree topology. The nodes within the cluster exchange data or sensing results and form a consensus statistic that is then transmitted to the fusion center, see [173]. The information sent by the cluster head may be a local decision statistic with side information or an estimate of the power spectral density.

Ad hoc configurations may also be used, in which there is no dedicated fusion center and the sensing information is distributed among the nodes, see Fig. 5.2. For ad hoc configurations in which the sensing information is distributed to all nodes, each node will have the same information needed to make the decision about whether the spectrum is occupied or not. Distributing the information to all nodes potentially over multiple hops may lead, however, to unacceptable delays in decision making, especially in cases where the spectrum is not persistently available, or the state of the spectrum may vary rapidly. Thus, configurations in which the information forwarding is limited to small local neighborhoods may have to be employed.

Let us now return to the parallel topology. As noted in Chapter 4, the detection problem is typically modeled as a binary hypothesis testing problem. Obviously one can use a variety of detection strategies. For example, if there is prior statistical knowledge about the primary user activity and persistence of the free spectrum, such information can be used in Bayesian detector design to improve performance. Such information is

5.2 Advanced spectrum sensing techniques

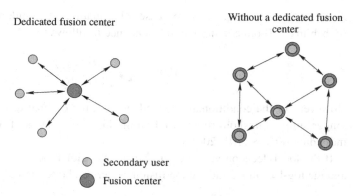

Figure 5.2 Parallel topology with a fusion center (FC) and ad hoc scenarios where decision statistics and side information are shared among the nodes.

also valuable in designing sensing and access policies, i.e., how the sensing task is allocated to the sensors for different subbands of interest over time and when and who gets access to the idle spectrum. The selected detection strategy is also intimately related to interference management and accessing the underutilized spectrum. If missed detections are not properly controlled, there will be collisions with the primary signal, and retransmissions are required from both the primary users and the secondary users. On the other hand, if there are too many false alarms, there will be fewer opportunities to access the spectrum and the improvements in spectrum efficiency obtained by cognitive radios are reduced. These considerations lead to the Neyman–Pearson scheme for distributed detection which can be stated as follows (see also Chapter 4): for a predefined global probability of false alarm level P_{FA}, find (optimum) local and global decision rules $\Gamma = \{\nu_0(\mathbf{u}), \nu_1(y_1), ..., \nu_{N_{\text{SU}}}(y_{N_{\text{SU}}})\}$, where $\nu_0(\cdot)$ is the global decision rule and $\nu_i(\cdot), i = 1, \ldots, N_{\text{SU}}$ are the local rules, that minimize the global probability of missed detection P_{MD}.

A crucial assumption that leads to a tractable solution to the distributed detection problem is conditional independence of sensor observations, conditioned on the hypothesis:

$$p(y_1, ..., y_{N_{\text{SU}}} \mid \mathcal{H}_l) = \prod_{i=1}^{N_{\text{SU}}} p(y_i \mid \mathcal{H}_l), \quad l = 0, 1, \qquad (5.3)$$

where $p(\cdot)$ is used here and elsewhere generically to denote probability densities or mass functions of its argument(s). If the assumption holds, the local mapping rules as well as the global decision rule at the fusion center become likelihood-ratio $\Lambda(\cdot)$ based threshold rules [194]. The likelihood ratio is given by

$$\Lambda(\mathbf{u}) = \frac{p(\mathbf{u} \mid \mathcal{H}_1)}{p(\mathbf{u} \mid \mathcal{H}_0)}. \qquad (5.4)$$

This hypothesis $\mathcal{H}_0$ at the fusion center is that no signal is present and the alternative hypothesis $\mathcal{H}_1$ is that a signal is present. The decision between these two is made by comparing $\Lambda(\mathbf{u})$ to a threshold value ψ_0. If $\Lambda(\mathbf{u}) > \psi_0$, we decide $\mathcal{H}_1$, and if

$\Lambda(\mathbf{u}) < \psi_0$, we decide $\mathcal{H}_0$; in case of equality, we randomly decide $\mathcal{H}_1$ with probability ε. From conditional independence it follows that $\Lambda(\mathbf{u})$ factors as follows:

$$\Lambda(\mathbf{u}) = \prod_{i=1}^{N_{\text{SU}}} \frac{p(u_i \mid \mathcal{H}_1)}{p(u_i \mid \mathcal{H}_0)}. \tag{5.5}$$

However, for the conditionally dependent case, the optimal tests at the sensors are no longer simple threshold rules based on the likelihood ratios of the observations at the individual sensors [101, 206].

If the local decision variables are binary, the global decision rule is a Boolean rule and the log-likelihood ratio at the fusion center can be written as

$$\log \Lambda(\mathbf{u}) = \sum_{i=1}^{N_{\text{SU}}} \left[u_i \log \frac{1 - P_{\text{MD},i}}{P_{\text{FA},i}} + (1 - u_i) \log \frac{P_{\text{MD},i}}{1 - P_{\text{FA},i}} \right], \tag{5.6}$$

where u_i is the binary local decision of the ith sensor (i.e., 0 or 1), $P_{\text{FA},i}$ is the probability of false alarm at the ith sensor, and $P_{\text{MD},i}$ is the probability of missed detection at the ith sensor. This fusion rule is known as the Chen–Varshney fusion rule [200, p. 63]. One sees immediately that the above expression is a weighted sum of local sensor decisions. The weights are functions of probabilities of false alarm and missed detection of individual sensor decisions. These local performance indices may not be known in practice and they have to be learned or estimated. Computationally simpler decision rules are obtained by using Boolean rules based on logical functions, such as the OR, AND, MAJORITY or more generally the K-out-of-N rule which decides in favor of $\mathcal{H}_1$ if at least K of the sensors do. Using such rules typically does not result in significant performance degradation. In fact, for identical local binary decision rules and identical distribution at the sensors, the optimal global decision rule is a K-out-of-N rule [200, Sec. 3.4]. Nevertheless, the detector is designed so that we control the global probability of false alarm $P_{\text{FA}} = P(u_0 = 1 \mid \mathcal{H}_0)$, and probability of missed detection $P_{\text{MD}} = P(u_0 = 0 \mid \mathcal{H}_1)$, where $u_0 = \nu_0(\mathbf{u})$ is the global decision rule. A detailed treatment of Boolean fusion rules and their optimization is given in [200].

In the Bayesian framework we have prior probabilities for the two hypotheses and costs for making the decisions (see Chapter 4 for more information). The corresponding global decision rule compares the log-likelihood ratio with a threshold value

$$\psi \triangleq \log \frac{\pi_0 (C_{10} - C_{00})}{\pi_1 (C_{01} - C_{11})}, \tag{5.7}$$

where C_{ij} is the cost of making a global decision $\mathcal{H}_i$ when $\mathcal{H}_j$ is true, and π_0 and π_1 denote the prior probabilities of the two hypotheses $\mathcal{H}_0$ and $\mathcal{H}_1$ associated with the channel under consideration being idle or occupied, respectively.

Decentralized sensing in cognitive radio systems

In the following we briefly review some main types of decentralized spectrum sensing methods needed for creating awareness about the state of the radio spectrum. The differences among the methods are in the topology of the network of sensors, type of local detector or estimator employed, and type of local information passed to the fusion center. In addition, there are differences in the way sensors cooperate or compete, and the fusion rule or combining method employed at the fusion center. In cooperative sensing one also needs to design a sensing policy, i.e., decide how different sensors are allocated in sensing different parts of the spectrum at any given time instance. The local detector can be an energy detector, a coherent detector (i.e., a matched filter), or a feature-based detector such as a cyclostationarity-based detector, for example. These techniques were described in detail in Chapter 4. Local sensors may also estimate the power spectrum and characterize the level of interference at that location. The detectors may be designed using well-known Neyman–Pearson or Bayesian approaches, as described above. The detector may also be designed jointly with the access strategy taking into account the interference constraints and energy constraints. For example, if the channel is in a deep fade and the secondary user terminal is battery-operated, the secondary user might not want to access the channel even if it is found to be idle.

A crucial element of cooperative spectrum sensing is the combining technique used at the fusion center. The local detection algorithms may be used in conjunction with either soft- or hard-decision combining at the fusion center. As noted above, hard combining uses Boolean rules and binary decisions from the local detectors. Many widely used rules are K-out-of-N type of rules, with AND, OR, and MAJORITY rules being important special cases.

Soft decisions are known to improve the performance of decentralized sensing significantly and to exploit fully cooperation among the nodes. As noted above, under the conditional independence assumption the optimal global decision rule under both Bayesian and Neyman–Pearson criteria is a likelihood ratio test. However, in practice it may not always be possible to directly use a likelihood ratio test because some of the distribution parameters may be unknown. In such a case, a simple combining scheme of practical interest is combining the local generalized likelihood ratios or logarithms of them under a conditional independence assumption. The generalized likelihood ratio is obtained from the likelihood ratio by replacing the unknown parameter values with their estimates. The resulting test statistic is simply the sum of local generalized log-likelihoods. Clearly the local detector can be any detector that employs a generalized likelihood ratio test. We will later provide an example of a collaborative generalized likelihood ratio test using cyclostationarity-based local detectors and Neyman–Pearson design. Also belief propagation (BP) techniques [79] have been employed in combining. Belief propagation requires message passing and message updating among the nodes. It also requires forming a graph based on network topology. This approach is applicable to different local detector structures. Conditional independence is assumed which allows for combining the local information by using a product of local functions. Conditional independence can be captured by using Markov random fields or Bayesian networks as an underlying model. Dempster–Shafer theory for combining evidence from multiple sources has also been employed in distributed sensing for cognitive radios [157]. A degree of belief may be achieved based on all the evidence while allowing the radios to specify ignorance and uncertainty in a flexible manner. Both local sensing and combining of the evidence at the fusion center may be covered with the theory.

A simple way of fusing information gathered from several sensors is by combining linearly, possibly with equal weights, a suitable function of the observed samples. We assume here that there are N_{SU} sensors, with the kth one observing the signal

$$y_{k,n} = h_k x_n + i_{k,n} + w_{k,n}, \qquad k = 1, \ldots, N_{\text{SU}}, \tag{5.8}$$

where h_k denotes the fading affecting the path between the primary source and the kth sensor, which we assume to remain constant during the observation interval, x_n is the primary signal whose presence should be detected, $i_{k,n}$ are interference terms, and $w_{k,n}$ the additive white Gaussian noises. All terms on the right-hand side of (5.8) are assumed, as usual, to be independent.

Cooperative energy detection

The linear combination fusion technique has been advocated, in particular, for cooperative energy detection [50, 51, 127, 164, 187]. Each sensor k, $k = 1, \ldots, N_{\text{SU}}$, measures energy

$$\mathcal{E}_k \triangleq \sum_{n=0}^{N_{\text{S}}-1} |y_{k,n}|^2 \tag{5.9}$$

along an observation interval, where N_{S} is the number of observations. The fusion center performs the linear combination

$$\mathcal{E} = \sum_{k=1}^{N_{\text{SU}}} a_k \mathcal{E}_k, \tag{5.10}$$

where a_k, $k = 1, \ldots, N_{\text{SU}}$, are the linear weights, and makes a global decision by comparing $\mathcal{E}$ against threshold ψ:

$$\mathcal{E} \underset{\mathcal{H}_0}{\overset{\mathcal{H}_1}{\gtrless}} \psi. \tag{5.11}$$

Assuming conditional independence of observations and iid Gaussian signal and noise it follows that the optimal likelihood ratio test is equivalent to the linear weighting scheme in (5.10) with weights given by $a_k = |h_k|^2/(1 + |h_k|^2)$ [127]. Two other commonly employed linear weighting schemes are equal gain combining (EGC) [50, 51, 127] and maximal ratio combining (MRC) [127] in which the weights are given by $a_k = 1, k = 1, \ldots, N_{\text{SU}}$, and $a_k = |h_k|^2, k = 1, \ldots, N_{\text{SU}}$, respectively. The MRC-based fusion scheme is attractive in the low SNR regime while the EGC-based fusion scheme is considered to be a good choice for combination under limited knowledge of channel state information. Another combining method is selection combining which employs the maximum value for global detection and compares that to a threshold, i.e., $\mathcal{E} = \max\{\mathcal{E}_1, \ldots, \mathcal{E}_{N_{\text{SU}}}\}$ [50, 51].

In practice, one rarely has the instantaneous channel state information necessary to use the optimal or MRC weighting schemes. However, if the second-order statistics of the squared magnitudes of the fading and primary user signals are known, one can also seek to optimize the weight vector to explicitly minimize the probability of missed detection [164, 187]. The optimization problem may be formulated as follows. If we make the following two assumptions:

1. the sensor energy measurements $\mathcal{E}_k$ are transmitted to the fusion center through a noisy reporting channel in which the noise is iid zero-mean Gaussian, and
2. the number of observations N_S is sufficiently large that the chi-square distribution of the energy measurements may be approximated with a Gaussian distribution according to the central limit theorem [164],

then the components of the vector $\mathbf{z} \triangleq (\mathcal{E}_1, \ldots, \mathcal{E}_{N_{SU}})^T$ are conditionally jointly Gaussian: $\mathbf{z} \sim \mathcal{N}(\boldsymbol{\mu}_h, \boldsymbol{\Sigma}_h)$, with mean and covariance

$$\begin{cases} \boldsymbol{\mu}_h = \mathbb{E}[\mathbf{z} \mid \mathcal{H}_h], \\ \boldsymbol{\Sigma}_h = \mathbb{E}[\mathbf{z}\mathbf{z}^T \mid \mathcal{H}_h] - \boldsymbol{\mu}_h \boldsymbol{\mu}_h^T, \end{cases} \quad h = 0, 1. \tag{5.12}$$

Introducing the weight vector $\mathbf{a} \triangleq (a_1, \ldots, a_{N_{SU}})^T$, we have the following expressions for the probability of false alarm:

$$P_{\text{FA}} = P\left[\mathbf{a}^T \mathbf{z} > \psi \mid \mathcal{H}_0\right] = Q\left(\frac{\psi - \mathbf{a}^T \boldsymbol{\mu}_0}{\sqrt{\mathbf{a}^T \boldsymbol{\Sigma}_0 \mathbf{a}}}\right) \tag{5.13}$$

and of missed detection:

$$P_{\text{MD}} = P\left[\mathbf{a}^T \mathbf{z} < \psi \mid \mathcal{H}_1\right] = Q\left(\frac{\mathbf{a}^T \boldsymbol{\mu}_1 - \psi}{\sqrt{\mathbf{a}^T \boldsymbol{\Sigma}_1 \mathbf{a}}}\right), \tag{5.14}$$

where $Q(\cdot)$ denotes the tail probability of the standard Gaussian distribution.

False alarm and missed detection probabilities are connected by the value of the decision threshold ψ, which can be computed by choosing a target P_{FA} or P_{MD}:

$$\psi = \begin{cases} \mathbf{a}^T \boldsymbol{\mu}_0 + Q^{-1}(P_{\text{FA}})\sqrt{\mathbf{a}^T \boldsymbol{\Sigma}_0 \mathbf{a}}, \\ \mathbf{a}^T \boldsymbol{\mu}_1 - Q^{-1}(P_{\text{MD}})\sqrt{\mathbf{a}^T \boldsymbol{\Sigma}_1 \mathbf{a}}. \end{cases} \tag{5.15}$$

Alternatively, the ROC linking the probability of false alarm P_{FA} to the probability of missed detection P_{MD} are obtained from the following expression:

$$P_{\text{FA}} = Q\left(\frac{\mathbf{a}^T \mathbf{d} - Q^{-1}(P_{\text{MD}})\sqrt{\mathbf{a}^T \boldsymbol{\Sigma}_1 \mathbf{a}}}{\sqrt{\mathbf{a}^T \boldsymbol{\Sigma}_0 \mathbf{a}}}\right), \tag{5.16}$$

where $\mathbf{d} \triangleq \boldsymbol{\mu}_1 - \boldsymbol{\mu}_0$, or from

$$P_{\text{MD}} = Q\left(\frac{\mathbf{a}^T \mathbf{d} - Q^{-1}(P_{\text{FA}})\sqrt{\mathbf{a}^T \boldsymbol{\Sigma}_0 \mathbf{a}}}{\sqrt{\mathbf{a}^T \boldsymbol{\Sigma}_1 \mathbf{a}}}\right). \tag{5.17}$$

Following [164, 187], the weight vector optimization problem may be formulated in the form

$$\min_{\mathbf{a}} P_{\text{MD}}(\mathbf{a}) \triangleq Q\left(\frac{\mathbf{a}^T \mathbf{d} - Q^{-1}(P_{\text{FA}})\sqrt{\mathbf{a}^T \boldsymbol{\Sigma}_0 \mathbf{a}}}{\sqrt{\mathbf{a}^T \boldsymbol{\Sigma}_1 \mathbf{a}}}\right).$$

Since the function $Q(\cdot)$ is decreasing in its argument, this problem can be equivalently formulated as that of minimizing the function

$$g(\mathbf{a}) \triangleq \frac{Q^{-1}(P_{\mathrm{FA}})\sqrt{\mathbf{a}^{\mathrm{T}}\boldsymbol{\Sigma}_0\mathbf{a}} - \mathbf{a}^{\mathrm{T}}\mathbf{d}}{\sqrt{\mathbf{a}^{\mathrm{T}}\boldsymbol{\Sigma}_1\mathbf{a}}} \tag{5.18}$$

with respect to the weight vector $\mathbf{a} \in \mathbb{R}^{N_{\mathrm{SU}}}$.

We can now observe that $g(\mathbf{a})$ is not a convex function, and it is undefined for $\mathbf{a} = \mathbf{0}$. Nonconvexity makes the solution of the optimization problem harder, as illustrated by the complex algorithm advocated in [164]. However, inspection of (5.18) shows that the function $g(\mathbf{a})$ is positively homogeneous of degree 0, i.e., invariant to multiplications of its vector argument by a positive scalar. Homogeneity can be exploited to obtain a simpler, more efficient solution based on global optimization principles [187]. The algorithm proposed in [187] obtains the globally optimum solution by calculating one root (lying in a specific interval) of a polynomial equation whose degree is not greater than $2N_{\mathrm{SU}}$. This can be accomplished, for example, by employing the Newton–Raphson iterative algorithm [187]. Moreover, the algorithm proposed in [187] can also be used with non-Gaussian primary signals, known instantaneous channel state information, and primary user symbols.

Cooperative autocorrelation detection
Autocorrelation-based linear combining fusion with equal weights has been considered in [232]. The autocorrelation of signal (5.8) is

$$A_{y_k}(\tau) = \mathbb{E}[y_{k,n+\tau} y_{k,n}^\star], \qquad k = 1, \ldots, N_{\mathrm{SU}}, \tag{5.19}$$

where we have assumed that $y_{k,n}, k = 1, \ldots, N_{\mathrm{SU}}$, are zero-mean. Then we have, with obvious notation derived from (5.8),

$$A_{y_k}(\tau) = |h_k|^2 A_{x_k}(\tau) + A_{i_k}(\tau) + A_{w_k}(\tau), \qquad k = 1, \ldots, N_{\mathrm{SU}}. \tag{5.20}$$

The fusion center averages the received autocorrelations by computing

$$A_y(\tau) \triangleq \frac{1}{N_{\mathrm{SU}}} \sum_{k=1}^{N_{\mathrm{SU}}} A_{y_k}(\tau) \tag{5.21}$$

and uses autocorrelation-based detection; see Chapter 4. To do this, it should be assumed that the interfering signals received by different sensors are statistically independent and identically distributed, which makes it very likely that the sum $(1/N_{\mathrm{SU}}) \sum_{k=1}^{N_{\mathrm{SU}}} A_{i_k}(\tau)$ is vanishingly small for $\tau \neq 0$ for a sufficiently large number of sensors.

Cooperative cyclostationarity-based detection
In [122] and [123] cooperative cyclostationarity-based detection algorithms based on generalized likelihood ratios have been proposed. In particular, in [123], the local detectors are multicycle detectors given by

$$\mathcal{T} = N_{\mathrm{S}} \hat{\mathbf{r}}_y \hat{\boldsymbol{\Sigma}}_y^{-1} \hat{\mathbf{r}}_y^{\mathrm{T}}, \tag{5.22}$$

where N_S is the number of observations, $\hat{\mathbf{r}}_y$ is a $1 \times 2M$ vector consisting of stacked real and imaginary parts of estimated cyclic autocorrelations $\hat{R}_y^\alpha(\tau)$ at the cyclic frequencies of interest $\alpha \in \mathcal{A}$ for some set of time lags $\{\tau_{i,j}\}_{i=1}^{N_{\tau,j}}$ for each $\alpha_j \in \mathcal{A}$, and $\hat{\boldsymbol{\Sigma}}_y$ is the estimate of the asymptotic covariance matrix of $\hat{\mathbf{r}}_y$ (see Chapter 4 for estimating the cyclic autocorrelation). Thus, $M = \sum_{j=1}^{|\mathcal{A}|} N_{\tau,j}$.

The test statistic in (5.22) is in fact the generalized log-likelihood ratio in which the unknown parameters of the log-likelihood ratio have been replaced with their estimates obtained from the received data [123]. Assuming conditional independence, a collaborative test statistic at the fusion center is obtained by taking the sum of the local generalized log-likelihood ratios [123]:

$$\mathcal{T}_{N_{SU}} = \sum_{k=1}^{N_{SU}} \mathcal{T}_k, \tag{5.23}$$

where $\mathcal{T}_k$ is the generalized log-likelihood ratio of secondary user k. Under $\mathcal{H}_0$, when the primary user's signal is not present, $\mathcal{T}_{N_{SU}}$ is asymptotically chi-square distributed with $2MN_{SU}$ degrees of freedom, and under $\mathcal{H}_1$ it is noncentral chi-square distributed with $2MN_{SU}$ degrees of freedom and noncentrality parameter $\sum_{k=1}^{N_{SU}} N_S \mathbf{r}_{y,k} \hat{\boldsymbol{\Sigma}}_{y,k}^{-1} \mathbf{r}_{y,k}^T$, where each $\mathbf{r}_{y,k}$ is a vector of the real and imaginary parts of the corresponding true cyclic autocorrelations of the primary user signal at user k, $k = 1, \ldots, N_{SU}$, respectively [123]; see Chapter 4 for details on the chi-square distribution.

Other test statistics that are computationally less complex are obtained by assuming the cyclic correlation estimates at different cyclic frequencies to be uncorrelated. This assumption is valid under the null hypothesis if only wide-sense stationary noise is present [123]. Under the uncorrelatedness assumption the covariance matrix $\boldsymbol{\Sigma}_y$ is a block-diagonal matrix. Hence, the generalized log-likelihood ratio may be expressed as the sum of the generalized log-likelihood ratios calculated for each cyclic frequency $\alpha \in \mathcal{A}$ separately:

$$\mathcal{T}_S = \sum_{\alpha \in \mathcal{A}} \mathcal{T}(\alpha) \approx \mathcal{T}, \tag{5.24}$$

where $\mathcal{T}(\alpha)$ denotes the generalized log-likelihood ratio in (5.22) calculated for cyclic frequency α. Moreover, a selection combining-based statistic is obtained by taking the maximum over the generalized log-likelihood ratios calculated for the cyclic frequencies in $\mathcal{A}$ [123]:

$$\mathcal{T}_M = \max_{\alpha \in \mathcal{A}} \mathcal{T}(\alpha). \tag{5.25}$$

The corresponding collaborative sensing statistics are obtained by taking the sum over the local test statistics [123]:

$$\mathcal{T}_{S,N_{SU}} = \sum_{k=1}^{N_{SU}} \mathcal{T}_{S,k}, \tag{5.26}$$

$$\mathcal{T}_{M,N_{SU}} = \max_{\alpha \in \mathcal{A}} \sum_{k=1}^{N_{SU}} \mathcal{T}_{M,k}(\alpha), \tag{5.27}$$

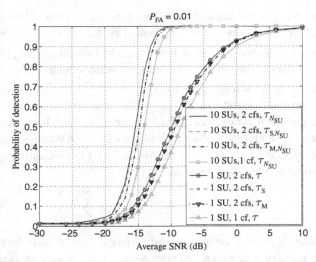

Figure 5.3 Probability of detection vs. average SNR (dB) for an OFDM signal in a Rayleigh fading channel for single-user sensing and cooperative detection of ten secondary users using cyclostationarity-based local detectors [123]. Cooperative sensing improves the detection performance significantly due to spatial diversity. Moreover, using multiple cyclic frequencies further improves the performance. (SU = secondary user, cf = cyclic frequency).

where k in the subscript denotes the kth user's local test statistic.

Figure 5.3 illustrates the single-user and collaborative detection performance of different cyclostationarity-based detectors in a Rayleigh fading channel.

Energy-efficient sensing

In wireless networks, secondary user terminals are often battery-operated. Consequently, low power consumption is an important design goal. There are many ways to lower the power consumption both in spectrum sensing and during exploitation of the identified idle spectrum. In spectrum sensing, the power consumption depends obviously on the employed sensing algorithms and their implementation, the duty cycle used in sensing, the employed sensing policy, as well as the power used in reporting the sensing results. In addition, deciding if and how the idle spectrum is accessed plays a key role in prolonging battery life. For example, if the secondary user is experiencing poor channel quality even if the channel is detected to be idle, it may make sense not to access the channel for transmitting with low rate while using high transmit power.

Censoring is a promising method to reduce power consumption in reporting sensing results to a fusion center [23, 166]. In order to limit the amount of data transmitted and save energy, only sufficiently informative decision statistics may be communicated to the fusion center or shared with other users. A secondary user sends its test statistic to the fusion center only when its test statistic, denoted here by $\log \Lambda_i$, is above a censoring threshold defined by the communication rate constraint specified by the designers:

$$P\left[\log \Lambda_i > \psi_i \mid \mathcal{H}_0\right] \leq \kappa_i, \ i = 1, \ldots, N_{\text{SU}}, \tag{5.28}$$

where $\kappa_i \leq 1$ is the communication rate of user i and ψ_i is the upper limit of the censoring (no-send) region of the user i. That is, each user will transmit its test statistic to the fusion center only if its value is above ψ_i, where ψ_i is chosen such that the probability of the user i transmitting the test statistic to the fusion center under $\mathcal{H}_0$ is κ_i. In the cooperative sensing context we may use the following censoring test statistic (L out of N_{SU} users transmit) [123]:

$$\mathcal{D}_{N_{\text{SU}}} = \sum_{i=1}^{L} \log \Lambda_i + \sum_{i=L+1}^{N_{\text{SU}}} d_i, \qquad (5.29)$$

where d_i is the average of the local log-likelihood ratio of the ith secondary user in the no-send region under the null hypothesis:

$$d_i = \mathbb{E}\left[\log \Lambda_i \mid \log \Lambda_i \leq \psi_i, \mathcal{H}_0\right]. \qquad (5.30)$$

This can be considered as a quantization to only one constant value. Consequently, the entire distribution of values in the no-send region is represented by a single quantity. By choosing the conditional mean as the value for d_i, the choice is optimal in the minimum mean-square error (MMSE) sense. In practice, there is no need to transmit d_i as it can be calculated at the fusion center since the limits ψ_i are determined by the communication rate constraints κ_i.

In [123] cyclostationarity-based local generalized likelihood ratio test detectors are used, and the secondary users communicate their generalized log-likelihood ratios exceeding the censoring threshold ψ_i to the fusion center. Under the null hypothesis the generalized log-likelihood ratio is asymptotically chi-square distributed [123]. However, censoring affects the distribution of the global test statistic at the fusion center where the statistics are combined. The distribution of the test statistic $\mathcal{D}_{N_{\text{SU}}}$ can be defined using conditional distributions as follows:

$$p(\mathcal{D}_{N_{\text{SU}}} \mid \mathcal{H}_0) = \sum_{k=0}^{N_{\text{SU}}} p(\mathcal{D}_{N_{\text{SU}}} \mid L = k, \mathcal{H}_0) P[L = k \mid \mathcal{H}_0], \qquad (5.31)$$

where the probabilities of different values of L are obtained by enumerating all possible combinations and computing their respective probabilities. If the same communication rate constraint is used for each secondary user, i.e., $\kappa = \kappa_i, \forall i$, then the probabilities are given by

$$P[L = k \mid \mathcal{H}_0] = \binom{N_{\text{SU}}}{k} \kappa^k (1-\kappa)^{N_{\text{SU}}-k}. \qquad (5.32)$$

The task at hand is then to determine the distribution of a sum of truncated chi-square distributed random variables. A closed-form solution is very difficult to obtain. However, the distribution may be numerically approximated accurately by using the fact that the cumulative distribution function may be obtained by inverting the characteristic function; see [123] for details. Censoring results in only a minor performance loss even with strict communication rate constraints. This is demonstrated in Figure 5.4 which plots the probability of detection vs. average SNR for an OFDM signal operating in a Rayleigh fading channel under different communication rate constraints κ.

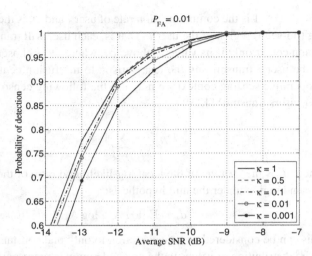

Figure 5.4 Probability of detection vs. average SNR (dB) for an OFDM signal in a Rayleigh fading channel for cooperative detection of ten secondary users using cyclostationarity-based local detectors under different communication rate constraints κ [123]. Only a minor performance loss is experienced even for very strict communication rate constraints.

The above test can be applied to any test statistic that is chi-square distributed under $\mathcal{H}_0$. Moreover, the equations for a normally distributed local test statistic can be found in [121, p. 101]. The normally distributed version can also be used as an approximation for a chi-square distributed test statistic for faster computation when the number of degrees of freedom is large.

Impact of quantization and reporting-channel errors
Using binary local decisions in a distributed decentralized detection system reduces the communication cost at the expense of a loss of information. Hard-decision-based sensing schemes are popular because of their simplicity. Using soft decisions such as log-likelihood ratios or their quantized versions typically leads to improvement in performance. It is commonly argued that soft-decision-based sensing approaches increase significantly the amount of data to be transmitted. This is not necessarily true since there may be significant overhead related to frame structure and different layers of the communication protocol stack used in transmitting the sensing results to the fusion center. That overhead will be present even if only binary decisions are transmitted. Consequently, the difference in the amount of data transmitted and bandwidth requirements compared to sending quantized soft-decision statistics may be small. Moreover, individual sensors may also convey information about interference levels, channel quality (SNR, SINR), and occupancy information. Such side information helps in managing interference, making access and scheduling decisions as well as utilizing the spectrum more efficiently.

The distribution of decision statistics for local quantized likelihood ratios is derived in [35]. Moreover, the distribution of the global decision statistic at the fusion center is established. Uniform quantization is used for simplicity. The analytical studies indicate that by using four or more bits for the quantization of the likelihood ratios, the performance loss is negligible. The analysis is done assuming an error-free reporting channel. The actual local detector is an autocorrelation-based detector exploiting the cyclic prefix of an OFDM transmission scheme. The impact of quantization on distributed detection has also been analyzed in [206].

It is common in the literature on distributed detection using the quantized values for decision statistics to make the assumption that the information transmitted by the user is received without any errors at the fusion center. This assumption may not be accurate in practice and the control (reporting) channel used for sending information to the fusion center may be subject to propagation effects, interference, and collisions. Using powerful error-correcting coding and increasing transmit power are obvious solutions to mitigate reporting-channel errors. However, in the case of simple battery-operated cognitive radio terminals with stringent energy constraints, these solutions are not always practical, and hence reporting-channel errors may cause a significant performance degradation in a decentralized detection scheme. Performance of hard-decision-based cooperative sensing in the presence of reporting-channel errors has been studied in the distributed detection research community [40] and in the cognitive radio literature [103]. The performance of hard-decision-based cooperative sensing using the OR fusion rule at the fusion center with nonideal reporting channels was studied in [144].

The problem of designing local sensor decision rules in the presence of independent nonideal reporting channels, i.e., the reporting channels may cause errors to the transmitted local decisions, is addressed in [40] (with further comments in [81]). Under the Bayesian criterion that minimizes the error probability at the fusion center the optimal decision rules turn out to be likelihood ratio tests under quite general conditions on the fusion rule and assuming conditional independence. The study is restricted to binary local sensor outputs. A design example using a simple binary symmetric channel model for the reporting channel is provided. The analysis in the paper is also applicable to fading channels. In [41] the analysis is extended to dependent nonideal reporting channels and the optimality of likelihood ratio tests at the local sensors is shown under both Bayesian and Neyman–Pearson criteria.

Reporting-channel errors also introduce fundamental performance limitations to decentralized distributed detection systems. These limitations have been studied in [37] by modeling the reporting channels as iid binary symmetric channels that cause bit errors with a certain bit-error probability (BEP). Figure 5.5 illustrates the system model. The secondary users sense the presence of primary users in a listening channel and transmit their local decisions to the fusion center through reporting channels that may cause errors. At the fusion center, the corrupted decisions are combined for cooperative sensing using the Neyman–Pearson criterion of constraining the false alarm rate while maximizing the probability of detection. A performance limitation in the form of a *BEP wall* is shown to exist for hard-decision-based cooperative sensing. That is, if the BEP of the reporting channel exceeds the BEP wall value, then irrespective of the received signal quality on the listening channel or the sensing time at the secondary users, constraints on the detector performance (the probability of false alarm and/or the

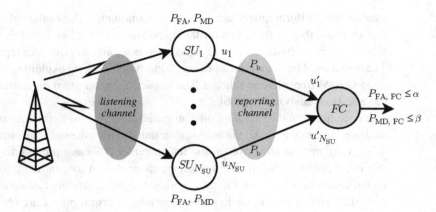

Figure 5.5 Spectrum sensors detect the primary user activity in a listening channel and pass their decision statistics u_i to the fusion center through a reporting channel with bit error probability P_b. The fusion center makes the global decision. However, due to reporting-channel errors the fusion center may not be able to satisfy the constraints α and β on the probability of false alarm and missed detection, respectively, regardless of the sensing time or the listening channel SNR [37].

probability of missed detection) cannot be met at the fusion center [37]. In addition, expressions for the BEP wall have been derived in [37] for the K-out-of-N rules in terms of the error probabilities at the fusion center and the number of users cooperating. The case of hard-decision-based systems and nonidentical reporting channels has been considered in [38]. Moreover, a BEP wall exists also for soft-decision-based systems [39]. Soft-decision-based systems are, however, significantly more robust to reporting-channel errors than hard-decision-based systems.

5.2.2 Sequential and quickest detection

Sequential and quickest detection techniques have applications in a number of fields, such as target detection in radar systems, pattern recognition, fault detection, finance, and clinical trials, among others [96, 162]. In cognitive radio systems sequential detection techniques are important for finding available spectral opportunities in minimal average time. In addition, sequential detection facilitates vacating a specific frequency band as soon as possible after the primary user becomes active.

One of the most critical performance indices in cognitive radio systems is the sensing time. Cognitive radio systems should aim at minimizing the time spent for sensing, and thus, maximizing the time available for transmission as well as simultaneously minimizing the energy resources spent for sensing. Sequential detection aims at minimizing the detection time given specified levels of error probabilities. The main difference between a sequential and a nonsequential test is in the length of the data sequence used for decision making. In the nonsequential case this length is fixed, whereas in the case of sequential test it varies according to the amount and quality of information obtained from the observations. In cognitive radio systems sequential detection may be used for both single-user and collaborative distributed detection tasks. Moreover, in collaborative distributed spectrum sensing sequential detection algorithms may be used either only at the local sensors, only at the fusion center, or both at the local sensors and at the fusion center.

In this section we will consider two sequential analysis problems. The classical sequential detection problem aims at distinguishing between two hypotheses from a sequence of identically distributed random observations. The objective of the classical sequential detection problem is to make a decision as quickly as possible given specified error levels. The second sequential analysis problem we consider is the quickest detection problem, in which the objective is to detect a change in the distribution of the observations as quickly as possible. The distribution of the random observations changes at some unknown point, and the objective is to detect the change point with minimal detection delay.

Sequential detection

Let $y_1, y_2, \ldots$, be a sequence of iid random observations with a common distribution F_0 or F_1. The hypothesis-testing problem of testing $\mathcal{H}_0$ against $\mathcal{H}_1$ may be formulated as

$$\mathcal{H}_0: \quad y_n \sim F_0, \; n = 1, 2, \ldots,$$
$$\mathcal{H}_1: \quad y_n \sim F_1, \; n = 1, 2, \ldots. \tag{5.33}$$

For simplicity we assume there are probability density functions p_0 and p_1 associated with F_0 and F_1, respectively. Sequential detection aims at deciding between these two hypotheses in a way that minimizes the number of observations given constraints on the error probabilities.

The sequential probability ratio test (SPRT) of Wald [207] requires the minimal average number of observations under both hypotheses among all tests with equal (or smaller) error probabilities [208]. The stopping time of the SPRT is given by

$$N_S = \inf\{n \geq 1 \mid \Lambda_n \leq B \text{ or } \Lambda_n \geq A\}, \tag{5.34}$$

where $\Lambda_n = \prod_{k=1}^{n} p_1(y_k)/p_0(y_k)$ is the likelihood ratio and A and B are upper and lower stopping boundaries, respectively. Thus, after each sample the SPRT accepts $\mathcal{H}_1$ if $\Lambda_n \geq A$, or accepts $\mathcal{H}_0$ if $\Lambda_n \leq B$, or if neither of the conditions is fulfilled, i.e., $B < \Lambda_n < A$, it takes an additional observation. The SPRT is illustrated in Fig. 5.6. The stopping boundaries A and B may in practice be chosen as

$$A = \frac{1-\beta}{\alpha} \text{ and } B = \frac{\beta}{1-\alpha}, \tag{5.35}$$

where α and β are the constraints on the probability of false alarm and the probability of missed detection, respectively. This guarantees that the actual probabilities of false alarm P_{FA} and missed detection P_{MD} are bounded by

$$P_{\text{FA}} \leq \frac{\alpha}{1-\beta} \tag{5.36}$$

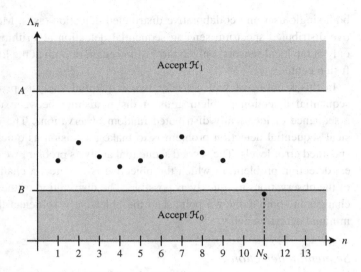

Figure 5.6 Illustration of an SPRT.

and

$$P_{\text{MD}} \leq \frac{\beta}{1-\alpha}. \quad (5.37)$$

Moreover, it follows that $P_{\text{FA}} + P_{\text{MD}} \leq \alpha + \beta$. (See, e.g., [161, p. 104].)

However, although the SPRT requires the minimal average number of observations among all tests with given error probabilities, this applies only for testing a simple null hypothesis versus a simple alternative hypothesis, i.e., when the distributions are completely specified. Hence, the distributions must be fully known under both hypotheses for the SPRT to be optimal. In cognitive radio applications, in particular, exact knowledge of the distribution of the primary user's signal and its parameters is difficult to obtain. Although various standards describe the primary user waveforms in great detail, the number of unknown variables, such as transmit powers, propagation conditions, adaptive modulation and coding schemes, and different primary user modes, makes it quite difficult to obtain accurate knowledge of the actual distribution parameters. To combat this problem the SPRT may be employed with assumed distribution parameters, e.g., corresponding to a certain primary user signal SNR. That is, consider a composite hypothesis-testing problem of testing $\mathcal{H}_0 : \theta \leq \theta_0$ versus $\mathcal{H}_1 : \theta \geq \theta_1 (> \theta_0)$ where θ is a real parameter of the parametric likelihood ratio. The SPRT can be used to test $\mathcal{H}_0 : \theta = \theta_0$ versus $\mathcal{H}_1 : \theta = \theta_1$, however, the maximum expected number of observations for such a test may be far from optimal when the true distribution parameter θ is different from the assumed value θ_0 or θ_1. Thus, there have been numerous efforts to design sequential detection tests for the case of composite hypotheses, such as the 2-SPRT [119], invariant SPRTs [59, 93], sequential generalized likelihood ratio (GLR) tests [94, 95, 99], and minimax tests [27, 28]. For further information, see [27, 28, 96] and the references therein. Moreover, it may sometimes be necessary to make a decision within a certain time. Hence, a truncated test [186, 207] that guarantees a decision will be made before a certain number of observations is reached may have to be employed. A truncated test uses a final decision rule to decide between the two hypotheses if a pre-defined upper limit on the number of observations is reached before a decision has been made by the SPRT. Truncation, however, changes the probabilities of false alarm and missed detection. This needs to be taken into account when deriving a truncated sequential detection test. For more information, see [59, 180] where comprehensive treatments of classical sequential detection can be found.

A Bayesian framework for distributed detection with sequential testing at the fusion center in a parallel topology has been introduced in [202]. The optimal Bayesian sequential detector, consisting of a stopping time N_S and a terminal decision rule δ_{N_S}, minimizes the Bayesian risk $P_e(\delta_{N_S}) + c\mathbb{E}[N_S]$, where $P_e(\delta_{N_S})$ is the average probability of error, $\mathbb{E}[N_S]$ is the average number of observations used by the detector, and $c > 0$ is the cost per observation for decision making. Different system configurations regarding feedback from the fusion center to the sensors and the sensors' local memory capabilities are considered. The optimal solution for the quasi-classical system structure in which there is full feedback but local memory for only the past sensor decisions is shown to be tractable. In [132] asymptotically Bayes-optimal distributed sequential detection methods for systems with local sensors having full and limited memory are proposed. The asymptotic optimality is shown in the sense that the Bayesian risk tends to that of an optimal centralized SPRT as the cost per time step for decision making goes to zero. Feedback from the fusion center to the sensors does not improve the asymptotic performance if the sensors have full local memory of their past observations. However, in the case of limited local memory even a one-time one-bit feedback can significantly improve the asymptotic performance. In [72] a decentralized SPRT (D-SPRT) scheme in which both the local sensors and the fusion center employ SPRTs is proposed. The local sensors employ repeated SPRTs to determine the transmit instances and binary decisions transmitted to the fusion center. After making a local decision (i.e., the SPRT stops), the local detector transmits the binary decision to the fusion center and starts a new local SPRT on its subsequent observations. The fusion center employs an SPRT on the received local decisions from the sensors. In [53] an analogous continuous time D-SPRT is proposed. Moreover, the asymptotic first-order optimality of the discrete-time D-SPRT and the asymptotic second-order optimality of the continuous time D-SPRT are shown in the non-Bayesian setting as the error probabilities go to zero.

The SPRT has been considered for cognitive radio systems in [36, 175, 185]. In [36] a sequential detection test for the fusion center based on the SPRT and an autocorrelation-based detector for OFDM signals is proposed. The secondary users transmit their local log-likelihood ratios (LLRs) of a fixed-sample-size autocorrelation-based detector to the fusion center that employs an SPRT. Figure 5.7 illustrates the benefit of sequential testing in comparison to fixed-sample-size testing by plotting the number of secondary user statistics used at the fusion center vs. SNR. The probability of false alarm constraint is set to $\alpha = 0.05$ and the probability of missed detection constraint to $\beta = 1 - P_D = 0.05$. Transmission is assumed to take place over an AWGN channel.

In [185], a collaborative sequential detection scheme based on energy detection and the SPRT is proposed. The secondary users transmit their energy measurements to each

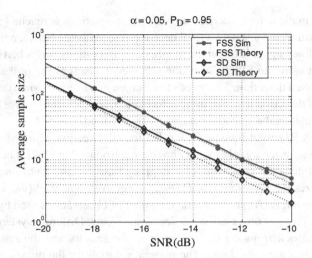

Figure 5.7 Comparison of the theoretical average number of secondary user decision statistics for sequential detection (SD) using the SPRT at the fusion center and fixed-sample-size (FSS) detection needed for equal performance [36]. The sequential method requires significantly fewer decision statistics than FSS detection for the same error probabilities.

other and each user employs an SPRT. A collaborative distributed detection scheme based on energy detection and the SPRT at both the local sensors and fusion center is proposed in [175]. An iterative log-likelihood ratio update for a sequential energy detector assuming white Gaussian noise is proposed in [89].

In [238] a collaborative sequential detection scheme is proposed in which each user transmits its generalized log-likelihood ratio (GLLR) for each sample to the fusion center. The fusion center uses a sequential GLLR test in which the unknown distribution parameters are replaced with their maximum likelihood estimates. This facilitates adapting to the varying radio channel conditions and reducing the average sensing time by making it dependent on the state of the radio channel.

In [46] a single-cycle cyclostationarity-based sequential detection test is proposed. The proposed test is based on a generalized M-ary SPRT (MSPRT) [203] in which the multiple hypotheses correspond to a set of different signal phases. This structure is employed since the signal phase is in practice unknown.

A collaborative sequential detection scheme based on energy detection and a truncated SPRT is proposed in [111]. The secondary users employ energy detection and a truncated SPRT to obtain local decisions. The local decisions are then transmitted to the fusion center which employs an SNR weighted K-out-of-N fusion rule. In [224] a simple energy-detection-based truncated sequential detection test is proposed that does not require any knowledge about the primary user signals. The test is called the sequential shifted chi-square test (SSCT) because the test statistic is a sum of shifted chi-square random variables, i.e., constant shifted energy measurements. The constant depends on the desired minimum detection SNR. In [64] a randomly truncated distributed sequential detection algorithm based on a repeated significance test (RST) is proposed. The local sensors use censoring to refrain from transmitting their data to the fusion center when it is not sufficiently informative. The fusion center employs the RST in addition to a truncation to avoid exceeding a communication constraint. That is, if the amount of data transmitted exceeds the communication constraint before the RST stops, the null hypothesis $\mathcal{H}_0$ is accepted.

An interesting and important twist to the collaborative spectrum sensing problem is obtained when some of the secondary users may be malicious. Such users may falsify sensing data for their own purposes. A similar situation arises when a secondary user's sensor is malfunctioning and thus produces erroneous data. Sequential detection algorithms derived from the SPRT that are robust against data falsification have been proposed in [42, 57, 237]. For example, in [42] a collaborative weighted SPRT (WSPRT), in which the likelihood ratios of the users are weighted using weights based on the credibility of previous sensing results is proposed. The WSPRT takes into account all the previous samples to estimate the credibility of the individual sensors. A windowed credibility weighting scheme for WSPRT is proposed in [57]. In [237] two further tests are proposed: enhanced weighted SPRT (EWSPRT) and enhanced weighted sequential zero/one test (EWSZOT). Both tests use weighting based on the credibility of previous sensing results to weight the current sensing results.

Quickest detection

Let $y_1, y_2, \ldots$, be a sequence of independent random observations with an unknown change point m, such that $y_1, \ldots, y_{m-1}$ have a common distribution F_0 and $y_m, y_{m+1}, \ldots$ have another common distribution F_1. Let p_0 and p_1 denote probability density functions of F_0 and F_1, respectively.

Both Bayesian and non-Bayesian quickest detection approaches have received considerable attention in the research community. Page's cumulative sum (CUSUM) test [151] has been the most influential and most researched algorithm in the development of non-Bayesian quickest detection methods. The stopping time of the CUSUM test for detecting a change is given by

$$N_S = \inf \{n \geq 1 \mid U_n \geq \psi\}, \quad (5.38)$$

where $U_n = \max_{1 \leq k \leq n} \prod_{i=k}^{n} p_1(y_i)/p_0(y_i)$ and $\psi \geq 0$ is the test threshold. The CUSUM test statistic U_n may be updated recursively via $U_n = \max\{1, U_{n-1}\} p_1(y_n)/p_0(y_n), n \geq 1, U_0 = 1$. Let ψ be chosen such that the mean time between false alarms is $\mathbb{E}_\infty [N_S] = \zeta$, where $\mathbb{E}_\infty [\cdot]$ denotes the expectation when there is no change in the distribution of the observations. Then the CUSUM test is optimal in the minimax sense, i.e., it minimizes the worst-case expected detection delay among all tests satisfying $\mathbb{E}_\infty [N_S] \geq \zeta$ [136, 162].

Page's CUSUM scheme considers the change point m as an unknown constant. An alternative Bayesian formulation is obtained by treating the change point as a random parameter with a prior distribution. Shiryaev [177] proposed a Bayesian change detection formulation in which the stopping point is given by the solution of the following optimization problem:

$$\inf_{N_S} \left\{ P\left[N_S < m\right] + c\, \mathbb{E}\left[(N_S - m + 1)^+\right] \right\}, \tag{5.39}$$

where c is a positive constant and $(N_S - m + 1)^+ = \max\{0, N_S - m + 1\}$. The constant c controls the relative importance of minimizing the probability of false alarm $P\left[N_S < m\right]$ and the expected detection delay $\mathbb{E}\left[(N_S - m + 1)^+\right]$.

Let us assume a geometric prior distribution for the change point:

$$P[m = n] = \begin{cases} \omega, & n = 0, \\ (1-\omega)q(1-q)^{n-1}, & n = 1, 2, \ldots, \end{cases} \tag{5.40}$$

where ω is the probability that the change has already occurred before the first observation is obtained and q is the conditional probability that the sequence transfers to the post-change state at any given time. Both ω and q are constants in $(0,1)$.

Under the above assumptions, Shiryaev [177] showed that the stopping point that minimizes the Bayes risk in (5.39) is given by

$$N_S = \inf \{n \geq 0 \mid \omega_n \geq \omega^*\}, \tag{5.41}$$

where ω^* is a certain appropriately chosen constant threshold in $[0,1]$ and $\omega_n = P[m \leq n \mid y_1, \ldots, y_n]$ is the posterior probability that a change has already happened given the observations up to that point. In addition Shiryaev [177] showed that if the threshold ω^* is chosen to satisfy a given false alarm constraint $P\left[N_S < m\right] = \alpha$ then the stopping rule in (5.41) minimizes the expected detection delay among all rules satisfying the false alarm constraint $P\left[N_S < m\right] \leq \alpha$. Shiryaev's quickest change detection test is illustrated in Fig. 5.8.

The Bayes optimal stopping time may be written also as

$$N_S = \inf \{n \geq 0 \mid U_{n,q} \geq \psi\}, \tag{5.42}$$

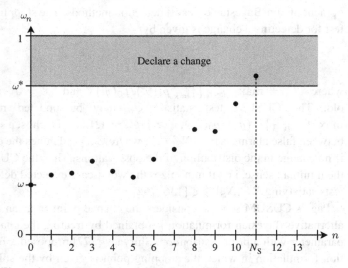

Figure 5.8 Illustration of a Bayesian quickest change detection test.

where $\psi = \omega^*/((1-\omega^*)q) > \omega/((1-\omega)q)$ and

$$U_{n,q} = \frac{\omega}{(1-\omega)q}\prod_{i=1}^{n}\left(\frac{1}{1-q}\frac{p_1(y_i)}{p_0(y_i)}\right) + \sum_{k=1}^{n}\prod_{i=k}^{n}\left(\frac{1}{1-q}\frac{p_1(y_i)}{p_0(y_i)}\right). \quad (5.43)$$

Note that $P[m \leq n \mid y_1, \ldots, y_n] = U_{n,q}/(U_{n,q} + q^{-1})$. The form of the test statistic allows the following recursive update:

$$U_{n,q} = (1 + U_{n-1,q})\frac{1}{1-q}\frac{p_1(y_n)}{p_0(y_n)}, \; n \geq 1, \; U_{0,q} = \frac{\omega}{(1-\omega)q}. \quad (5.44)$$

Now, choosing $\omega = 0$ and letting $q \to 0$, we obtain the Shiryaev–Roberts (SR) [168] stopping rule

$$N_S = \inf\{n \geq 0 \mid U_n \geq \psi\}, \quad (5.45)$$

where $U_n = \sum_{k=1}^{n}\prod_{i=k}^{n} p_1(y_i)/p_0(y_i)$, which allows a recursive update $U_n = (1 + U_{n-1})p_1(y_n)/p_0(y_n), n \geq 1, U_0 = 0$. Note that the SR rule does not assume any specific prior distribution for the change point m, i.e., as $q \to 0$ with $\omega = 0$ the prior distribution approaches an improper uniform distribution. Pollak and Tartakovsky [159] showed that the discrete-time SR rule is optimal in the sense of minimizing $\sum_{m=0}^{\infty} \mathbb{E}_m[(N_S - m)^+]$ for every $\zeta > 1$ among all rules satisfying $\mathbb{E}_\infty[N_S] \geq \zeta$.

Moustakides et al. [137] introduced a generalization of the SR rule that does not necessarily start from zero but from any point $r \geq 0$. The test is called the SR-r test and it is obtained from the Shiryaev test by choosing $\omega = rq$ and then letting $q \to 0$. The stopping point of the SR-r rule is given by

$$N_S = \inf\{n \geq 0 \mid U_n^r \geq \psi\}, \quad (5.46)$$

where $U_n^r = r\prod_{i=1}^{n} p_1(y_i)/p_0(y_i) + \sum_{k=1}^{n}\prod_{i=k}^{n} p_1(y_i)/p_0(y_i)$. The SR-$r$ rule can be updated recursively via $U_n^r = (1 + U_{n-1}^r)p_1(y_n)/p_0(y_n), n \geq 1$, and $U_0^r = r$. The asymptotic third-order optimality in the minimax sense of the SR-r rule with an appropriately chosen r has been shown in [189] (see also [137, 160]). Moreover, comprehensive treatments of quickest detection can be found in [15, 162, 178, 188].

As already discussed, in cognitive radio applications it is unrealistic to assume exact knowledge of the distributions under both hypotheses. Moreover, it is well known that the optimum quickest detection algorithms are sensitive to uncertainty in distribution parameters. Many of the approaches dealing with unknown distribution parameters are based either on nonparametric approaches [26, 158] or on generalized likelihood ratio (GLR)-based algorithms [96]. Moreover, other approaches involving, e.g., different prior distributions for the unknown pre- and post-change distribution parameters or least favorable pre- and post-change distributions have been proposed as well [98, 131, 133, 198]. In [158] a nonparametric change point detection method is proposed that satisfies a prespecified lower bound on the ratio of average run length and false alarm regardless of the true distribution. In addition, if the true distribution belongs to a suspected family the method is asymptotically first-order optimal. In [198] robust quickest change detection methods are proposed. Least favorable distributions (LFDs) under certain uncertainty classes are employed to obtain minimax optimal robust CUSUM, Shiryaev's Bayesian, and Shiryaev–Roberts–Pollak (SRP) rules (SRP is an SR rule with a random starting point). The robust Shiryaev's Bayesian and SRP rules require a fully known pre-change distribution. However, the robust CUSUM rule may be applied in the case of unknown pre- and post-change distributions. Moreover, the robust CUSUM rule allows a recursive implementation unlike the GLR change detection algorithm. However, the optimal robust stopping rules obtained with LFDs may perform significantly worse for other distributions in the uncertainty class than the optimal stopping rules with exactly known distributions. In [98, 131, 133] quickest change detection problems involving pre- or post-change distributions with unknown parameters are considered. In [131] a required detection delay at a given post-change distribution is specified and the frequency of false alarms is minimized for every possible pre-change distribution. In [133] a mixture distribution is employed for the pre-change distribution with unknown parameters. In [98] Shiryaev's model is extended by employing an exponential family of pre- and post-change distributions with conjugate priors for the unknown parameters. A sliding window type rule is proposed and shown to be asymptotically optimal in the Bayesian sense.

Single-user quickest detection methods for cognitive radio systems have been proposed in [91, 105, 235, 236]. In [91] quickest detection methods for different cases of prior knowledge in cognitive radio applications are proposed. The CUSUM, the GLR change detection algorithm, and a nonparametric rank-based change detection algorithm are proposed for the cases of known signal and noise power, known noise power and unknown signal power, and unknown signal and noise power, respectively. In [236] a quickest detection algorithm based on the output energy of a stochastic resonance system is proposed. In [235] single-user quickest change detection over multiple frequency bands is considered. The problem corresponds to quickest detection of an idle period in multiple iid on-off processes. A Bayesian formulation for the quickest change detection problem is proposed and optimal decision rules based on a threshold structure are obtained. Another quickest detection problem over multiple frequency bands is considered in [105]. In this problem the goal is to detect the emergence of primary users and to select the best frequency band to sense. A framework for obtaining a tradeoff between the minimization of false alarms and detection delays is proposed. Dynamic programming is employed to obtain a control policy for selecting which frequency band to sense and when to declare the emergence of a primary user.

Collaborative quickest detection schemes applicable to cognitive radio systems have been proposed in [13, 65, 71, 74, 75, 108–110, 190, 192, 201, 231]. Moreover, general treatments of distributed quickest detection can be found in [22, 191]. In [201] a Bayesian formulation of a decentralized quickest detection problem in parallel topology is proposed. The optimal Bayesian solution is obtained for the quasi-classical information structure in which there is full feedback from the fusion center to the sensors and local memory is restricted to the past decisions of all sensors. Moreover, a suboptimal solution without any feedback is proposed and shown to perform close to optimally in the considered simulation examples. In [65] a decentralized quickest detection scheme is considered in which the local sensors use the CUSUM algorithm for quickest change detection. Figure 5.9 illustrates the scheme. In this scheme the sensors communicate

5.2 Advanced spectrum sensing techniques

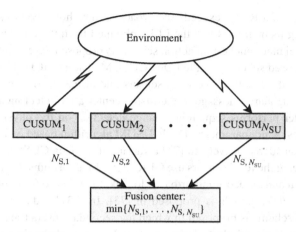

Figure 5.9 Decentralized quickest detection system corresponding to a minimum of N_{SU} CUSUMs [65]. Each sensor communicates with the fusion center only after it has detected a change in the distribution. The fusion center declares a change immediately after receiving the first message.

with the fusion center only after detecting a change in the distribution. The fusion center declares a change after receiving at least one message from the sensors. Thus, this scheme corresponds to a minimum of N_{SU} CUSUMs. It is shown to be asymptotically optimal as the mean time between false alarms tends to infinity. In [192] asymptotically optimal decentralized Bayesian and minimax quickest detection schemes are proposed for two cases in which the local sensors transmit binary quantized observations and local decisions, respectively. Full knowledge of the pre-change and post-change distributions is assumed. Binary quantization-based tests were found to perform better than local decision-based tests. In [190] a distributed quickest detection approach, called the M-BQ-CUSUM test, is proposed for the case when the post-change distribution has an unknown parameter. The approach is based on using $M \geq 2$ putative values for the unknown parameter. The fusion center runs M binary quantized (BQ) CUSUM statistics in parallel using optimally quantized data sent by the local sensors. The M-BQ-CUSUM scheme is shown to be asymptotically optimal at the putative values. In [71] a collaborative quickest detection scheme for cognitive radios is proposed where the secondary users make local decisions for each observation and send them to the fusion center only when a change is detected locally. The fusion center employs a CUSUM test to detect a change in primary user occupancy. The CUSUM change detector is designed to maximize the throughput of the secondary users given a constraint on the average interference time. In [109, 110] two-node and multi-node collaborative quickest detection methods based on the CUSUM algorithm are proposed for cognitive radio ad hoc networks. The proposed methods take into account the communication delay of the sensing result exchanged among the secondary users. In [108] a collaborative quickest detection scheme based on random broadcast is proposed for cognitive radio ad hoc networks. Without any explicit coordination to avoid collisions while broadcasting the local test statistics, each secondary user randomly chooses a time slot after sensing to broadcast its LLR if it exceeds the broadcast threshold. Each user employs a CUSUM test using its own LLR and the LLRs obtained from the other users. In [13] a collaborative distributed quickest detection scheme is proposed based on the CUSUM algorithm. The proposed scheme is called DualCUSUM. The DualCUSUM scheme employs CUSUM algorithms both at the local sensors and at the fusion center. Each local sensor transmits its alarm message to the fusion center after detecting a change locally. The fusion center performs sequential change detection on the physical layer fused sensor messages. A nonparametric DualCUSUM algorithm based on constant shifted observations is considered as well. In [74] a variant of the DualCUSUM scheme is proposed. In this scheme the local sensors use GLR CUSUM algorithms. Moreover, a DualCUSUM algorithm and autocorrelation-based log-likelihood ratio for collaborative quickest detection of OFDM signals is proposed in [75]. In [231] a distributed quickest change detection scheme is proposed in which the secondary users transmit their quantized observations over a wireless Gaussian multiple-access channel to a fusion center that performs change detection on the physical layer fused observations. The control parameters of the distributed detection network, such as the secondary user transmit powers, are optimized through feedback to minimize the error variance of the observations at the fusion center. An energy constrained formulation and solution are provided as well.

Quickest detection of newly appearing users in code division multiple access (CDMA) systems has been considered in [30, 134, 150]. In [150] the matrix CUSUM, an online minimax sequential multiple-hypothesis detection algorithm for the detection of a user switching from passive to active mode in a CDMA system is proposed. In [30] an online Bayesian sequential multiple-hypothesis detection algorithm – the Shiryaev sequential probability radio test (SSPRT) – for the detection of a user switching from passive to active mode in a CDMA system is proposed. In [134] distribution-free detectors for the presence of a new user in a CDMA system are proposed.

In addition to detecting a change in distribution, ideally a cognitive radio system is able to also recognize the signal or system causing the post-change distribution. Quickest detection combined with signal recognition methods have been proposed in [45, 47]. In [47] a Bayesian formulation of a sequential change diagnosis problem is proposed. Sequential change diagnosis is the joint problem of quickest detection and identification of a change in the distribution of a random sequence. An optimal decision strategy for this problem is derived in [47]. In [45] a quickest detection and primary signal recognition approach based on hidden Markov models (HMMs) and the CUSUM algorithm is proposed.

5.3 Optimized spectrum exploration and exploitation: sensing and access policy design

Sensing and access policies provide answers to the questions: what should the radios sense and access, and when? Moreover, in centrally controlled multiuser systems spectrum sensing and access policies allocate and divide the sensing, learning, and interference management tasks among the network nodes and different spectrum areas. That is, the spectrum of interest may be very wide and noncontiguous. Thus, from the perspective of a single secondary user, sensing the whole spectrum of interest in each sensing instant may not be practical due to the difficulties it presents to the sensing receiver's radio frequency (RF) front-end design as well as due to the high energy consumption of such an approach. Consequently, a more practical solution involves dividing the spectrum of interest into smaller frequency subbands that can be sensed one at a time. Hence, the problem becomes deciding which frequency band or bands to sense and subsequently access and by which user or users in order to optimize the cognitive radio network's performance. Obviously the secondary users should sense and access frequency bands in which there exist persistent spectral opportunities and high average throughput. Moreover, in a cognitive radio network it becomes essential to distribute the frequency bands for sensing and access among the network members in the most efficient and effective manner that results in a high probability of finding available spectrum as well as in low interference to both primary and secondary users of the spectrum. In order to make such decisions most effectively, the cognitive radio network must obtain situational awareness. Thus, various approaches to modeling, learning, and predicting the time-, frequency-, and location-varying spectrum state and the behavior of other spectrum users are of special interest.

The nature and constraints of the optimization problem and the design of sensing and access policies depend on whether the cognitive radio network model is a single-user or multiuser network model, whether the spectrum model is a single-band or multiband model, whether the users sense and access the spectrum synchronously or asynchronously, and whether the secondary users in a multiuser model cooperate or compete with each other, among other things. In general, in an infrastructure-based network cooperation among the secondary users is more naturally enforced than in an ad hoc network in which the users are more inclined to compete with each other. However, this by no means rules out the possibility of competition in an infrastructure-based network or cooperation among the users in an ad hoc network. On the contrary, such approaches may offer substantial benefits by simplifying the system design and/or by providing improved performance. In this section we will describe various different approaches and cognitive radio network models that are competitive, cooperative, or a combination of the two. We will mostly focus on the design of spectrum sensing and access policies in cognitive radio networks, and introduce various approaches such as bandit problems, reinforcement learning, and game theory that may be employed to address these issues. However, we will start this section by looking at some more direct optimization techniques for optimized spectrum exploration and exploitation in cognitive radio networks.

5.3.1 Optimization techniques

In this subsection various techniques and approaches for optimizing the performance of cognitive radio networks are studied. We begin by briefly introducing an optimization technique called *dynamic programming*. Dynamic programming is a collection of optimization techniques that solve complicated sequential decision problems by breaking them into smaller easily solved subproblems. Moreover, dynamic programming is paramount for understanding the nature of learning techniques such as reinforcement learning that will be studied later in this chapter. In addition, in this section we will look at a few interesting dynamic programming and other optimization approaches for maximizing the cognitive radio throughput by means of optimizing the frequency band sensing order or the sensing parameters, such as the detection thresholds, sensing and transmission times, and the allocation of sensing resources to different frequency bands in multiband cognitive radio scenarios.

Dynamic programming

Dynamic programming is an optimization approach in which an original complex problem is broken down into smaller, simpler subproblems that are then solved in a recursive manner. Dynamic programming applies to many different types of problems. However, here we focus on solving finite Markov decision processes (MDPs) to illustrate the main idea of dynamic programming. A finite MDP consists of the following:

- a sequence of discrete time steps $n = 0, 1, 2, \ldots$;
- a finite set of possible states of the environment $s \in S$;
- a finite set of possible actions in each state $a \in A$;
- a state transition function $\varphi : S \times A \times S \to [0, 1]$ which defines the transition probability $P[s_{n+1} \mid s_n, a_n]$;
- a reward function $r : S \times A \times S \to \mathbb{R}$ which gives a reward for taking action a_n in state s_n resulting in new state s_{n+1}.

Let us consider the maximization of the expected sum of discounted rewards given an initial state $s_n = s$:

$$\max \mathbb{E}\left[\sum_{k=0}^{N_S} \xi^k r_{n+k+1}\right], \qquad (5.47)$$

where r_n is the reward at time n and ξ, $0 \leq \xi \leq 1$, is the discount rate. Note that the model allows also the infinite horizon case if $N_S = \infty$; however, in this case we must assume $\xi < 1$. The goal of the decision problem is to find an optimal policy π^* that determines in each state $s \in S$ the optimal action $a \in A$, so that the above discounted sum of rewards is maximized. The value of a policy π starting from state s is defined by

$$\mathcal{V}^\pi(s) = \mathbb{E}_\pi\left[\sum_{k=0}^{N_S} \xi^k r_{n+k+1} \,\bigg|\, s_n = s\right], \qquad (5.48)$$

where $\mathbb{E}_\pi[\cdot]$ denotes the expectation given that the policy π is followed.

The value function can be broken down as follows:

$$\mathcal{V}^\pi(s) = \mathbb{E}_\pi\left[\sum_{k=0}^{N_S} \xi^k r_{n+k+1} \,\bigg|\, s_n = s\right]$$

$$= \mathbb{E}_\pi\left[r_{n+1} + \sum_{k=1}^{N_S} \xi^k r_{n+k+1} \,\bigg|\, s_n = s\right]$$

$$= \mathbb{E}_\pi \left[r_{n+1} + \xi \sum_{k=0}^{N_S-1} \xi^k r_{(n+1)+k+1} \,\bigg|\, s_n = s \right]$$

$$= \mathbb{E}_\pi \left[r_{n+1} + \xi \mathcal{V}^\pi(s_{n+1}) \,\bigg|\, s_n = s \right]$$

$$= \sum_{s' \in S} P[s' \mid s, a] \left[r(s, a, s') + \xi \mathcal{V}^\pi(s') \right], \quad (5.49)$$

where it can be seen that the successive states have a recursive relationship, thus allowing the problem to be broken down into simpler subproblems. The last two forms in (5.49) are forms of *the Bellman equation* [16, 19].

The optimal value function $\mathcal{V}^*$ satisfies *the Bellman optimality equation*:

$$\mathcal{V}^*(s) = \max_a \mathbb{E}\left[r_{n+1} + \xi \mathcal{V}^*(s_{n+1}) \mid s_n = s, a_n = a \right]$$

$$= \max_a \sum_{s' \in S} P[s' \mid s, a] \left[r(s, a, s') + \xi \mathcal{V}^*(s') \right]. \quad (5.50)$$

Dynamic programming offers a computationally more efficient method of finding the optimal solution compared to brute-force techniques. However, dynamic programming still suffers from *the curse of dimensionality*: the computational and memory requirements grow exponentially as the number of states and actions increases. Hence, dynamic programming may not be the best overall solution in many cognitive radio applications, especially, those involving mobile, battery-powered user equipment with limited computational and memory resources. However, dynamic programming provides insight into the optimal solution that can then be employed to obtain near-optimal reduced-complexity algorithms and methods to be employed in practice.

Finally, we would like to point out that dynamic programming requires a model for the state transitions and rewards. In Section 5.3.3 we will introduce model-free reinforcement learning methods that can be used to solve the same problems as dynamic programming (although we note that in general reinforcement learning methods can only approximate the optimal solutions). However, in the following we will briefly review a few dynamic programming as well as other optimization approaches for optimizing the performance of cognitive radio systems. For comprehensive treatments of dynamic programming we refer the reader to [16] and [19].

Optimizing the sequential order of sensing multiple frequency bands
In [76] the optimization of the frequency band sensing order in a multiband cognitive radio system is considered. That is, in each time slot, given n_B frequency bands, the secondary user senses the frequency bands one at a time according to a sequential order until it finds a frequency band that is both available and has acceptable channel quality. Then the secondary user proceeds to transmit on that frequency band for the rest of that time slot. The goal is to optimize the frequency band sensing order so that the expected throughput of the secondary user is maximized. The problem may be formulated as an optimal stopping problem as follows.

Let $b_1, b_2, \ldots, b_{n_B}$ denote a sensing order, thus being a permutation of the set $\{1, 2, \ldots, n_B\}$. Moreover, let $\omega_i \in [0,1]$ denote the probability that the frequency band i is available and let $r(\gamma_i)$ denote the achievable transmission rate, where $r(\cdot)$ is a non-decreasing function and γ_i is the SNR of the secondary user on frequency band i. The probabilities ω_i are assumed to be different for all frequency bands. The γ_is are iid and fixed within a time slot but change randomly between different time slots.

The secondary user senses the frequency bands according to the sensing order $b_1, \ldots, b_{n_B}$. If frequency band b_i is occupied, the secondary user proceeds to sense frequency band b_{i+1}. On the other hand, if b_i is available the secondary user will transmit on that frequency band provided that the expected throughput is larger than the expected throughput is if the user proceeds to sense the next frequency band in the order. The expected throughput of frequency band b_i is given by [76]

$$r_i = \begin{cases} \eta_i r(\gamma_{b_i}), & \text{if } i = n_B \text{ (transmit on } b_i\text{)}, \\ \eta_i r(\gamma_{b_i}), & \text{if } i < n_B \text{ and } \eta_i r(\gamma_{b_i}) > R_{i+1} \\ & \text{(transmit on } b_i\text{)}, \\ R_{i+1}, & \text{otherwise (proceed to sense } b_i\text{)}, \end{cases} \quad (5.51)$$

where η_i is the transmission time $T - iT_s$ relative to the time slot duration T, i.e., $\eta_i = 1 - iT_s/T$, where T_s is the sensing time per frequency band, and R_{i+1} is the expected throughput if the user proceeds to frequency band b_{i+1} given by

$$R_{i+1} = \begin{cases} \omega_{b_{i+1}} \mathbb{E}[r_{i+1}] + (1 - \omega_{b_{i+1}})R_{i+2}, & \text{if } i < n_B - 1, \\ \omega_{b_{i+1}} \mathbb{E}[r_{i+1}], & \text{if } i = n_B - 1. \end{cases} \quad (5.52)$$

Starting from R_{n_B} to R_1 the sequence $R_1, \ldots, R_{n_B}$ can be obtained recursively.

Dynamic programming is employed to obtain the optimal sensing order in both cases of known and unknown availability probabilities ω_i. The optimal solution in the case of unknown ω_i balances between short-term and long-term gains. Dynamic programming suffers from the curse of dimensionality, and thus obtaining the optimal solution via dynamic programming may become computationally prohibitive if the number of frequency bands is large. Hence, computationally more efficient, yet optimal sensing orders are obtained for certain special cases of availability probabilities and achievable rates. Moreover, it is shown in [76] that the intuitive sensing order selecting the frequency bands in the order of their availability probabilities ω_i is in general not optimal if adaptive modulation is employed, and thus the achievable rates are not equal on different frequency bands.

Other approaches to optimizing the sensing order in multiband cognitive radio systems have been proposed in [83, 126] and in [52], where a collaborative spectrum search strategy is proposed for maximizing the expected number of frequency bands identified as being available.

Optimizing the secondary system throughput given interference constraints
In [163, 165] an optimal energy-detection-based joint multiband detection approach for cognitive radio systems is proposed. The throughput of the secondary system is

maximized by finding optimal detection thresholds for the joint energy detection problem over multiple frequency bands. Let $i = 1, 2, \ldots, n_B$ denote the different frequency bands. The single secondary user optimization problem is formulated as:

$$\max_{\psi} \quad R(\psi) = \mathbf{r}^T[1 - \mathbf{P}_{\text{FA}}(\psi)] \tag{5.53}$$

$$\text{s.t.} \quad \sum_{i \in S_j} c_i P_{\text{MD}}^{(i)}(\psi_i) \leq \epsilon_j, \; j = 1, 2, \ldots, J, \tag{5.54}$$

$$\mathbf{P}_{\text{MD}}(\psi) \leq \boldsymbol{\beta}, \tag{5.55}$$

$$\mathbf{P}_{\text{FA}}(\psi) \leq \boldsymbol{\alpha}, \tag{5.56}$$

where $\mathbf{r}$ is a column vector of the achievable rates on the different frequency bands, ψ is a vector of the energy detector thresholds for the different frequency bands, $\mathbf{P}_{\text{FA}}(\psi)$ is a vector of probabilities of false alarm at the different frequency bands given ψ, $\mathbf{P}_{\text{MD}}(\psi)$ is a vector of probabilities of missed detection at the different frequency bands given ψ, c_i is a priority factor for the frequency band i, S_j is the subset of frequency bands occupied by the primary user j, ϵ_j is a constraint for the interference caused to the primary user j, $\boldsymbol{\beta}$ is a vector of probability of missed detection constraints for the different frequency bands, and $\boldsymbol{\alpha}$ is a vector of probability of false alarm constraints for the different frequency bands. The constraints in (5.55) and (5.56) are defined item-wise. The above generally nonconvex optimization problem is reformulated in [163, 165] in a convex form with convex constraints under certain conditions by exploiting the properties of the energy detector and its distributions under both hypotheses. That is, the fact that the Q-function is monotonically nonincreasing allows transforming the constraints (5.55) and (5.56) into linear constraints. Moreover, an alternative formulation in which the priority weighted probability of missed detection, i.e., $\mathbf{c}^T \mathbf{P}_{\text{MD}}(\psi)$, is minimized given constraints on the minimum secondary throughput as well as on the probabilities of missed detection and false alarm on the different frequency bands is proposed. In addition, a collaborative spectrum sensing optimization problem controlled by a fusion center is formulated. The secondary users transmit their local summary energy measurements to the fusion center that performs a linear combination to obtain the global test statistics for the different frequency bands (see also (5.10)):

$$\mathcal{E}_i = \sum_{k=1}^{N_{\text{SU}}} a_{k,i} \mathcal{E}_{k,i}, \; i = 1, \ldots, n_B, \tag{5.57}$$

where N_{SU} is the number of collaborating secondary users, $a_{k,i}$ is a weighting factor for the frequency band i of the secondary user k, and $\mathcal{E}_{k,i}$ is the summary energy detection test statistic for the frequency band i of the secondary user k. An optimization problem, similarly to (5.53)–(5.56), is formulated in which both the weights $a_{k,i}$ and the thresholds ψ_i are optimized to maximize the secondary system throughput. Two near-optimal solutions to this generally nonconvex problem are proposed. In [172] genetic algorithms, a global optimization approach, have been considered for solving this problem as well.

In [112] the sensing time within a fixed time slot is optimized to maximize the throughput of the secondary user given a constraint on the probability of detection:

$$\max_{T_s} \quad R_0(\psi, T_s), \tag{5.58}$$

$$\text{s.t.} \quad P_D(\psi, T_s) \geq 1 - \beta, \tag{5.59}$$

where $R_0(\psi, T_s)$ is the throughput of the secondary user under $\mathcal{H}_0$ when the primary user is not present for a detection threshold ψ and sensing time T_s, $P_D(\psi, T_s)$ is the probability of detection, and β is the constraint on the probability of missed detection to limit the interference caused to the primary users. For the energy detector and circularly symmetric complex Gaussian noise $R_0(\psi, T_s)$ is given by [112]

$$R_0(\psi, T_s) = C_0 P\left[\mathcal{H}_0\right] \left(1 - \frac{T_s}{T_p}\right) \left(1 - Q\left(\sqrt{2\gamma + 1}Q^{-1}(1-\beta) + \sqrt{T_s f_s}\gamma\right)\right), \tag{5.60}$$

where C_0 is throughput of the secondary user when the primary user is not present, T_p is a fixed time slot duration (thus the transmission time is given by $T_t = T_p - T_s$), $Q(\cdot)$ denotes the tail probability of the standard Gaussian distribution (i.e., $Q(\cdot)$ is the Gaussian tail function), γ is the primary user SNR, and f_s is the sampling frequency. Moreover, $P_D(\psi, T_s)$ has been chosen as $P_D(\psi, T_s) = 1 - \beta$ since this value maximizes throughput. The above throughput maximization framework has been extended to collaborative detection of multiple secondary users in [112] and [155]. However, the approaches in [112, 155, 163, 165] do not consider how to divide the limited sensing resources over multiple frequency bands. Such an approach has been proposed in [102].

In [102] a framework for optimizing the spectrum sensing and transmission times of a centrally controlled cognitive radio network over multiple frequency bands is proposed. The goal of this approach is to maximize the secondary throughput given constraints on the interference caused to the primary systems and the amount of sensing resources (i.e., the number of sensors sensing simultaneously). The proposed framework is based on optimizing the sensing efficiency of single-user single-band sensing. The sensing efficiency measures the transmission time relative to the sum of the transmission and sensing times. That is, the optimum sensing and transmission times T_s^* and T_t^*, respectively, for each frequency band are found by maximizing the sensing efficiency η given a constraint on the interference caused to the primary users:

$$\max_{T_s, T_t} \quad \eta = \frac{T_t}{T_s + T_t}, \tag{5.61}$$

$$\text{s.t.} \quad T_I \leq T_c, \tag{5.62}$$

where T_I is the expected fraction of the transmission time of the primary users interrupted by the secondary user transmissions and T_c is a given interference constraint. Compared to the approach in (5.58) the time slot duration $T_p = T_s + T_t$ is not fixed and thus both the sensing and transmission times are optimized. In order to accomplish that, a maximum a posteriori (MAP) probability detection model based on an energy detector and a two-state birth–death process for the primary user traffic with known death rate a and birth rate b is employed. Hence, state transitions follow a Poisson arrival process with exponentially distributed state lengths. An analytical model for the expected interference ratio is obtained:

$$T_\mathrm{I} = \frac{a}{b}\left[e^{-\mu T_\mathrm{t}}\alpha + (1 - e^{-\mu T_\mathrm{t}})\frac{b}{a+b}\right], \qquad (5.63)$$

where $\mu \triangleq \max\{a, b\}$, α is the false alarm probability of the employed energy detector, and the parameters of the detector are to be selected such that the posterior detection and false alarm probabilities are related as $P_\mathrm{on} - P_\mathrm{D} = P_\mathrm{FA}$, where $P_\mathrm{on} = b/(a+b)$. It follows that the single-user single-band optimization problem in (5.61) is equivalent to finding an optimal α that maximizes the sensing efficiency. See [102] for the details of the proposed iterative method.

The single-band optimization method is extended over multiple frequency bands with limited sensing resources (i.e., the number of frequency bands may exceed the number of sensors) by formulating the problem as a maximization of the expected transmission rate [102]:

$$\max_{\mathbf{x}} \sum_{i=1}^{n_\mathrm{B}} \eta_i\, C_i\, P_{\mathrm{off},i}\, x_i \qquad (5.64)$$

$$\mathrm{s.t.} \sum_{i=1}^{n_\mathrm{B}} \frac{T_{\mathrm{s},i}^*}{T_{\mathrm{s},i}^* + T_{\mathrm{t},i}^*}\, x_i \le N_\mathrm{SU}, \qquad (5.65)$$

where η_i is the sensing efficiency of frequency band i, C_i is the transmission rate the frequency band i can support (i.e., spectral efficiency multiplied by the bandwidth), $P_{\mathrm{off},i} = a_i/(a_i + b_i)$, N_SU is the number of sensors sensing, and $\mathbf{x} = [x_i]$ is a binary vector indicating whether each frequency band $i = 1, \ldots, n_\mathrm{B}$ is sensed ($x_i = 1$) or not ($x_i = 0$). Thus, the problem is a linear binary integer programming problem with linear constraints that can be solved, e.g., using branch-and-bound algorithms [222]. However, the fact that the optimal time slot durations may be different for each frequency band may render practical implementation impossible due to the interference caused by the overlapping of the sensing and transmission slots between different frequency bands. Hence, a scheduling algorithm that takes into account the cost of synchronizing the time slot durations on all frequency bands is proposed in [102]. In addition, an extension to cooperative sensing is proposed in [102] as well. Another sensing time optimization approach allowing different time slot durations for different frequency bands has been proposed in [83].

All of the above approaches are based on fixed sample size tests and the optimization of the parameters of these tests such as the sensing time and the test threshold. In [84, 86] sequential multiband detection algorithms for OFDM systems that optimize the throughput of the secondary system given a constraint on the probability of interfering with the primary users are proposed. The problem is formulated as an optimal stopping problem given a fixed time slot duration T that constrains the maximum sensing time. An optimal dynamic programming framework is developed and suboptimal reduced-complexity solutions are proposed. A permit to access is given when a channel is deemed to be idle. The access decision in each band is based on likelihood ratios computed using an energy detector. Channel knowledge is assumed to be obtained in an acquisition phase and the channels are assumed to be quasi-stationary during the operational phase. Hence, the proposed scheme is mainly applicable to flexible use of TV bands where the nodes are not mobile or have low mobility. Extensions of the proposed framework and algorithms to cooperative sensing among secondary users are proposed in [86] using both periodically transmitted raw observations and quantized binary observations. The collaborative sequential detection algorithm is employed at a fusion center.

5.3.2 Bandit problems

Bandit problems are a form of sequential decision problem in which at each time instant there are a number of different actions that may be selected. Bandit problems get their name from the correspondence to a traditional slot machine that has a lever (an arm) that the gambler pulls to operate the machine. These slot machines are also called one-armed bandits because of the single lever and the fact that the machine can "rob" the gambler of his/her money by not returning any of the placed bet. In cognitive radio applications, bandit problems are well suited for modeling multiband spectrum sensing and access problems and designing optimal sensing and access policies. For example, spectrum sensing and access problems can be modeled as multiarmed bandit problems in which the arms of the multiarmed bandit correspond to different frequency bands. Thus, choosing an arm to play corresponds to either sensing or accessing the corresponding frequency band.

In bandit problems selecting an action, i.e., an arm to play, results in a stochastic reward. The goal of the agent selecting the actions is to balance between maximizing the immediate reward and trying out the other arms to possibly improve the long-term reward by finding arms that produce better rewards. In other words, the agent employs a policy to tradeoff between exploitation and exploration to maximize its reward.

Multiarmed bandit problems

The classical multiarmed bandit problem may be described as follows. There are L arms and an agent controlling the selection of actions. At each stage, the agent selects one arm to play and receives a reward. Each arm is described by two stochastic processes s_n^k and $r_n^k = r_n^k(s_n^k, a_n^k), k = 1, \ldots, L, n = 0, 1, \ldots$, where s_n^k denotes the state of the kth arm, r_n^k is the reward for the kth arm, and $a_n^k \in \{0, 1\}$ is the agent's action for the kth arm at time n. Action a_n^k is 1 if arm k is played at time n and 0 otherwise. The processes for the different arms are independent of each other. Moreover, the agent does not receive a reward for the non-played arms, i.e., $r_n^k(s, 0) = 0, \forall n, k, s$. In addition, the state of each arm s_n^k changes only when the arm is played. The state of the nonplayed arms does not change. In general, the new state depends on all previous states. However, in the following we will focus on Markovian multiarmed bandit problems where the conditional state transition probability depends only on the current state. That is, the state transition probabilities are defined by

5.3 Optimized spectrum exploration and exploitation

$$P[s_{n+1} \mid s_n, a_n] = \begin{cases} P_{s_{n+1}s_n}, & \text{if } a_n = 1, \\ \delta_{s_{n+1}s_n}, & \text{if } a_n = 0, \end{cases} \quad (5.66)$$

where $\delta_{s_{n+1}s_n}$ is the Kronecker delta function, i.e., $\delta_{ss'} = 1$ if $s = s'$ and 0 otherwise.

A commonly employed goal for the agent in bandit problems is to optimize the expected sum of discounted rewards:

$$\mathcal{R} = \mathbb{E}\left[\sum_{n=0}^{\infty}\left(\xi^n \sum_{k=1}^{L} r_n^k(s_n^k, a_n^k)\right)\right], \quad (5.67)$$

where ξ, $0 \leq \xi < 1$, is the discount rate. This criterion takes into account all future rewards. It places a higher importance on immediate rewards and gradually smaller importance on future rewards. This can be interpreted as an interest rate or as an encoding factor taking into account the higher uncertainty of the future rewards. Other possibilities for the performance criterion include the expected reward over a finite horizon N_S:

$$\mathcal{R} = \mathbb{E}\left[\sum_{n=1}^{N_S} \sum_{k=1}^{L} r_n^k(s_n^k, a_n^k)\right], \quad (5.68)$$

and the expected average reward over the infinite horizon:

$$\mathcal{R} = \mathbb{E}\left[\lim_{N_S \to \infty} \frac{1}{N_S} \sum_{n=1}^{N_S} \sum_{k=1}^{L} r_n^k(s_n^k, a_n^k)\right]. \quad (5.69)$$

In addition, system regret has been employed as a performance criterion as well. The regret measures the expected loss of the employed policy compared to an optimal policy:

$$\mathcal{L} = \sum_{n=1}^{N_S} \mu_n^* - \sum_{n=1}^{N_S} \mathbb{E}\left[\sum_{k=1}^{L} r_n^k(s_n^k, a_n^k)\right], \quad (5.70)$$

where $\mu_n^* \triangleq \max\{\mu_n^1, \ldots, \mu_n^L\}$ is the expected reward of the best arm. An asymptotically order-optimal policy has a logarithmic order of regret $\mathcal{L} = O(\log N_S)$ [7, 8, 97].

Gittins and Jones [61, 62] showed that the optimal policy maximizing the expected sum of discounted rewards admits an index structure. That is, for each state of each arm of the multiarmed bandit a priority index, called the *Gittins index*,[1] that depends only on that process may be calculated. The optimal policy then reduces to selecting the arm with the largest Gittins index at each time. Hence, the L-dimensional optimization problem reduces to determining the Gittins index for L independent one-armed bandit processes. This reduces the computational complexity of the problem exponentially. The Gittins index for the kth arm is given by

---

[1] Gittins called the index initially *dynamic allocation index* (DAI) but later the name *Gittins index* has become more familiar in honor of its inventor.

$$v_k(s_0^k) = \max_{\tau > 0} \frac{\mathbb{E}\left[\sum_{n=0}^{\tau-1} \xi^n r_n^k(s_n^k, 1) \mid s_0^k\right]}{\mathbb{E}\left[\sum_{n=0}^{\tau-1} \xi^n \mid s_0^k\right]}, \quad (5.71)$$

where the maximization is over the set of all stopping times $\tau > 0$. Several algorithms have been proposed for efficient calculation of Gittins indices [20, 82, 140, 182, 199]. However, in general, calculating the Gittins indices requires full knowledge of the rewards and state transition probabilities for the arms.

There are several variants of the classical multiarmed bandit problem.

- *Multiarmed bandit problems with dependent arms* are multiarmed bandit problems in which the different arms are not independent [153].
- *Arm-acquiring multiarmed bandit problems* are variants of the classical multiarmed bandit problem in which the number of arms may increase as time advances [219].
- *Multiarmed bandit problems with switching penalties* are multiarmed bandit problems in which switching from one arm to another incurs a switching cost and/or a switching delay [1, 9, 14, 77].
- *Multiarmed bandit problems with multiple plays* are a form of multiarmed bandit problems in which the agent may at each time instant choose $M \leq L$ arms to play [2, 7, 8, 152].
- *Restless multiarmed bandit problems* are a generalization of the multiarmed bandit problem with multiple plays in which the state of all the arms changes at each time according to a stochastic process even if the arms are not played [220].
- *Multiarmed bandit problems with side observations* are a form of multiarmed bandit problem in which in addition to the reward sequence there is a sequence of side observations that are correlated with the rewards [100, 210, 211]. These problems are also called *contextual bandit problems*.
- *Multiarmed bandit problems with availability constraints* are a variation of the classical multiarmed bandit problem in which the arms are not always available [48, 225].

General treatments of multiarmed bandit problems discussing many of the above variants and their properties can be found in [17, 130].

From the cognitive radio point-of-view, the most important multiarmed bandit problem is the restless multiarmed bandit problem. The restless multiarmed bandit problem fits well the spectrum sensing and exploitation problems in dynamic environments in which the arms correspond to different frequency bands and primary user activity in each frequency band (and thus the arm state) may change even if the corresponding frequency band is not sensed or accessed. Thus, the restless multiarmed bandit problem provides the general framework for cognitive radio applications. In addition, there may be elements from other multiarmed bandit problems as well. The number of frequency bands available for secondary spectrum use may vary during the operation (*arm-acquiring multiarmed bandit problems* and *multiarmed bandit problems with availability constraints*). The cognitive radio or radios may be able to sense or access multiple frequency bands simultaneously (*multiarmed bandit problems with multiple plays*). The secondary user throughputs and primary user occupancies on different frequency bands may be correlated (*multiarmed bandit problems with dependent arms*). Switching the cognitive radio's operating frequency band may result in a delay (*multiarmed bandit problems with switching penalties*). There may be some additional information available correlated with the spectrum availabilities or throughputs (*multiarmed bandit problems with side observations*). The above lists only a part of the possible scenarios. Consequently, the list of various possibilities and combinations is numerous. In the following we focus on the restless multiarmed bandit problem.

Restless multiarmed bandit problems
The restless multiarmed bandit problem is a generalization of the classical multiarmed bandit problem where the agent may play multiple $M \leq L$ arms simultaneously and the nonplayed arms may change state and produce rewards as well [220]. This complicates the problem considerably. Consequently, the Gittins index policy is no longer optimal. Whittle proposed a heuristic index policy using a Lagrangian multiplier approach for the restless multiarmed bandit problems [220]. The *Whittle index* policy is optimal under a relaxed constraint that the average number of played arms is equal to M for the average reward over the infinite horizon criterion. Moreover, the Whittle index policy is asymptotically optimal under certain conditions as M and L tend to infinity with M/L fixed [217, 218]. However, the Whittle index is not always meaningful, i.e., the ordering of the arms is not consistent. The restless multiarmed bandits for which the Whittle index is meaningful are called *indexable* [220]. Establishing indexability and calculating the Whittle indices may be difficult in practice. For indexability of restless multiarmed bandit problems and numerical algorithms for establishing indexability and calculating the Whittle index, we refer the reader to [63, 138, 139, 141–143].

In [115] a single-user multiband spectrum sensing and access problem is formulated as a restless multiarmed bandit problem. The design of an optimal policy for choosing M frequency bands to sense and subsequently access if they are vacant is considered. The goal is to maximize the long-run reward. Sensing is assumed to be perfect, i.e., the detector does not make any errors. The states of the frequency bands are assumed to follow Markov chains with two states (good (1) and bad (0)) and different transition probabilities (P_{01}^i, P_{11}^i) for each band $i = 1, \ldots, n_\text{B}$. The user receives a reward for each frequency band sensed available and subsequently accessed. The rewards for the different frequency bands are denoted by $B_i, i = 1, \ldots, n_\text{B}$. The frequency bands that are either not sensed or identified to be occupied do not produce any reward. The user is able to observe the frequency band states only after sensing. Hence, it needs to infer or predict the state from its past decisions and observations. The conditional probability that a frequency band is in state 1 given all past decisions and observations is a sufficient statistic. The vector of conditional probabilities is referred to as the belief vector $\Omega_n = [\omega_n^1, \ldots, \omega_n^{n_\text{B}}]$. The belief states $\omega_n^i, i = 1, \ldots, n_\text{B}$, can be updated recursively [115]:

$$\omega_{n+1}^i = \begin{cases} P_{11}^i, & a_n^i = 1, s_n^i = 1, \\ P_{01}^i, & a_n^i = 1, s_n^i = 0, \\ \omega_n^i P_{11}^i + (1 - \omega_n^i) P_{01}^i, & a_n^i = 0. \end{cases} \qquad (5.72)$$

Consequently, the spectrum sensing and access problem is a restless multiarmed bandit problem in which the frequency bands correspond to the arms of the bandit and the state of the ith arm at time n is given by the belief state ω_n^i.

The indexability of the restless multiarmed bandit formulation of the single-user multiband spectrum sensing and access problem is established in [115]. Moreover, the Whittle index is obtained in closed form for both the discounted and average infinite-horizon reward criteria. In [142] a similar restless multiarmed bandit formulation for centralized multiuser, multiband spectrum sensing and access with sensing errors is proposed. Indexability is shown and Whittle's marginal productivity (MP) index [139, 141] policy is obtained. In general, calculating the Whittle index or Whittle's MP index requires full knowledge of the transition probabilities P_{11}^i and P_{01}^i as well as the rewards B_i. However, for stochastically identical arms (i.e., all arms have equal transition probabilities and rewards) the Whittle index policy has been shown in [115] to be equivalent to a myopic (greedy) policy. The myopic policy chooses at each time the action that maximizes the immediate reward [3, 4, 117, 233]. The myopic action $\hat{a}_n = [\hat{a}_n^1, \ldots, \hat{a}_n^i]$, $\hat{a}_n^i \in \{0, 1\}$, is given by

$$\hat{a}_n = \arg\max_{a_n} \sum_{i=1, a_n^i \neq 0}^{n_B} \omega_n^i B_i, \quad \text{s.t.} \quad \sum_{i=1}^{n_B} a_n^i = M, \quad (5.73)$$

where the first sum is over the sensed frequency bands in action a_n. The myopic policy has a simple queue structure for stochastically identical arms [117, 233]. That is, the frequency bands can be maintained in an ordered queue and at each time the M frequency bands at the head of the queue are chosen. Moreover, implementing the myopic policy in practice requires only the ordering of P_{11} and P_{01} to be known. Figure 5.10 illustrates the structure of the myopic policy for iid frequency bands. In [233] the structure of the myopic policy for $M = 1$ has been established and shown to be optimal for $n_B = 2$ when the arms are iid. The optimality of the myopic sensing policy has been extended for positively correlated ($P_{11} \geq P_{01}$) iid frequency bands for $M = 1$ and $n_B > 2$ in [4].

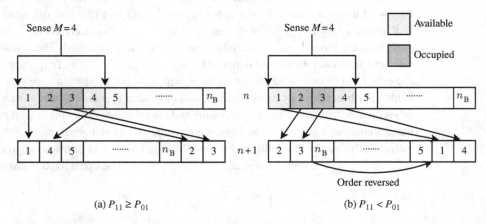

Figure 5.10 The structure of the myopic sensing policy for iid frequency bands when (a) $P_{11} \geq P_{01}$ and (b) $P_{11} < P_{01}$ [117].

Moreover, the optimality is shown for the case of $n_B = 2$ and $n_B = 3$ iid bands with negatively correlated ($P_{11} < P_{01}$) state transitions as well. In [3] the optimality of the myopic policy is extended to the case of multiple plays $M \geq 2$ for positively correlated frequency bands, whereas a more general class of objective functions and the general conditions for the optimality of the myopic policy have been considered in [214]. In [117] the myopic policy and the case of imperfect frequency band state detection (i.e., sensing errors) are considered. Moreover, in [197] myopic policies are proposed for collaborative multiband spectrum sensing and access with both known and unknown frequency band availability statistics. In addition, a more general framework has been proposed in [44, 234] by formulating the multiband sensing and access problem in cognitive radios as a partially observable Markov decision process (POMDP) taking into account the potential correlations among the frequency bands. However, the optimal solution has exponential complexity, and the state transition probabilities are assumed to be known.

In [90] another restless multiarmed bandit formulation of the multiband spectrum sensing and access problem is proposed. The proposed network model is slightly more simplified than the Markov model in [115]. That is, the frequency band availability is assumed to be determined by a single Bernoulli random variable. Thus, with a probability θ_i the frequency band i ($i = 1, \ldots, n_B$) is available in a given time slot. The availability is independent among different time slots and frequency bands. Both single-user and multiuser scenarios are considered. Scenarios where users are able to sense one or multiple frequency bands simultaneously are considered separately. Moreover, scenarios with both known and unknown frequency band availability probabilities are considered. Optimal and asymptotically optimal sensing and access strategies for the case of known and unknown frequency band availability probabilities are developed, respectively. We consider here the multiuser scenario with unknown θ_i where each user is able to sense one frequency band in each time slot. The goal of each user is to design a sensing and access policy that minimizes the regret over a finite time interval N_S. The users employ a generalized carrier sense multiple access with collision avoidance (CSMA/CA) protocol to access the available frequency bands after sensing. The proposed multiuser sensing and access policy consist of three stages [90].

1 *Initialization* Each user $k = 1, \ldots, N_{SU}$ senses each frequency band once and tries to access the sensed frequency band if it is sensed free. Moreover, each user maintains the following statistics: $X_{k,i}$ which denotes the number of times user k has sensed frequency band i free and $Y_{k,i}$ which denotes the total number of times user k has sensed frequency band i. Set $X_{k,i} = 1$ regardless of the sensing result.
2 $n < \ln N_S$ Each user estimates the probabilities for frequency band availability

$$\hat{\theta}_{k,i} = \frac{X_{k,i}}{Y_{k,i}}. \tag{5.74}$$

Moreover, each user senses the frequency band i with probability $\hat{\theta}_{k,i} / \sum_{i=1}^{n_B} \hat{\theta}_{k,i}$.
3 $n \geq \ln N_S$ The $\hat{\theta}_i$ are updated similarly as in the second stage. Each user senses frequency band i with probability

$$\hat{P}_i = \left\{ 1 - \left(c/\hat{\theta}_i\right)^{1/(N_{\text{SU}}-1)} \right\}^+, \qquad (5.75)$$

where c is a constant such that $\sum_{i=1}^{n_B} \hat{P}_i = 1$, and $\{x\}^+ = \max\{0, x\}$.

Thus, each user employs the same prespecified strategy without any additional coordination or information exchange among the users. This strategy is shown to achieve a linear order of regret. The multiuser algorithms may be extended to the more general Markov model for the frequency band availability [92].

The single-user multiband spectrum access problem with unknown frequency band availability statistics using a multiarmed bandit problem formulation has been considered also in [135]. A sensing and access policy based on Gittins indices and estimated frequency band availability statistics is proposed. Moreover, a heuristic restless multiarmed bandit policy, called the restless upper confidence bound (RUCB), has been proposed in [114]. In the RUCB policy time is divided into exploration and exploitation epochs. In an exploration epoch each arm is played multiple times (each arm is played an equal number of times) in order to find the best arm according to a heuristic index rule. The best arm is then played exclusively in the exploitation epoch. The RUCB policy may be applied to single-user multiband spectrum sensing and access problems in cognitive radio systems. A similar approach, but with a slightly different structure, is proposed in [193]. Both of these algorithms are based on the upper confidence bound algorithm UCB1 proposed for the multiarmed bandit in [11]. The UCB1 first plays each arm once and then chooses at time n the arm i that maximizes the following index:

$$\bar{r}_i + \sqrt{\frac{2 \ln n}{n_i}}, \qquad (5.76)$$

where $\bar{r}_i$ is the average reward received from arm i and n_i is the number of times the arm i has been played until time n. Improved versions of the original UCB1 algorithm with smaller regrets have been recently proposed in [10, 12].

Distributed restless multiarmed bandit formulations and policies for multiuser spectrum sensing and access have been proposed also in [5, 6, 116]. In [116] a distributed competitive multiuser spectrum sensing and access policy achieving optimal logarithmic order of system regret with N_{SU} users competing for n_B independent arms with iid rewards with different unknown means that are common to all users is proposed. The proposed distributed policy is based on time-division fair sharing (TDFS) of the best N_{SU} arms with any order optimal single-user policy. In the TDFS policy the users sense the N_{SU} best arms in a round-robin fashion with different time offsets. The time offsets are adjusted based on collisions with other users. The TDFS policy ensures fairness among users without assuming any pre-agreement or information exchange among users. In [5, 6] a similar approach is proposed. The main difference is that in [5, 6] the users employ a policy that aims only at finding orthogonal frequency bands for the users while in [116] the users share the frequency bands in time as well. Both schemes achieve asymptotically optimal logarithmic regret but the policy in [116] provides potentially a fairer sharing of the available resources.

In [56] a centralized multiarmed bandit problem formulation of multiuser, multiband spectrum access problem is proposed. In this formulation the spatially dispersed secondary users may experience different unknown spectrum availability statistics. A combinatorial multiarmed bandit problem where each arm corresponds to a matching of the secondary users to frequency bands is formulated. This results in dependent arms. A learning policy achieving regret that is logarithmic in time slots and polynomial in the number of frequency bands and secondary users is proposed.

5.3.3 Reinforcement learning

Machine learning methods may be broadly divided into three different classes: supervised learning, unsupervised learning, and reinforcement learning. In supervised learning the learning process involves a teacher that tells the agent the correct action in each state. At the opposite end of the categorization is unsupervised learning. Unsupervised learning is performed completely without any indication or feedback about how good or bad the actions selected by the agent are. Reinforcement learning operates between these two opposites. In reinforcement learning the agent learns through experiment and experience. After taking an action, the agent receives a scalar reward that measures the quality of the action and the resulting state transition. This feedback allows the agent to reinforce the taken actions that result in high rewards and thus make them more likely to be selected in similar situations in the future. Of the three classes, reinforcement learning methods are the most suitable for cognitive radio tasks, such as spectrum sensing and access. The correct action in each state is in general unknown or difficult to find out due to the dynamically changing environment. Hence, supervised algorithms are difficult to employ in practice. However, the natural rewards for reinforcement learning methods in these tasks are related to the throughput of the secondary users and collisions encountered with other users of the spectrum. This information may be easily obtained during the online operation of the secondary system. Hence, in the following we focus on reinforcement learning techniques.

Reinforcement learning techniques may be employed to solve the bandit problems introduced in the previous subsection. Thus, reinforcement learning may be applied to sensing and access policy design in cognitive radio networks. However, reinforcement learning can be applied to more general problems as well that are not well described by the bandit problems. Nonetheless, in reinforcement learning problems we are faced with the same exploitation vs. exploration dilemma as in bandit problems. That is, at each stage the agent must decide whether to exploit the currently best action or to explore other actions in the hope of a better future reward. In the following we will briefly review single-agent and multiagent reinforcement learning and their application to spectrum sensing and access policy design in cognitive radio systems.

Single-agent reinforcement learning
A Markov decision process (MDP) formulation of a single-agent reinforcement learning problem consists of

- a sequence of discrete time steps $n = 0, 1, 2, \ldots$;
- an agent i;
- a set of possible states of the environment $s \in S$;
- a set of possible actions for the agent in each state $a \in A$;
- a state transition function $\varphi : S \times A \times S \to [0, 1]$ which defines the transition probability $P[s_{n+1} \mid s_n, a_n]$;
- a reward function $r : S \times A \times S \to \mathbb{R}$ which gives the agent a reward for taking action a_n in state s_n resulting in new state s_{n+1}.

In every state s_n the agent chooses an action a_n according to some policy π. A policy π is a stochastic or deterministic function that determines how the agent selects its action a_n in each state s_n. Taking the action a_n in state s_n results in a new state s_{n+1} and a scalar reward r_{n+1} for the agent. The goal of the agent is to try to act in a way that maximizes not just the immediate reward but also the future rewards. A commonly employed goal in reinforcement learning is to optimize the expected sum of discounted rewards

$$\mathcal{R}_n = \mathbb{E}\left[\sum_{k=0}^{\infty} \xi^k r_{n+k+1}\right], \tag{5.77}$$

where r_n is the reward at time n and ξ, $0 \leq \xi < 1$, is the discount rate. This model may not be the most appropriate model in all applications. In particular, in episodic tasks, for which the episode length is known beforehand, models with finite horizon are typically more appropriate. However, the infinite-horizon discounted model is mathematically more tractable than most other models. Other options for defining the goal in reinforcement learning include the ones already introduced for the bandit problems in Section 5.3.2.

A convenient way of optimizing the sum of expected discounted future rewards is to learn an optimal action-value function. The action-value function $\mathcal{Q}(s, a)$ evaluates the value of each action in a given state. The action-value function is defined as the expected return (i.e., the sum of discounted rewards) of taking an action a in state s and then following policy π:

$$\mathcal{Q}^\pi(s, a) = \mathbb{E}_\pi [\mathcal{R}_n \mid s_n = s, a_n = a, \pi]. \tag{5.78}$$

The optimal action-value function $\mathcal{Q}^*$ satisfies the Bellman optimality equation [183, p. 76]

$$\begin{aligned}\mathcal{Q}^*(s, a) &= \max_\pi \mathcal{Q}^\pi(s, a) \\ &= \sum_{s' \in S} P[s' \mid s, a] [r(s, a, s') + \xi \max_{a'} \mathcal{Q}^*(s', a')],\end{aligned} \tag{5.79}$$

where $r(s, a, s')$ is the reward function.

In the following two commonly employed powerful model-free temporal-difference reinforcement learning algorithms, Sarsa [169] and $\mathcal{Q}$-learning [215, 216], for optimizing the action-value functions are introduced. Sarsa is an on-policy temporal-difference algorithm. An on-policy algorithm means that the action-value updates are based on the actually taken actions by the policy. The action-value updates of the Sarsa-algorithm are defined by

$$\mathcal{Q}_{n+1}(s_n, a_n) = \mathcal{Q}_n(s_n, a_n) + \alpha_n \left[r_{n+1} + \xi \mathcal{Q}_n(s_{n+1}, a_{n+1}) - \mathcal{Q}_n(s_n, a_n) \right], \quad (5.80)$$

where α_n ($0 < \alpha_n \leq 1$) is a step size parameter (learning rate). Thus, the action-value function update is given by the temporal difference between the $\mathcal{Q}$-values of the employed policy at time steps n and $n+1$ multiplied by the step size α_n.

$\mathcal{Q}$-learning is an off-policy temporal-difference algorithm. Thus, the temporal-difference update of $\mathcal{Q}$-learning does not necessarily employ the $\mathcal{Q}$-values of the employed policy as on-policy algorithms, such as Sarsa, do. The action-value updates of $\mathcal{Q}$-learning are given by

$$\mathcal{Q}_{n+1}(s_n, a_n) = \mathcal{Q}_n(s_n, a_n) \\ + \alpha_n \left[r_{n+1} + \xi \max_{a' \in A} \mathcal{Q}_n(s_{n+1}, a') - \mathcal{Q}_n(s_n, a_n) \right]. \quad (5.81)$$

The temporal difference update of $\mathcal{Q}$-learning depends on the maximum $\mathcal{Q}$-values of the possible actions. Note that for a greedy policy that always chooses the action with maximum $\mathcal{Q}$-value the Sarsa and $\mathcal{Q}$-learning algorithms are identical.

The convergence of the action-value function $\mathcal{Q}_n$ to the optimum $\mathcal{Q}^*$ has been established for one-step Sarsa in [181] and for $\mathcal{Q}$-learning in [73, 196, 216]. The convergence is guaranteed with probability one if the agent employs a lookup-table to store the $\mathcal{Q}$-values, visits every state-action pair infinitely many times, and $\sum_{n=0}^{\infty} \alpha_n = \infty$ and $\sum_{n=0}^{\infty} \alpha_n^2 < \infty$. Moreover, to ensure the convergence of one-step Sarsa, the learning policy must become greedy in the limit.

The convergence requires infinite visits to every state-action pair. Intuitively this is easy to understand. In order to find out whether the current policy is optimal the agent must explore other actions than the ones given by the current policy as well. This presents the agent with the tradeoff between exploitation and exploration. That is, whether to exploit the current policy and choose the greedy action or to explore other actions instead to possibly find out some other action that provides a higher return than the greedy action. The ϵ-greedy action selection is a simple method that balances between exploitation and exploration by selecting the action that maximizes the action-value function, i.e., $a^* = \max_a \mathcal{Q}(s, a)$, with a probability $1 - \epsilon$, or a random action, uniformly, with a probability ϵ regardless of the action-value function estimates. Another simple, yet effective, action selection method is the softmax action selection method. In the softmax method the action a is selected with probability

$$\pi(s, a) = \frac{\exp(\mathcal{Q}(s, a)/\tau)}{\sum_{a' \in A} \exp(\mathcal{Q}(s, a')/\tau)}, \quad (5.82)$$

where $a \in A$ and τ is a positive parameter called the temperature that controls the weighting of different actions. Low temperatures increase the differences in the action selection probabilities while high temperatures cause all actions to be almost equiprobable. In the limit, when $\tau \to 0$, the softmax method corresponds to greedy action selection, and when $\tau \to \infty$, the actions are selected randomly from a uniform distribution. In addition, adaptive algorithms for balancing between exploitation and exploration have been proposed in [195, 204].

We note that the above convergence results apply to a stationary situation only. That is, the convergence results state the conditions under which one-step Sarsa and Q-learning converge to an optimal stationary policy. In dynamic nonstationary environments, such as the one encountered in cognitive radio applications, one is interested in tracking a nonstationary problem. In such cases one of the most commonly employed approaches is to use a constant step size $\alpha_n = \alpha, \forall n$. A constant step size does not satisfy the above conditions for convergence. However, it guarantees that the more recent samples are weighted more heavily than the ones in the distant past which is a desired feature in nonstationary problems. Moreover, in nonstationary environments it is important to continue to explore. This may be accomplished, for example, by employing the ϵ-greedy algorithm with a constant, sufficiently large ϵ.

Multiagent reinforcement learning

A multiagent reinforcement learning problem can be formulated as a stochastic game [31, 113]. A stochastic game formulation of a multiagent reinforcement learning problem consists of [31]

- a sequence of discrete time steps $n = 0, 1, 2, \ldots$;
- a group of agents $\mathcal{N} = \{1, \ldots, N_{\text{SU}}\}$;
- a set of possible states of the environment $s \in S$;
- a set of possible actions for each agent in each state $a_i \in A_i$, $i \in \mathcal{N}$; the combined action space is $A = A_1 \times \cdots \times A_{N_{\text{SU}}}$;
- a state transition function $\varphi : S \times A \times S \to [0, 1]$ which defines the transition probability $P[s_{n+1} \mid s_n, a_{1,n}, \ldots, a_{N_{\text{SU}},n}]$;
- reward functions $r_i : S \times A \times S \to \mathbb{R}$, $i \in \mathcal{N}$, which give the agents a reward for the joint action $a_{1,n}, \ldots, a_{N_{\text{SU}},n}$ of the agents in state s_n resulting in new state s_{n+1}.

The rewards in a multiagent reinforcement learning problem depend on the joint actions of all the agents. This complicates the learning problem considerably. The agents have to adapt their own policies not only to the environment but to the other agents' policies as well. Defining the goal of a multiagent reinforcement learning task is a challenging problem. A good learning goal must include components that provide both stability to the learning process of the individual agents and adaptation to the other agents' dynamic behavior [31]. The learning goal depends also on the nature of the learning problem. Multiagent reinforcement learning tasks may be categorized into fully competitive, fully cooperative, and mixed tasks [31]. Mixed tasks are neither fully competitive nor fully cooperative. Each one of these classes requires different goals for the learning algorithm. Moreover, the level of coordination among the agents has a significant effect on both the stability of the learning process and the adaptation to the other agents' policies. The level of coordination depends on the application and its restrictions as well as on the chosen reinforcement learning algorithm and its goal. In some tasks there may be explicit coordination, e.g., in the form of pre-established preferences toward certain joint actions, negotiation of action selections among agents, or actions may be selected in turn and communicated to other agents. On the other hand, in the absence of any explicit coordination mechanism the agents may be able to observe the behavior of the other agents and try to learn a model of their policies.

For more advanced discussions on reinforcement learning, its challenges, benefits, and different learning algorithms, the reader is referred to [24, 31, 32, 78, 183, 184, 226] and the references therein.

Reinforcement learning-based spectrum sensing and access policies
Reinforcement learning-based methods have been proposed for spectrum sensing and access in cognitive radio networks. Both single- and multiagent reinforcement learning methods have been proposed. In the following we will give an overview of the state-of-the-art in reinforcement learning in cognitive radio systems.

Energy-efficient cooperative multiband sensing policy for cognitive radio systems
In [148, 149] a two-stage reinforcement learning-based collaborative multiband sensing policy for centralized cognitive radio systems is proposed. The fusion center operates as the learning agent of the cognitive radio network consisting of a group of secondary users. The actions determine which frequency bands are sensed and which frequency band each secondary user senses. The actions are chosen in two stages. In the first stage the frequency bands to be sensed are chosen. In the second stage the secondary users are assigned to sense the frequency bands chosen in the first stage. Thus, in the first stage the action corresponds to choosing the frequency bands and in the second stage the action corresponds to choosing the secondary users for each frequency band. The policy employs ϵ-greedy action selection and single-state $\mathcal{Q}$-learning with discount rate $\xi = 0$. Thus, the update of the action $\mathcal{Q}$-values is given by

$$\mathcal{Q}_{n+1}(a) = \mathcal{Q}_n(a) = \alpha_n[r_{n+1} - \mathcal{Q}_n(a)]. \qquad (5.83)$$

A separate $\mathcal{Q}$-value is kept for each frequency band. In addition, for each secondary user a separate $\mathcal{Q}$-value is kept for each frequency band. The reward r is the obtained throughput for the frequency bands, and for the secondary users the reward is defined as follows:

$$r_{n+1,k} = \begin{cases} u_{n+1,k}, & u_{n+1,\text{FC}} = 1, \\ \mathcal{Q}_n(a), & u_{n+1,\text{FC}} = 0, \end{cases} \qquad (5.84)$$

where $u_{n+1,k}$ is the local decision of the kth secondary user and $u_{n+1,\text{FC}}$ is the global decision at the fusion center for that frequency band. Here, decision $u = 1$ denotes that the frequency band is occupied by a primary system. The secondary user $\mathcal{Q}$-values are updated only when the decision at the fusion center is considered to be correct with high certainty, i.e., the global test statistic exceeds the decision threshold with a certain safety margin.

In the exploitation phase, the first stage of the sensing policy chooses greedily the L best frequency bands to sense. In the second stage the following sensing assignment problem (SAP) is solved to assign the secondary users to sense each of the L frequency bands:

$$\min_{\mathbf{A}} \sum_{i=1}^{n_B} \sum_{k=1}^{N_{SU}} g_k a_{ki} \tag{5.85}$$

$$\text{s.t.} \quad \hat{P}_{MD,FC}^i(\mathbf{A}) \leq \beta_i, \ i=1,\ldots,n_B, \tag{5.86}$$

$$\sum_{i=1}^{n_B} a_{ki} \leq L_k, \ k=1,\ldots,N_{SU}, \tag{5.87}$$

where n_B and N_{SU} are the number of frequency bands and the number of secondary users, respectively, L_k is the number of frequency bands secondary user k can sense simultaneously, g_k is the weight of user k which can be used to favor or unfavor certain users, e.g., due to a high or low level of remaining battery power, $\mathbf{A}$ is an $N_{SU} \times L$ binary sensing assignment matrix where the components $[\mathbf{A}]_{ki} = a_{ki}$ are 0 or 1, $\hat{P}_{MD,FC}^i(\mathbf{A})$ is the estimate of the probability of missed detection for the frequency band b at the fusion center given $\mathbf{A}$, and β_i is the probability of missed detection constraint for the frequency band i. The estimates $\hat{P}_{MD,FC}^i(\mathbf{A})$ are formed from the Q-values of the secondary users for each band.

The SAP is a binary integer programming problem. It optimizes energy efficiency by minimizing the number of secondary users sensing given the probability of missed detection constraints. This is accomplished by assigning the users with the best propagation channel to each primary user to sense the corresponding frequency band. Moreover, during exploration a pseudorandom frequency hopping policy [145, 147] is employed that guarantees a given diversity order, i.e., the number of secondary users sensing each frequency band simultaneously; see Fig. 5.11. Figure 5.12 depicts the throughput and probability of missed detection of the proposed scheme in a scenario with ten frequency bands and six secondary users.

Single-agent reinforcement learning-based spectrum sensing and access
Single-agent reinforcement learning-based spectrum sensing and access policies have been proposed in [18, 54, 55, 67, 227]. In [55] a model-based reinforcement learning

SU1	1	2	3	1	2	3	1	2	3
SU2	1	2	3	2	3	1	2	3	1
SU3	2	3	1	1	2	3	2	3	1
SU4	2	3	1	2	3	1	1	2	3

Figure 5.11 An example pseudorandom frequency hopping table indicating the index of the frequency band sensed by each secondary user (SU) in each time slot [145, 147]. In this example, there are four secondary users, three frequency bands, and the employed sensing diversity order is two.

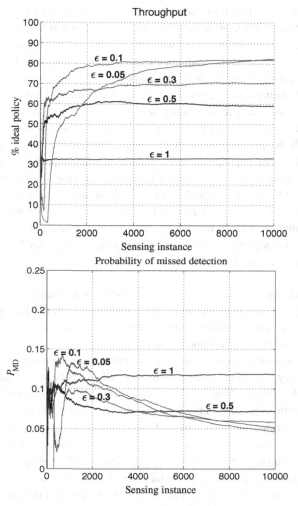

Figure 5.12 Throughput and probability of missed detection as a function of the sensing instance for the two-stage multiuser Q-learning-based sensing policy with ϵ-greedy action selection [148]. With an appropriately chosen ϵ the proposed algorithm is able to achieve high throughput while minimizing the probability of missed detection. The ideal policy accesses always the unoccupied frequency bands providing the best throughput.

algorithm, called the tiling algorithm, is proposed for balancing between exploitation and exploration in single-user spectrum sensing. A single frequency band case is considered in which the secondary user's policy determines how long the secondary user waits until it senses the frequency band again, depending on the previous sensing result. Sensing is assumed to incur a fixed cost. In the proposed scheme the secondary user first performs an exploration phase to learn the model parameters. Then in the second phase a fixed sensing policy based on the learned parameters is employed to exploit the spectrum. In [54] a single secondary user multiband sensing and access policy is proposed. State transition probabilities of the primary users are assumed to be known and value iteration with a discretized state space is employed to optimize the policy. An actor-critic reinforcement learning-based multiband sensing policy is proposed in [18] for deciding which frequency band to sense as well as for balancing between sensing, transmission, and switching between different frequency bands. In [227] a single-agent Q-learning-based approach for dynamic frequency band selection for secondary access is proposed. The secondary user receives a negative reward whenever its transmission collides with a primary user. Hence, the secondary user learns to avoid collisions with the primary users. A single-user spectrum sensing and access scheme based on Q-learning is proposed in [67]. At each time step the secondary user makes a decision based on Q-learning whether to explore for new spectral opportunities or to exploit its home band. The secondary users are assumed to have license to use the home band at all times. The secondary user may decide to search for new spectral opportunities if the quality of its home band degrades too much. During the exploration phase the secondary user (or a group of secondary users) senses frequency bands in a sequential order until it finds an available frequency band.

Multiagent Q-learning for Aloha-like multiband spectrum access
In [107] a Q-learning multiagent, multiband Aloha-like spectrum access policy is proposed. The secondary users in the cognitive radio network are assumed to be employing an Aloha-like transmission scheme without any explicit collision avoidance. There is no communication or negotiation among the secondary users. Thus, two or more secondary users transmitting simultaneously on the same frequency band will collide with each other resulting in a failed transmission for all involved parties.

In the cognitive radio scenario considered in [107] there are N_{SU} secondary users and frequency bands. Moreover, the spectral opportunities (and thus the primary user occupancy statistics) are assumed to be identical for each secondary user. Each secondary user employs a reinforcement learning-based sensing and access policy individually without any information exchange with the other secondary users. The other secondary users considered as part of the environment. Thus, in addition to learning the primary user occupancy statistics, the secondary users have to learn how to avoid other secondary users as well. However, the users are assumed to be synchronized in the time domain and using a slotted time structure with alternating sensing and access slots.

Both fully observable and partially observable spectrum occupancy scenarios are considered. The more interesting scenario is the partially observable scenario in which each secondary user is able to sense only one frequency band in each time slot. In this case, the state of the environment as observed by the secondary user i is defined as

$$\mathbf{s}_{i,n} = [\tau_{i,b,n}, \ldots, \tau_{i,n_\text{B},n}, u_{i,1,n}, \ldots, u_{i,n_\text{B},n}], \qquad (5.88)$$

where $\tau_{i,b,n}$, $b = 1, \ldots, n_\text{B}$, denotes the number of time slots that have passed since the user i sensed frequency band b the last time. And $u_{i,b,n}$, $b = 1, \ldots, n_\text{B}$, denotes the corresponding occupancy state of the frequency band b the last time it was sensed by user i. Each user employs independently the Q-learning algorithm with softmax action selection to maximize the expected sum of its discounted throughputs. It is shown in [107] that the Q-values of the secondary users converge to stationary values under certain conditions regardless of the initial strategies.

Other multiagent reinforcement learning-based spectrum sensing and access without any information exchange
Multiagent reinforcement learning-based approaches for spectrum sensing and access without any communication or information exchange among the secondary users have been proposed in [120, 176, 223]. In [223] a Q-learning approach is proposed for multiagent, multiband spectrum access in cognitive radio networks. The state of a single agent is the index of the current frequency band accessed by the secondary user and the corresponding transmit power value. Kanerva coding-based function approximation [183] is proposed to reduce the size of the state-space. There is no information exchange or explicit coordination of actions among the secondary users. Each secondary user employs the Q-learning algorithm independently and tries to learn how to avoid other secondary users and primary users based on the negative reward received from collisions. In [120] a modified regret-matching learning algorithm is proposed for distributed dynamic spectrum access in cognitive radio networks. A group of secondary users compete for spectral opportunities in a multiband dynamic spectrum access scenario using a modified regret-matching learning algorithm that leads to a correlated equilibrium. In the modified regret-matching algorithm [69] the secondary users know only their own realized rewards and actions. There is no communication among the users. In [176] a multiagent learning scheme for multiband dynamic spectrum sensing and access is proposed. The proposed scheme is based on a noncooperative framework without any information exchange among the secondary users. The secondary users learn through collisions with other users to select the best frequency band to access to minimize collisions. The secondary users estimate belief vectors for themselves and for the other secondary users. The belief vector gives the probability of accessing each frequency band in each time slot.

Collaborative distributed multiagent reinforcement learning-based sensing policy with linear function approximation
In [124, 125] collaborative distributed multiagent, multiband reinforcement learning-based sensing policies are proposed for cognitive radio networks. The policies employ a one-step Sarsa-algorithm with linear function approximation. The secondary users in the cognitive radio network aim to maximize the amount of free spectrum found by adjusting locally the number of secondary users sensing each frequency band simultaneously. This allows controlling of the probability of false alarm for a given constraint on the probability of missed detection. The secondary users try to optimize their joint action, i.e., the frequency bands to be sensed, through multiagent reinforcement learning. The secondary users collaborate through local interactions by sharing their local test statistics or decisions as well as their current and future actions with their neighboring users. The use of the linear function approximation method for approximating the state-action space provides a computation- and memory-efficient approach even with large numbers of secondary users and frequency bands. Figure 5.13 illustrates the sensing

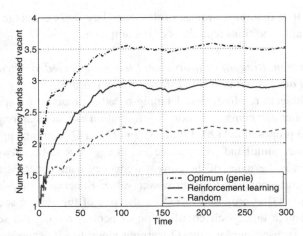

Figure 5.13 Average number of frequency bands sensed as vacant as a function of time for a Sarsa-algorithm-based sensing policy and a random sensing policy in a scenario with 40 spatially dispersed secondary users and eight primary user frequency bands [124]. Compared to the random policy the reinforcement learning-based policy improves the performance significantly by allowing the secondary users to find more unoccupied spectrum. The optimum, genie-aided policy knows at every time instant which frequency bands are available for secondary use.

performance compared to a random policy in a scenario with 40 spatially dispersed secondary users and eight primary user frequency bands. The random policy chooses the sensed frequency bands randomly from a uniform distribution. However, similarly to the Sarsa-algorithm-based policy, it employs action coordination to try to guarantee that each user is sensing different frequency bands.

Other multiagent reinforcement learning-based spectrum sensing and access with information exchange

Multiagent reinforcement learning-based spectrum sensing and access policies with coordination and communication among the secondary users have been proposed in [49, 106, 228, 229]. In [106] collaborative filtering-based multiagent machine learning policies are proposed for learning the primary user occupancy probabilities and for maximizing the data rate of spectrum access. The secondary users collaborate with their neighboring secondary users by sharing their local estimates to improve the accuracy of their own estimates or to find more free spectrum. Two collaboration schemes are proposed: a linear prediction-based collaboration scheme using the correlation factors among the secondary users to weight the local estimates of the primary user probabilities, and a reward oriented collaboration scheme maximizing the sum of available channels for the collaborating secondary users. The goal is to have secondary users whose spectrum occupancy statistics are highly correlated collaborating with each other. A distributed multiagent Q-learning-based approach for dynamic frequency band selection for secondary access has been proposed in [228]. In this approach the secondary users coordinate their actions by transmitting control messages to their neighbors in order to indicate that they are exploring instead of exploiting. Only one user in a neighborhood is allowed to explore at a time. This provides stability to the network. In [229] a distributed multiagent learning-based approach for dynamic frequency band selection for secondary access is proposed. In this approach the users employ payoff propagation [88] by broadcasting their payoff messages to their neighbors. The payoff message of a given user consists of the local action-value function of the user's current action and its one-hop neighbors' actions. This coordination mechanism allows joint action selection optimization. In [49] a multiagent reinforcement learning-based multi-band sensing and access policy is proposed for cognitive wireless mesh networks for balancing between sensing, transmission, and switching the operating frequency band. The wireless mesh networks consist of clusters and each cluster head (i.e., the mesh router) employs the proposed reinforcement learning method to optimize the cluster's performance. A distance-based weighted cooperative framework in which the cluster heads share their local frequency band value estimates with others is proposed as well.

5.3.4 Game-theoretic approaches

Game theory is a collection of mathematical methods and tools for modeling, analyzing, and predicting the behavior of multiple interacting entities. In cognitive radio applications, game theory is important for modeling and analyzing the interaction of individual spectrum users, as well as designing algorithms for efficient and effective sharing of the scarce spectrum resources.

The simplest model for the interaction of multiple decision makers is a *strategic game* (or a *normal form game*). A strategic game consists of

- a set of players $\mathcal{N} = \{1, \ldots, N_{\text{SU}}\}$;
- a set of strategies (actions) for each player $A_i, i \in \mathcal{N}$. A strategy $a_i \in A_i$ is a complete plan of action for each situation in the game. Moreover, let a_{-i} denote the strategies of all the other players except player i. The combined strategy space is the set of strategy profiles $A = A_1 \times \cdots \times A_{N_{\text{SU}}}$;
- payoff (utility) functions $r_i : A \to \mathbb{R}, i \in \mathcal{N}$, which give the players a payoff (utility) for the joint strategies $a_1, \ldots, a_{N_{\text{SU}}}$.

Game theory can be broadly categorized into noncooperative and cooperative game theory. In the following the main concepts of both noncooperative and cooperative games are presented. Noncooperative games apply naturally to cognitive radio systems in which the secondary users act selfishly and compete with each other, while cooperative games apply to systems in which the users cooperate with each other and act jointly. In addition, we will take a brief look at auction games and finally an overview of stochastic games that provide an excellent fit to the dynamic environment encountered in cognitive radio applications will be given.

Noncooperative games

In a noncooperative game the players choose their strategies independently to achieve their own goals. Note, however, that this does not mean that the players are not allowed to cooperate in a noncooperative game. It only means that this cooperation cannot be enforced. That is, each player decides individually whether to cooperate with the other players or not. *Nash equilibrium* is a fundamental concept in noncooperative game theory. Nash equilibrium is a strategy profile $a^* \in A$ such that

$$r_i(a_i^*, a_{-i}^*) \geq r_i(a_i, a_{-i}^*), \ \forall i, a_i \in A_i. \tag{5.89}$$

That is, no player can unilaterally improve his/her payoff by changing his/her strategy if the other players employ the Nash equilibrium strategies. Thus, the Nash equilibrium defines each player's best response strategy given the other players' strategies. However, in general, a Nash equilibrium is not necessarily the optimal equilibrium providing the highest cumulative payoff due to the noncooperative nature of the game. Moreover, a game may have multiple Nash equilibria with different payoffs. Hence, different concepts such as Pareto optimality, evolutionary equilibrium, and correlated equilibrium as well as different techniques such as pricing have been proposed for choosing the optimal equilibrium and improving and refining inefficient equilibria. For more information on these concepts, see, e.g., [118, 209].

In [174] the spectrum access in cognitive radio networks is formulated as a strategic noncooperative game with power and interference constraints. Each secondary user aims at maximizing its own transmit rate given local constraints on the transmit power and interference caused to the primary users. Both null and soft interference constraints are considered. That is, the projection of the transmitted signal power along the subspaces corresponding to the primary users (in the spatial, frequency, or time domain) is either zero (null constraint) or below a given threshold (soft constraint). Distributed asynchronous MIMO iterative waterfilling algorithms (IWFAs) are proposed for the cognitive radio network under different interference constraints. The Nash equilibrium is shown to exist for all of the different game formulations. Moreover, sufficient conditions for its uniqueness and the convergence of the proposed distributed algorithms to the Nash equilibrium are established.

Spectrum access in cognitive radio networks under global interference constraints has been considered in [154]. Each secondary user seeks to maximize its own transmit rate over multiple frequency bands given a local constraint on the transmit power and global constraints on maximum per-carrier and total aggregate interference caused to the primary users. A pricing mechanism is proposed that introduces a penalty in the payoff function depending on the interference levels. The noncooperative game is formulated using the framework of finite-dimensional variational inequalities (VIs). Conditions for the uniqueness of the Nash equilibrium are established and distributed algorithms with minimum signaling from the primary users to the secondary users (i.e., to broadcast the updated prices) converging to the Nash equilibrium are proposed. Similar pricing-based interference management concept is employed in [70] and distributed IWFAs are proposed for both base station-based and ad hoc cognitive radio networks.

The algorithms in [70, 154, 174] assume perfect knowledge of the channel state information (CSI) between the secondary transmitters and primary receivers. In [213] a noncooperative strategic game formulation of spectrum access in cognitive radio networks with robust interference constraints is proposed. Each secondary user is assumed to know only the nominal channels to the primary receivers. A nominal channel is defined as a corrupted version of the actual channel corrupted with an error belonging to an elliptical uncertainty region. Thus, each secondary user should satisfy the worst-case interference constraints defined by the uncertainty region. The robust noncooperative strategic game in which the secondary users aim at maximizing their own transmit rates given the robust interference constraints is formulated as a VI problem and asynchronous distributed algorithms are proposed for achieving the Nash equilibrium.

In [230] a collaborative spectrum sensing framework is proposed in which the secondary users try to maximize individual revenue functions that depend on the achievable rate as well as on delay and energy consumption due to cooperative sensing. The problem is formulated as a noncooperative game in which the secondary users decide whether to participate in cooperative sensing or not. Distributed algorithms based on information exchange among the secondary users are proposed for obtaining a Nash equilibrium with both identical and nonidentical detection and false alarm probabilities among the users.

Cooperative games

In general, cooperation among the players provides an opportunity to improve the overall cumulative payoff compared to that obtained in noncooperative games. In cognitive radio systems cooperation may be realized in various forms: the users may share their local sensing results to find more free spectrum, as well as helping the other users in routing, packet forwarding, and interference management, for example.

An important class of cooperative games comprises *coalitional games* [171]. Coalitional games are cooperative games in which the players form coalitions (i.e., groups of players) that may enforce cooperation within the coalition. Thus, a coalitional game may be viewed as a game between coalitions instead of between the individual players. A coalitional game consists of

- a set of players $\mathcal{N} = \{1, \ldots, N_{\text{SU}}\}$;
- a coalition value function v that quantifies the worth of a coalition $\mathcal{S}$. The value of a coalition $\mathcal{S}$ is denoted by $v(\mathcal{S})$.

A widely used and important subclass of coalitional games is composed of coalitional games that are in *characteristic form*. In coalitional games that are in characteristic form the value of a coalition $\mathcal{S}$ depends only on the members of $\mathcal{S}$. Moreover, a coalitional game may have either transferable or nontransferable utility. For a coalitional game in characteristic form with transferable utility the value $v(\mathcal{S})$ of a coalition $\mathcal{S}$ is a real number. The value $v(\mathcal{S})$ can be thought of as a total payoff that may be divided among the members of $\mathcal{S}$. However, the value of a coalition may not always be represented as a single real number or be divided in a meaningful manner among the members of the coalition. Such games are called coalitional games with non-transferable utility.

Let us consider coalitional games in characteristic form that are superadditive, i.e., the value of a larger coalition formed out of disjoint coalitions is always at least equal to the value obtained by the disjoint coalitions separately. Hence, a *grand coalition*, i.e., a coalition involving all the players, is the optimal coalition. Such a game is called a *canonical coalitional game* [171]. *The core* of a canonical coalitional game with transferable utility is defined by

$$\mathcal{C} \triangleq \left\{ x : \sum_{i \in \mathcal{N}} x_i = v(\mathcal{N}) \text{ and } \sum_{i \in \mathcal{N}} x_i \geq v(\mathcal{S}), \forall \mathcal{S} \subseteq \mathcal{N} \right\}, \quad (5.90)$$

where x_i are the rational payoffs of the players satisfying $\sum_{i \in \mathcal{N}} x_i = v(\mathcal{N})$ and $x_i \geq v(\{i\}), \forall i \in \mathcal{N}$. Thus, no group of players has incentive to leave the grand coalition whose payoff allocation is in the core of the game. Consequently, the concept of the core in coalitional games is similar to the Nash equilibrium in noncooperative games. If it exists, the core of a coalitional game with transferable utility is obtained by solving the following linear program:

$$\min_{\mathbf{x}} \sum_{i \in \mathcal{N}} x_i, \text{ s.t. } \sum_{i \in \mathcal{S}} x_i \geq v(\mathcal{S}), \forall \mathcal{S} \subseteq \mathcal{N}. \quad (5.91)$$

The core can be defined for other coalitional games as well. A comprehensive treatment of coalitional games and their applications in communication networks can be found in [171].

In [167] a multiuser, multiband spectrum sensing and sharing problem is formulated as a coalitional game in which the secondary users form coalitions to cooperatively sense the frequency bands. The work done by each user is rewarded by allocating a portion of the available spectrum to each user based on the quantity and quality of its sensing work. The more frequency bands the user senses and the more informative its sensing results are the larger is its allocated portion of the available spectrum. This is accomplished by defining the worth of each coalition as [167]

$$v(\mathcal{S}) \triangleq |\mathcal{S}| \sum_{j=1}^{n_{\text{B}}} \left(\frac{1 - H\left(\left|\max_{\forall i \in \mathcal{S}}(P_{\text{D},ij} u_j)\right|\right)}{c(j)} \right), \quad (5.92)$$

where $\mathcal{S}$ is a coalition among the users, $|\mathcal{S}|$ denotes the cardinality of the set $\mathcal{S}$, n_{B} is the number of frequency bands, $H(\cdot)$ is the binary entropy function, $P_{\text{D},ij}$ is the reported probability of detection by user i on frequency band j, u_j is the spectrum decision for frequency band j (1 when the frequency band is occupied and -1 when it is vacant), and $c(j)$ is the total number of users sensing frequency band j. Moreover, if the probability of detection $P_{\text{D},ij}$ reported by the user i contradicts the fusion center's spectrum decision the corresponding user is not rewarded by setting its $P_{\text{D},ij}$ equal to 0.5. Fair and stable spectrum allocations are obtained by using allocations based on one-point solutions such as Shapley, tau, or nucleolus values. The formulation of the game ensures that the core of the game is nonempty and that the one-point solutions lie within the core.

In [170] collaborative single-band spectrum sensing in cognitive radio networks is modeled as a coalitional game with nontransferable utility. Distributed algorithms are proposed for forming the coalitions. That is, the secondary users autonomously form coalitions using merge and split rules to maximize their payoff in terms of probability of detection and a cost for false alarms. Due to the cost for false alarms a grand coalition is rarely the optimal structure, hence, the class of games formed belong to the class of coalition formation games [171]. Moreover, a coalitional voting game for guaranteeing a desired probability of detection is proposed and used to complement the merge and split-based algorithm. In the proposed coalitional games each coalition head acts as the fusion center in the collaborative detection. The user with the highest probability of detection is chosen as the coalition head in each coalition.

Auction games

An auction is a sales process in which the prospective buyers compete by bidding according to the rules of the auction to buy the goods or services under auction. Auctions are most often used in situations when the seller does not have a clear idea of how the prospective buyers are valuing the goods under auction. A particularly interesting form of auction, especially from the cognitive radio perspective, is *the second-price sealed-bid auction* in which the bidder placing the highest bid wins the sealed-bid auction but instead of paying the price equal to his/her own bid the bidder pays the price equal to the second highest bid. The appealing feature of a second-price sealed-bid auction is that it gives incentive to the bidders to bid truthfully since bidding truthfully is the optimal strategy [205].

In [68] spectrum access in cognitive radio networks is formulated as a repeated auction game with entry and monitoring fees. The auction is coordinated by a secondary coordinator. The entry fee accounts for the energy consumption due to the spectrum access and channel estimation needed for bidding and the monitoring fee accounts for the cost of spectrum sensing and the subscription to the common control channel to obtain past auction history information. The entry fee is paid only when the user participates in the auction while the monitoring fee is paid in each time period. The repeated auction of spectral opportunities follows the second-price sealed-bid auction format in each time period. A learning scheme based on Dirichlet processes is proposed to optimize the bidding of the distributed secondary users. The Dirichlet process is employed to model the distribution of the winning bid given the past auction history.

Stochastic games

The concept of stochastic games was already introduced in Section 5.3.3 when we defined the multiagent reinforcement learning problem as a stochastic game. Stochastic games are an extension of MDPs in which multiple players interact in noncooperative or cooperative manner. We repeat the formulation of a stochastic game from the reinforcement learning section for convenience using more game-theoretic terminology. That is, a stochastic game consist of

- a sequence of discrete time steps $n = 0, 1, 2, \ldots$;
- a set of players $\mathcal{N} = \{1, \ldots, N_{\text{SU}}\}$;
- a set of possible states $s \in S$;
- a set of possible actions for each player in each state $a_i \in A_i, i \in \mathcal{N}$. The combined action space is $A = A_1 \times \cdots \times A_{N_{\text{SU}}}$;

- a state transition function $\varphi : S \times A \times S \to [0,1]$ which defines the transition probability $P\left[s_{n+1} \mid s_n, a_{1,n}, \ldots, a_{N_{\text{SU}},n}\right]$;
- payoff functions $r_i : S \times A \times S \to \mathbb{R}$, $i \in \mathcal{N}$, which give the players a payoff for the joint action $a_{1,n}, \ldots, a_{N_{\text{SU}},n}$ of the players in state s_n resulting in new state s_{n+1}.

Stochastic games differ from the other games discussed earlier in the sense that there are multiple states that change stochastically during the course of the game. Hence, stochastic games provide an excellent fit to the constantly changing dynamic environment in cognitive radio applications. The consequence is that in a stochastic game the players' strategies depend on the current state. The mapping of the states to actions is called a policy. Multiagent reinforcement learning is one of the most prominent approaches for optimizing the behavior of the players in a stochastic game (i.e., learning a policy that maximizes a given objective function). See Section 5.3.3 for an introduction to multiagent reinforcement learning, solving stochastic games, and their applications to cognitive radios.

In this section we have provided a short introduction to game theory and its applications in cognitive radio networks along with a review of the most recent literature. For a more comprehensive treatment of game theory in cognitive radio networks, we refer the reader to [118, 209].

5.3.5 Location awareness and geolocation

Location awareness and geolocation are at the core of situational awareness in cognitive radio systems. Or, more precisely, awareness of one's location relative to the other spectrum users is at the core of situational awareness since, depending on the application, it may not be necessary for a cognitive radio user to know its exact geographical location; rather a location relative to the other users of the spectrum may suffice. Nevertheless, location information plays an important role in spectrum access and interference management. The more accurately a secondary user knows the locations of the other users of the spectrum and the network topologies, the more accurately it can estimate the level of interference caused to them by its transmissions and subsequently adjust its transmission parameters accordingly. Location information is also beneficial in routing and scheduling problems. The capability to understand propagation phenomena and detection distances, primary user communication distances, and interference distances add to the location awareness and allow for modeling areas of harmful interference, achieving high data rates, as well as satisfying the interference constraints. The question then arises: how can a cognitive radio user obtain this location information?

Outdoors, a mobile secondary user may employ satellite positioning techniques such as global positioning system (GPS) to obtain its own geolocation. In some cognitive radio applications this may already provide the secondary user a wealth of information for efficient and effective spectrum exploitation. For example, the reuse of digital TV frequency bands is such an application. The TV broadcast towers are located in fixed, static positions that are typically publicly available. Hence, a cognitive radio user knowing its own geolocation can easily determine the nearby TV broadcast towers and the channels employed. However, in general, the primary transmitter geolocations may not be available in advance. Moreover, if the user is indoors or no GPS or other satellite positioning system is available the user may have difficulty in determining its own geolocation. Finding out an unknown location of a wireless transmitter is a challenging problem requiring advanced signal processing algorithms. Moreover, reliable and accurate localization requires distributed cooperative techniques. Various proposed approaches include received signal strength- (RSS) [85], time-of-arrival- (TOA) [33, 34, 58, 80, 87], direction-of-arrival- (DOA) [80], and sensing result-based [128] techniques.

The above techniques can also be applied to determine the locations of the other secondary users. However, such information may be obtained through other means as well, such as information exchange or mutual ranging [66], provided that the secondary users in the cognitive radio network cooperate with each other.

Finally, we would like to point out that the interference in a wireless communication system is experienced at the receiver and not at the transmitter, hence, knowing the location of the primary receivers is much more beneficial than knowing the location of the primary transmitters. However, this is notoriously difficult if the receivers are passive, as, for example, TV receivers are. In principle, even passive receivers may be detected by exploiting the local oscillator leakage power emitted by the RF front-end of a wireless receiver when a signal is received [29, 221]. However, the typical leakage power is very low, limiting the detection range to only a few meters at best. Thus, in applications involving passive receivers, one may have to be content with knowing only the locations of the primary transmitters.

5.4 Summary

In this chapter, the optimization of the spectrum exploration and exploitation processes in cognitive radio networks has been considered. Spectrum exploration is the process of obtaining local awareness of the spectrum state through spectrum sensing. The goal of spectrum exploration is to find idle spectrum that can then be exploited. Optimization of spectrum exploration includes optimization of the whole spectrum sensing process that determines which frequency bands are sensed, when they are sensed and for how long and by which users, and how are the sensing results from multiple users combined. This involves trading off quantities such as diversity, detection speed, and performance. Spectrum exploration is coupled with spectrum exploitation. Spectrum exploitation addresses the questions: what happens after idle spectrum has been found; and how is the idle spectrum subsequently exploited? Spectrum exploitation optimization involves optimizing the spectrum access process that determines which idle frequency bands to access, for how long, and by which users. It involves choosing the transmit powers and waveforms to be employed, as well. The goal of spectrum exploitation combined with spectrum exploration is to maximize the throughput of the secondary network and provide a desired quality of service for the secondary network while guaranteeing that the level of interference caused to the primary users is below the given interference constraints. In the first part of the chapter, advanced spectrum sensing techniques, such as distributed detection and sequential and quickest detection, have been reviewed. Distributed detection among spatially dispersed secondary users allows improving the sensing reliability by increasing the probability of detection for a given probability of false alarm. Or alternatively the sensing time can be reduced without sacrificing the sensing performance. Sequential and quickest detection aim at making the decision as quickly as possible for the given error levels. Quickest detection techniques facilitate also the secondary users rapid evacuation from a frequency band when a primary user becomes active. Both distributed detection and sequential and quickest detection techniques are thus important for more efficient and effective spectrum exploration.

The second part of the chapter focused on the design of optimum spectrum sensing and access policies for optimized spectrum exploration and exploitation. Various approaches for modeling, analyzing, and learning the behavior of other spectrum users have been considered. Introductions to dynamic programming, bandit problems, reinforcement learning, game theory, and their applications in cognitive radio systems have been provided. Dynamic programming is an optimization approach in which the original problem is broken into recursively solved simpler subproblems. Although this reduces the complexity of finding the optimal solution, the computational complexity of dynamic programming still limits its usefulness in practical cognitive radio systems. In high-dimensional problems the computational complexity of obtaining an optimal solution through dynamic programming may be prohibitive in particular for battery-operated terminals with limited computing power. Hence, computationally efficient techniques that may be used to obtain a close-to-optimal or asymptotically optimal solution are of interest. Reinforcement learning, bandit problems, and game theory offer such techniques. A truly cognitive radio should possess the ability to learn from the environment and the outcomes of its previous actions. The goal of reinforcement learning, bandit problems, and game-theoretic learning methods is to learn to act optimally in a given situation. Hence, these techniques are particularly suitable for cognitive radio systems. Whereas bandit problems focus on a limited class of problems that can be cast into the framework of (restless) multiarmed bandits, reinforcement learning is a vast field comprising a multitude of different methods and possible applications. In fact, bandit problems are the simplest form of reinforcement learning problems. However, bandit problems offer a sufficient complexity to model problems such as multiband spectrum sensing and access in cognitive radio systems. More complicated, full-blown reinforcement learning problems and algorithms may be needed, in particular, if the actions of the secondary users are allowed to be also something other than sensing or accessing a certain frequency band. Nevertheless, both bandit problems and more general reinforcement learning problems deal with the exploitation vs. exploration dilemma. That is, the agent or agents must in each state decide whether to exploit the currently best action or to explore other actions in the hope of a better future reward. The learning is performed through trial and error and by reinforcing actions leading to good rewards, not through a teacher telling whether the chosen action was correct or not. Hence, these techniques are particularly suitable for cognitive radio applications in which the radio environment is unknown, highly dynamic, and nonstationary. Game theory, on the other hand, provides a collection of mathematical methods and tools for modeling, analyzing, and predicting the behavior of multiple interacting entities. In cognitive radio systems, game theory offers a framework for modeling and analyzing the interaction of multiple competing and/or cooperating spectrum users. The majority of existing game theory deals with static single-shot or repeated games. This means that the players face the same game each time. Thus, the players' strategies do not depend on the current state of the environment. This does not provide a very good fit to cognitive radio applications in which the radio environment may be highly dynamic. However, a subfield of game theory called stochastic games provides an excellent fit to the dynamic environment encountered in cognitive radio applications. In stochastic games there are multiple states that change stochastically during the course of the game. This also means that the players' strategies will depend on the current state. In addition, stochastic games provide a connection between game theory and multiagent reinforcement learning since multiagent reinforcement learning problems can be modeled as stochastic games.

The main focus in this chapter has been on optimizing spectrum exploration and exploitation in interweave cognitive radio networks. For the vast majority of introduced techniques and algorithms the employed system framework assumes one or multiple primary systems and a single cognitive radio network consisting of identical (homogeneous) secondary users. However, in practice, there may be multiple heterogeneous secondary systems simultaneously trying to find and exploit the same underutilized spectrum. The coexistence of such heterogeneous secondary systems and networks has not been considered in this chapter.

What are the implications of heterogeneous secondary systems coexisting on the same frequency bands? In principle, two heterogeneous secondary systems may employ completely different waveforms, modulation, coding, and access schemes from one another, for example. How then can two or more completely different secondary systems coexist efficiently and effectively in a licensed primary frequency band while controlling the interference caused to the primary users? Can the other secondary systems be treated only as noise? On the other hand, how do the other asynchronous secondary systems affect the sensing process since quiet periods for sensing cannot be assumed anymore? Obviously the employed sensing algorithms should be able to distinguish among the secondary systems and the primary systems if underutilized spectrum is to be exploited as efficiently as possible. However, how much does the interference from the other secondary systems affect the reliability of sensing? How about spectrum access then? How do the interference and collisions with other secondary systems affect the efficiency of the spectrum access algorithms? These are all open questions that need to be answered before free coexistence of heterogeneous secondary systems can become reality.

5.5 Further reading

In this chapter various techniques and approaches for optimizing the spectrum exploration and exploitation processes in cognitive radio networks have been introduced. The first part of the chapter introduced advanced detection techniques for cognitive radio networks, such as distributed detection as well as sequential and quickest detection.

A more detailed treatment of distributed detection and fusion can be found in [200]. Comprehensive discussions of classical sequential detection can be found in [59, 180] while detailed treatments of quickest detection can be found in [15, 162, 178, 188].

The second part of the chapter focused on spectrum sensing and access policy design for optimized spectrum exploration and exploitation. Techniques such as dynamic programming, bandit problems, reinforcement learning, and game theory, were briefly introduced and their application to cognitive radio systems were discussed. For more detailed discussions of dynamic programming we refer the reader to [16, 19, 32]. General treatments of multiarmed bandit problems, including a detailed discussion of many of the variants listed in Section 5.3.2 and their properties, can be found in [17, 130]. Comprehensive introductions to reinforcement learning as well as more advanced material can be found in [24, 32, 78, 183, 184] and in [31, 226] where the emphasis is on multiagent reinforcement learning. Finally, more detailed discussions of game theory for cognitive radio networks and wireless communications can be found in [118, 129, 171, 209] while for more general treatments of game theory we refer the reader to [25, 60, 104, 156, 179].

References

[1] R. Agrawal, M. V. Hedge, and D. Teneketzis, "Asymptotically efficient adaptive allocation rules for the multiarmed bandit problem with switching cost," *IEEE Trans. Autom. Control*, **33**, no. 10, 899–906, Oct. 1988.

[2] R. Agrawal, M. V. Hedge, and D. Teneketzis, "Multi-armed bandit problems with multiple plays and switching cost," *Stochast. Stochast. Rep.*, **29**, 437–459, 1990.

[3] S. Ahmad and M. Liu, "Multi-channel opportunistic access: a case of restless bandits with multiple plays," *Proc. Allerton Conf. Communication, Control, and Computing*, Monticello, IL, Oct. 2009, pp. 1361–1368.

[4] S. H. Ahmad, M. Liu, T. Javidi, Q. Zhao, and B. Krishnamachari, "Optimality of myopic sensing in multi-channel opportunistic access," *IEEE Trans. Inf. Theory*, **55**, no. 9, 4040–4050, Sep. 2009.

[5] A. Anandkumar, N. Michael, and A. K. Tang, "Opportunistic spectrum access with multiple users: learning under competition," *Proc. IEEE Conf. Computer Communications (INFOCOM 2010)*, San Diego, CA, USA, Mar. 15–19, 2010.

[6] A. Anandkumar, N. Michael, A. K. Tang, and A. Swami, "Distributed algorithms for learning and cognitive medium access with logarithmic regret," *IEEE J. Sel. Areas in Commun.*, **29**, no. 4, 731–745, Apr. 2011.

[7] V. Anantharam, P. Varaiya, and J. Walrand, "Asymptotically efficient allocation rules for the multiarmed bandit problem with multiple plays – part I: i.i.d. rewards," *IEEE Trans. Autom. Control*, **AC-32**, no. 11, 968–977, Nov. 1987.

[8] V. Anantharam, P. Varaiya, and J. Walrand, "Asymptotically efficient allocation rules for the multiarmed bandit problem with multiple plays – part II: Markovian rewards," *IEEE Trans. Autom. Control*, **AC-32**, no. 11, 977–982, Nov. 1987.

[9] M. Asawa and D. Teneketzis, "Multi-armed bandits with switching penalties," *IEEE Trans. Autom. Control*, **41**, no. 3, 328–348, Mar. 1996.

[10] J.-Y. Audibert, R. Munos, and C. Szepesvári, "Exploration–exploitation tradeoff using variance estimates in multi-armed bandits," *Theor. Comput. Sci.*, **410**, no. 19, 1876–1902, Apr. 2009.

[11] P. Auer, N. Cesa-Bianchi, and P. Fischer, "Finite-time analysis of the multiarmed bandit problem," *Mach. Learn.*, **47**, no. 2–3, 235–256, 2002.

[12] P. Auer and R. Ortner, "UCB revisited: improved regret bounds for the stochastic multi-armed bandit problems," *Period. Math. Hungar.*, **61**, no. 1–2, 55–65, 2010.

[13] T. Banerjee, V. Sharma, V. Kavitha, and A. K. Jayaprakasam, "Generalized analysis of a distributed energy efficient algorithm for change detection," *IEEE Trans. Wireless Commun.*, **10**, no. 1, 91–101, Jan. 2011.

[14] J. Banks and R. Sundaram, "Switching costs and the Gittins index," *Econometrica*, **62**, no. 3, 687–694, May 1994.

[15] M. Basseville and I. V. Nikiforov, *Detection of Abrupt Changes – Theory and Application*, Prentice-Hall, 1993.

[16] R. Bellman, *Dynamic Programming*, Princeton University Press, 2010.

[17] D. A. Berry and B. Fristedt, *Bandit Problems: Sequential Allocation of Experiments*, Chapman and Hall, 1985.

[18] U. Berthold, F. Fu, M. van der Schaar, and F. K. Jondral, "Detection of spectral resources in cognitive radios using reinforcement learning," *Proc. 3rd IEEE Int. Symp. New Frontiers in Dynamic Spectrum Access Networks (DySPAN 2008)*, Chicago, IL, Oct. 14–17, 2008.

[19] D. P. Bertsekas, *Dynamic Programming and Optimal Control* (3rd edn.), Vols. 1 and 2, Athena Scientific, 2007.

[20] D. Bertsimas and J. Niño-Mora, "Conservations laws, extended polymatroids and multiarmed bandit problems; a polyhedral approach to indexable systems," *Math. Oper. Res.*, **21**, no. 2, 257–306, May 1996.

[21] E. Biglieri, R. Calderbank, A. Constantinides, A. Goldsmith, A. Paulraj, and H. V. Poor, *MIMO Wireless Communications*, Cambridge University Press, 2007.

[22] R. S. Blum, S. A. Kassam, and H. V. Poor, "Distributed detection with multiple sensors: Part II – advanced topics," *Proc. IEEE*, **85**, no. 1, 64–79, Jan. 1997.

[23] R. S. Blum and B. M. Sadler, "Energy efficient signal detection in sensor networks using ordered transmissions," *IEEE Trans. Signal Process.*, **56**, no. 7, 3229–3235, July 2008.

[24] M. Bowling and M. Veloso, "Multiagent learning using a variable learning rate," *Artif. Intell.*, **136**, no. 2, 215–250, Apr. 2002.

[25] R. Branzei, D. Dimitrov, and S. Tijs, *Models in Cooperative Game Theory*, 2nd edition, Springer-Verlag, 2008.

[26] B. E. Brodsky and B. S. Darkhovsky, *Nonparametric Methods in Change-Point Problems*, Kluwer Academic Publishers, 1993.

[27] B. E. Brodsky and B. S. Darkhovsky, "Minimax sequential tests for many composite hypothesis. I," *Theory Probab. Appl.*, **52**, no. 4, 565–579, 2008.

[28] B. E. Brodsky and B. S. Darkhovsky, "Minimax sequential tests for many composite hypothesis. II," *Theory Probab. Appl.*, **53**, no. 1, 1–12, 2009.

[29] T. X. Brown, "An analysis of unlicensed device operation in licensed broadcast service bands," *Proc. 1st IEEE Int. Symp. New Frontiers in Dynamic Spectrum Access Networks (DySPAN 2005)*, Baltimore, MD, Nov. 8–11, 2005, pp. 11–29.

[30] T. N. Bui, V. Krishnamurthy and H. V. Poor, "Online Bayesian activity detection in DS/CDMA network," *IEEE Trans. Signal Process.*, **53**, no. 1, 371 - 375, Jan. 2005.

[31] L. Buşoniu, R. Babuška, and B. De Schutter, "A comprehensive survey of multiagent reinforcement learning," *IEEE Trans. Syst., Man, Cybern. C, Appl. Rev.*, **38**, no. 2, 156–172, Mar. 2008.

[32] L. Buşoniu, R. Babuška, B. De Schutter, and D. Ernst, *Reinforcement Learning and Dynamic Programming Using Function Approximators*, CRC Press, 2010.

[33] H. Celebi and H. Arslan, "Adaptive positioning systems for cognitive radios," *Proc. 2nd IEEE Int. Symp. New Frontiers in Dynamic Spectrum Access Networks (DySPAN 2007)*, Dublin, Ireland, Apr. 17–20, 2007, pp. 78–84.

[34] H. Celebi and H. Arslan, "Cognitive positioning systems," *IEEE Trans. Wireless Commun.*, **6**, no. 12, 4475–4483, Dec. 2007.

[35] S. Chaudhari and V. Koivunen, "Effect of quantization and channel errors on collaborative spectrum sensing," *Proc. 43rd Asilomar Conf. Signals, Systems, and Computers*, Pacific Grove, CA, Nov. 1–4, 2009, pp. 528–533.

[36] S. Chaudhari, V. Koivunen, and H. V. Poor, "Autocorrelation-based decentralized sequential detection of OFDM signals in cognitive radios," *IEEE Trans. Signal Process.*, **57**, no. 7, 2690–2700, July 2009.

[37] S. Chaudhari, J. Lundén, and V. Koivunen, "BEP walls for collaborative spectrum sensing," *Proc. 36th IEEE Int. Conf. Acoustics, Speech, and Signal Processing (ICASSP 2011)*, Prague, Czech Republic, May 22–27, 2011.

[38] S. Chaudhari, J. Lundén, and V. Koivunen, "Performance limitations for cooperative spectrum sensing with reporting channel errors," *Proc. 22nd IEEE Int. Symp. Personal, Indoor, and Mobile Radio Communications (PIMRC 2011)*, Toronto, Canada, Sep. 11–14, 2011.

[39] S. Chaudhari, J. Lundén, V. Koivunen, and H. V. Poor, "Cooperative sensing with imperfect reporting channels: hard decisions or soft decisions?," *IEEE Trans. Signal Process.*, **60**, no. 1, 18–28, Jan. 2012.

[40] B. Chen and P. K. Willett, "On the optimality of the likelihood-ratio test for local sensor decision rules in the presence of nonideal channels," *IEEE Trans. Inf. Theory*, **51**, no. 2, 693–699, Feb. 2005.

[41] H. Chen, B. Chen, and P. K. Varshney, "Further results on the optimality of the likelihood-ratio test for local sensor decision rules in the presence of nonideal channels," *IEEE Trans. Inf. Theory*, **55**, no. 2, 828–832, Feb. 2009.

[42] R. Chen, J.-M. Park, and K. Bian, "Robust distributed spectrum sensing in cognitive radio networks," *Proc. 27th IEEE Conf. Computer Communications (INFOCOM 2008)*, Phoenix, AZ, Apr. 13–18, 2008.

[43] Y. Chen, "Optimum number of secondary users in collaborative spectrum sensing considering resource usage efficiency," *IEEE Commun. Lett.*, **12**, no. 12, 877–879, Dec. 2008.

[44] Y. Chen, Q. Zhao, and A. Swami, "Joint design and separation principle for opportunistic spectrum access in the presence of sensing errors," *IEEE Trans. Inf. Theory*, **54**, no. 5, 2053–2071, May 2008.

[45] Z. Chen, Z. Hu, and R. C. Qui, "Quickest spectrum detection using hidden Markov model for cognitive radio," *Proc. IEEE Military Communications Conf. (MILCOM 2009)*, Boston, MA, Oct. 18–21, 2009.

[46] K. W. Choi, W. S. Jeon, and D. G. Jeong, "Sequential detection of cyclostationary signal for cognitive radio systems," *IEEE Trans. Wireless Commun.*, **8**, no. 9, 4480–4485, Sep. 2009.

[47] S. Dayanik, C. Goulding, and H. V. Poor, "Bayesian sequential change diagnosis," *Math. Oper. Res.*, **33**, no. 2, 475–496, May 2008.

[48] S. Dayanik, W. Powell, and K. Yamazaki, "Index policies for discounted bandit problems with availability constraints," *Adv. Appl. Prob.*, **40**, no. 2, 377–400, June 2008.

[49] M. Di Felice, K. R. Chowdhury, W. Meleis, and L. Bononi, "To sense or to transmit: a learning-based spectrum management scheme for cognitive radio mesh networks," *Proc. 5th IEEE Workshop Wireless Mesh Networks (WIMESH 2010)*, Boston, MA, June 21, 2010.

[50] F. F. Digham, M.-S. Alouini, and M. K. Simon, "On the energy detection of unknown signals over fading channels," *Proc. IEEE Int. Conf. Communications (ICC 2003)*, Anchorage, USA, May 11–15, 2003, pp. 3575–3579.

[51] F. F. Digham, M.-S. Alouini, and M. K. Simon, "On the energy detection of unknown signals over fading channels," *IEEE Trans. Commun.*, **55**, no. 1, 21–24, Jan. 2007.

[52] S. Fazeli-Dehkordy, K. N. Plataniotis, and S. Pasupathy, "Wide-band collaborative spectrum search strategy for cognitive radio networks," *IEEE Trans. Signal Process.*, **59**, no. 8, 3903–3914, Aug. 2011.

[53] G. Fellouris and G. V. Moustakides, "Decentralized sequential hypothesis testing using asynchronous communication," *IEEE Trans. Inf. Theory*, **57**, no. 1, 534–548, Jan. 2011.

[54] S. Filippi, O. Cappé, F. Clérot, and E. Moulines, "A near optimal policy for channel allocation in cognitive radio," *Recent Advances in Reinforcement Learning: 8th European Workshop (EWRL 2008)*, Villeneuve d'Ascq, France, June 30–July 4, 2008, pp. 69–81.

[55] S. Filippi, O. Cappé, and A. Garivier, "Optimally sensing a single channel without prior information: the tiling algorithm and regret bounds," *IEEE J. Sel. Topics Signal Process.*, **5**, no. 1, 68–76, February 2011.

[56] Y. Gai, B. Krishnamachari, and R. Jain, "Learning multiuser channel allocations in cognitive radio networks: a combinatorial multi-armed bandit formulation," *Proc. IEEE Int. Symp. Dynamic Spectrum Access Networks (DySPAN 2010)*, Singapore, Apr. 6–9, 2010.

[57] F. Gao, W. Yuan, W. Liu, W. Cheng, and S. Wang, "A robust and efficient cooperative spectrum sensing scheme in cognitive radio networks," *Proc. IEEE Int. Conf. Communications (ICC 2010)*, Cape Town, South Africa, May 23–27, 2010.

[58] S. Gezici, H. Celebi, H. V. Poor, and H. Arslan, "Fundamental limits on time delay estimation in dispersed spectrum cognitive radio systems," *IEEE Trans. Wireless Commun.*, **8**, no. 1, 78–83, Jan. 2009.

[59] B. K. Ghosh, *Sequential Tests of Statistical Hypothesis*, Addison-Wesley, 1970.

[60] R. Gibbons, *A Primer in Game Theory*, Pearson Education Limited, 1992.

[61] J. C. Gittins, "Bandit processes and dynamic allocation indices," *J. Roy. Statist. Soc. Ser. B*, **41**, no. 2, 148–177, 1979.

[62] J. C. Gittins and D. M. Jones, "A dynamic allocation index for the sequential design of experiments," in J. Gani (ed.) *Progress in Statistics*, North Holland, pp. 241–266, 1974.

[63] K. D. Glazebrook, D. Ruiz-Hernandez, and C. Kirkbride, "Some indexable families of restless bandit problems," *Adv. Appl. Prob.*, **38**, no. 3, 643–672, 2006.

[64] M. Guerriero, V. Pozdnyakov, J. Glaz, and P. Willett, "A repeated significance test with applications to sequential detection in sensor networks," *IEEE Trans. Signal Process.*, **58**, no. 7, 3426–3435, July 2010.

[65] O. Hadjiliadis, H. Zhang, and H. V. Poor, "One shot schemes for decentralized quickest change detection," *IEEE Trans. Inf. Theory*, **55**, no. 7, 3346–3359, July 2009.

[66] A. Haghparast, T. Abrudan, and V. Koivunen, "OFDM ranging in multipath channels using time reversal method," *Proc. 10th IEEE Int. Workshop Signal Processing Advances for Wireless Communications (SPAWC 2009)*, Perugia, Italy, June 21–24, 2009, pp. 568–572.

[67] B. Hamdaoui, P. Venkatraman, and M. Guizani, "Opportunistic exploitation of bandwidth resources through reinforcement learning," *Proc. IEEE Global Communications Conf. (GLOBECOM 2009)*, Honolulu, HI, Nov. 30–Dec. 4, 2009.

[68] Z. Han, R. Zheng, and H. V. Poor, "Repeated auctions with Bayesian nonparametric learning for spectrum access in cognitive radio networks," *IEEE Trans. Wireless Commun.*, **10**, no. 2, 890–900, Mar. 2011.

[69] S. Hart and A. Mas-Colell, "A reinforcement procedure leading to correlated equilibrium," in G. Debreu, W. Neuefeind, and W. Trockel (eds.), *Economics Essays*, pp. 181–200, Springer, 2001.

[70] M. Hong and A. Garcia, "Equilibrium pricing of interference in cognitive radio networks," *IEEE Trans. Signal Process.*, **59**, no. 12, 6058–6072, Dec. 2011.

[71] T.-C. Hsu, T.-Y. Wang, and Y.-W. P. Hong, "Collaborative change detection for efficient spectrum sensing in cognitive radio networks," *Proc. 71st IEEE Vehicular Technology Conf. (VTC 2010-Spring)*, Taipei, Taiwan, May 16–19, 2010.

[72] A. M. Hussain, "Multisensor distributed sequential systems," *IEEE Trans. Aerosp. Electron. Syst.*, **30**, no. 3, 698–708, July 1994.

[73] T. Jaakkola, M. I. Jordan, and S. P. Singh, "On the convergence of stochastic iterative dynamic programming algorithms," *Neural Comput.*, **6**, 1185–1201, 1994.

[74] A. K. Jayaprakasam and V. Sharma, "Sequential detection based cooperative spectrum sensing algorithms in cognitive radio," *Proc. 1st IEEE UK-India Int. Workshop Cognitive Wireless Systems (UKIWCWS 2009)*, IIT Delhi, India, Dec. 11–12, 2009.

[75] A. K. Jayaprakasam, V. Sharma, C. R. Murthy, and P. Narayanan, "Cyclic prefix based cooperative sequential spectrum sensing algorithms for OFDM," *Proc. IEEE Int. Conf. Communication (ICC 2010)*, Cape Town, South Africa, May 2010.

[76] H. Jiang, L. Lai, R. Fan, and H. V. Poor, "Optimal selection of channel sensing order in cognitive radio," *IEEE Trans. Wireless Commun.*, **8**, no. 1, 297–307, Jan. 2009.

[77] T. Jun, "A survey on the bandit problem with switching costs," *De Economist*, **152**, no. 4, 513–541, 2004.

[78] L. P. Kaebling, M. L. Littman, and A. W. Moore, "Reinforcement learning: a survey," *J. Artif. Intell. Res.*, **4**, no. 1, 237–285, 1996.

[79] P. Kaewprapha, R. Wu, B. C. Ng, and T. J. Li, "Cooperative spectrum sensing for cognitive radios: bounds and algorithms," *Proc. IEEE Wireless Communications and Networking Conf. (WCNC 2010)*, Sydney, Australia, Apr. 18–21, 2010, pp. 1–6.

[80] S. Kandeepan, S. Reisenfeld, T. C. Aysal, D. Lowe, and R. Piesiewicz, "Bayesian tracking in cooperative localization for cognitive radio networks," *Proc. 69th IEEE Vehicular Technology Conf. (VTC 2009-Spring)*, Barcelona, Spain, Apr. 26–29, 2009.

[81] A. Kashyap, "Comments on the "On the optimality of the likelihood-ratio test for local sensor decision rules in the presence of nonideal channels"," *IEEE Trans. Inf. Theory*, **52**, no. 3, 1274–1275, Mar. 2006.

[82] A.-K. Katta and J. Sethuraman, "A note on bandits with a twist," *SIAM J. Discrete Math.*, **18**, no. 1, 110–113, 2004.

[83] H. Kim and K. G. Shin, "Efficient discovery of spectrum opportunities with MAC-layer sensing in cognitive radio networks," *IEEE Trans. Mobile Comput.*, **7**, no. 5, 533–545, May 2008.

[84] S.-J. Kim and G. B. Giannakis, "Rate-optimal and reduced-complexity sequential sensing algorithms for cognitive OFDM radios," *EURASIP J. Adv. Signal Process. – Special Issue on Dynamic Spectrum Access for Wireless Networking*, **2009**, Article ID 421540, 2009.

[85] S. Kim, H. Jeon, and J. Ma, "Robust localization with unknown transmission power for cognitive radio," *Proc. IEEE Military Communications Conf. (MILCOM 2007)*, Orlando, FL, Oct. 29–31, 2007.

[86] S.-J. Kim and G. B. Giannakis, "Sequential and cooperative sensing for multi-channel cognitive radios," *IEEE Trans. Signal Process.*, **58**, no. 8, 4239–4253, Aug. 2010.

[87] F. Kocak, H. Celebi, S. Gezici, K. A. Qarake, H. Arslan, and H. V. Poor, "Time delay estimation in dispersed spectrum cognitive radio systems," *EURASIP J. Adv. Signal Process. – Special Issue on*

Advanced Signal Processing for Cognitive Radio Networks, **2010**, Article ID 675959, 10 pages, 2010.

[88] J. R. Kok and N. Vlassis, "Collaborative multiagent reinforcement learning by payoff propagation," *J. Mach. Learn. Res.*, **7**, 1789–1828, Dec. 2006.

[89] N. Kundargi and A. Tewfik, "A performance study of novel sequential energy detection methods for spectrum sensing," *Proc. IEEE Int. Conf. Acoustics, Speech, and Signal Processing (ICASSP 2010)*, Dallas, TX, Mar. 14–19, 2010.

[90] L. Lai, H. El Gamal, H. Jiang, and H. V. Poor, "Cognitive medium access: exploration, exploitation, and competition," *IEEE Trans. Mobile Comput.*, **10**, no. 2, 239–253, Feb. 2011.

[91] L. Lai, Y. Fan, and H. V. Poor, "Quickest detection in cognitive radio: a sequential change detection framework," *Proc. IEEE Global Communications Conf. (GLOBECOM 2008)*, New Orleans, LA, Nov. 30–Dec. 4, 2008.

[92] L. Lai, H. Jiang, and H. V. Poor, "Medium access in cognitive radio networks: a competitive multi-armed bandit framework," *Proc. 42nd Asilomar Conf. Signals, Systems, and Computers*, Pacific Grove, USA, Oct. 26–29, 2008, pp. 98–102.

[93] T. L. Lai, "Asymptotic optimality of invariant sequential probability ratio tests," *Ann. Stat.*, **9**, no. 2, 318–333, 1981.

[94] T. L. Lai, "Nearly optimal sequential tests of composite hypotheses," *Ann. Stat.*, **16**, no. 2, 856–886, 1988.

[95] T. L. Lai, "On optimal stopping problems in sequential hypothesis testing," *Statist. Sinica*, **7**, pp. 33–51, 1997.

[96] T. L. Lai, "Sequential analysis: some classical problems and new challenges," *Statist. Sinica*, **11**, no. 2, 303–408, Apr. 2001.

[97] T. L. Lai and H. Robbins, "Asymptotically efficient adaptive allocation rules," *Adv. Appl. Math.*, **6**, no. 1, 4–22, Mar. 1985.

[98] T. L. Lai and H. Xing, "Sequential change-point detection when the pre- and post-change parameters are unknown," *Technical Report no. 2009-5*, Department of Statistics, Stanford University, Stanford, CA, Apr. 2009.

[99] T. L. Lai and L. Zhang, "A modification of Schwarz's sequential likelihood ratio tests in multivariate sequential analysis," *Sequential Anal.: Design Methods Appl.*, **13**, no. 2, 79–96, 1994.

[100] J. Langford and T. Zhang, "The epoch-greedy algorithm for contextual multi-armed bandits," in J.C. Platt, D. Koller, Y. Singer, and S. Roweis (eds.), *Advances in Neural Information Processing Systems 20*, pp. 817–824, MIT Press, 2008.

[101] G. S. Lauer and N. R. Sandell Jr., "Distributed detection of known signal in correlated noise," *Report ALPHATECH*, Burlington, MA, Mar. 1982.

[102] W.-Y. Lee and I. F. Akyildiz, "Optimal spectrum sensing framework for cognitive radio networks," *IEEE Trans. Wireless Commun.*, **7**, no. 10, 3845–3857, Oct. 2008.

[103] K. B. Letaief and W. Zhang, "Cooperative communications for cognitive radio networks," *Proc. IEEE*, **97**, no. 5, 878–893, May 2009.

[104] K. Leyton-Brown and Y. Shoham, *Essentials of Game Theory: A Concise Multidisciplinary Introduction*, Morgan & Claypool, 2008.

[105] H. Li, "Restless watchdog: selective quickest spectrum sensing in multichannel cognitive radio systems," *EURASIP J. Adv. Signal Process. – Special Issue on Dynamic Spectrum Access for Wireless Networking*, **2009**, Article ID 417457, Aug. 2009.

[106] H. Li, "Learning the spectrum via collaborative filtering in cognitive radio networks," *Proc. IEEE Int. Symp. New Frontiers in Dynamic Spectrum Access Networks (DySPAN 2010)*, Singapore, Apr. 6–9, 2010.

[107] H. Li, "Multiagent Q-learning for Aloha-like spectrum access in cognitive radio systems," *EURASIP J. Wireless Commun. Netw.*, **2010**, Article ID 876216, 2010.

[108] H. Li, H. Dai, and C. Li, "Collaborative quickest spectrum sensing via random broadcast in cognitive radio systems," *IEEE Trans. Wireless Commun.*, **9**, no. 7, 2338–2348, July 2010.

[109] H. Li, C. Li, and H. Dai, "Collaborative quickest detection in adhoc networks with delay constraint – Part I: Two-node network," *Proc. Conf. Information Sciences and Systems (CISS 2008)*, Princeton, NJ, Mar. 19–21, 2008, pp. 594–599.

[110] H. Li, C. Li, and H. Dai, "Collaborative quickest detection in adhoc networks with delay constraint – Part II: Multi-node network," *Proc. Conf. Information Sciences and Systems (CISS 2008)*, Princeton, NJ, Mar. 19–21, 2008, pp. 600–605.

[111] Z. Li, L. Liu, and C. Zhou, "Fast detection method in cooperative cognitive radio networks," *Int. J. Digit. Multimedia Broadcast.*, **2010**, Article ID 160462, 2010.

[112] Y.-C. Liang, Y. Zeng, E. C. Y. Peh, and A. T. Hoang, "'Sensing-throughput tradeoff for cognitive radio networks," *IEEE Trans. Wireless Commun.*, **7**, no. 4, 1326–1337, Apr. 2008.

[113] M. L. Littman, "Markov games as a framework for multi-agent reinforcement learning," *Proc. 11th Int. Conf. Machine Learning (ICML 1994)*, San Francisco, CA, 1994, pp. 157–163.

[114] H. Liu, K. Liu, and Q. Zhao, "Learning in a changing world: non-Bayesian restless multi-armed bandit," *Technical Report TR-10-03*, University of California, Davis, CA, 2010.

[115] K. Liu and Q. Zhao, "Indexability of restless bandit problems and optimality of Whittle index for dynamic multichannel access," *IEEE Trans. Inf. Theory*, **56**, no. 11, 5547–5567, Nov. 2010.

[116] K. Liu and Q. Zhao, "Distributed learning in multi-armed bandit with multiple players," *IEEE Trans. Signal Process.*, **58**, no. 11, 5667–5681, Nov. 2010.

[117] K. Liu, Q. Zhao, and B. Krishnamachari, "Dynamic multichannel access with imperfect channel state detection," *IEEE Trans. Signal Process.*, **58**, no. 5, 2795–2808, May 2010.

[118] K. J. R. Liu and B. Wang, *Cognitive Radio Networking and Security*, Cambridge University Press, 2011.

[119] G. Lorden, "2-SPRT's and the modified Kiefer-Weiss problem of minimizing an expected sample size," *Ann. Stat.*, **4**, no. 2, 281–291, 1976.

[120] Y. Lu, H. He, J. Wang, and S. Li, "Energy-efficient dynamic spectrum access using no-regret learning," *Proc. 7th Int. Conf. Information, Communications, and Signal Processing (ICICS 2009)*, Macau, China, Dec. 8–10, 2009.

[121] J. Lundén, *Spectrum Sensing for Cognitive Radio and Radar Systems*, D.Sc. Dissertation, Helsinki University of Technology, Espoo, Finland, 2009.

[122] J. Lundén, S. A. Kassam, and V. Koivunen, "Robust nonparametric cyclic correlation-based spectrum sensing for cognitive radio," *IEEE Trans. Signal Process.*, **58**, no. 1, 38–52, Jan. 2010.

[123] J. Lundén, V. Koivunen, A. Huttunen, and H. V. Poor, "Collaborative cyclostationary spectrum sensing for cognitive radio systems," *IEEE Trans. Signal Process.*, **57**, no. 11, 4182–4195, Nov. 2009.

[124] J. Lundén, V. Koivunen, S. R. Kulkarni, and H. V. Poor, "Reinforcement learning based distributed multiagent sensing policy for cognitive radio networks," *Proc. IEEE Int. Symp. New Frontiers in Dynamic Spectrum Access Networks (DySPAN 2011)*, Aachen, Germany, May 3–6, 2011.

[125] J. Lundén, V. Koivunen, S. R. Kulkarni, and H. V. Poor, "Exploiting spatial diversity in multiagent reinforcement learning based spectrum sensing," *Proc. 4th IEEE Int. Workshop Computational Advances in Multi-Sensor Adaptive Processing (CAMSAP 2011)*, San Juan, Puerto Rico, Dec. 13–16, 2011.

[126] L. Luo and S. Roy, "Analysis of search schemes in cognitive radio," *Proc. 2nd IEEE Workshop Networking Technologies for Software Defined Radio Networks*, San Diego, CA, USA, June 18–21, 2007, pp. 17–24.

[127] J. Ma, G. Zhao, and Y. Li, "Soft combination and detection for cooperative spectrum sensing in cognitive radio networks," *IEEE Trans. Wireless Commun.*, **7**, no. 11, 4502–4507, Nov. 2008.

[128] Z. Ma, W. Chen, K. B. Letaief, and Z. Cao, "A semi range-based iterative localization algorithm for cognitive radio networks," *IEEE Trans. Veh. Technol.*, **59**, no. 2, 704–717, Feb. 2010.

[129] A. B. MacKenzie and L. A. DaSilva, *Game Theory for Wireless Engineers*, Morgan & Claypool, 2006.

[130] A. Mahajan and D. Teneketzis, "Multi-armed bandit problems," in A. O. Hero III, D. Castañón, D. Cochran, and K. Kastella (eds.), *Foundations and Applications of Sensor Management*, pp. 121–151, Springer, 2008.

[131] Y. Mei, "Sequential change-point detection when unknown parameters are present in the pre-change distribution," *Ann. Stat.*, **34**, no. 1, 92–122, 2006.

[132] Y. Mei, "Asymptotic optimality theory for decentralized sequential hypothesis testing in sensor networks," *IEEE Trans. Inf. Theory*, **54**, no. 5, 2072–2089, 2008.

[133] Y. Mei, "Is average run length to false alarm always an informative criterion?," *Sequential Anal.*, **27**, no. 4, 354–376, Oct. 2008.

[134] U. Mitra and H. V. Poor, "Activity detection in a multi-user environment," *Wireless Pers. Commun. – Special Issue on Signal Separation and Cancellation for Personal, Indoor and Mobile Radio Communications*, **3**, Nos. 1–2, 149–174, Jan. 1996.

[135] A. Motamedi and A. Bahai, "Optimal channel selection for spectrum-agile low-power wireless packet switched networks in unlicensed band," *EURASIP J. Wireless Commun. Netw.*, **2008**, Article ID 896420, 2008.

[136] G. V. Moustakides, "Optimal stopping times for detecting changes in distributions," *Ann. Stat.*, **14**, no. 4, 1379–1387, 1986.

[137] G. V. Moustakides, A. S. Polunchenko, and A. G. Tartakovsky, "A numerical approach to performance analysis of quickest change-point detection procedures," *Statist. Sinica*, **21**, no. 2, 571–596, 2011.

[138] J. Niño-Mora, "Restless bandits, partial conservation laws and indexability," *Adv. Appl. Prob.*, **33**, no. 1, 76–98, 2001.

[139] J. Niño-Mora, "Dynamic priority allocation via restless bandit marginal productivity indices," *TOP*, **15**, no. 2, 161–198, 2007.

[140] J. Niño-Mora, "A $(2/3)n^3$ fast-pivoting algorithm for the Gittins index and optimal stopping of a Markov chain," *INFORMS J. Comput.*, **19**, no. 4, 596–606, 2007.

[141] J. Niño-Mora, "An index policy for dynamic fading-channel allocation to heterogeneous mobile users with partial observations," *Proc. 4th Euro-NGI Conf. Next Generation Internet Networks*, Kraków, Poland, Apr. 28–30, 2008.

[142] J. Niño-Mora, "A restless marginal productivity index for opportunistic spectrum access with sensing errors," in R. Núñez-Queija and J. Resing (eds.), *Network Control and Optimization, 3rd Euro-NF Conf., NET-COOP 2009 Eindhoven, The Netherlands, Nov. 23–25, 2009 Proceedings*, Lecture Notes in Computer Science, vol. 5894, Springer, 2009.

[143] J. Niño-Mora and S. S. Villar, "Multitarget tracking via restless bandit marginal productivity indices and Kalman filter in discrete time," *Proc. Joint 48th IEEE Conf. Decision and Control and 28th Chinese Control Conf.*, Shanghai, China, Dec. 16–18, 2009.

[144] D.-C. Oh, H.-C. Lee, and Y.-H. Lee, "Linear hard decision combining for cooperative spectrum sensing in cognitive radio systems," *Proc. 72nd IEEE Vehicular Technology Conf. (VTC 2010-Fall)*, Ottawa, ON, Canada, Sep. 6–9, 2010, pp. 1–5.

[145] J. Oksanen, V. Koivunen, J. Lundén, and A. Huttunen, "Diversity-based spectrum sensing policy for detecting primary signals over multiple frequency bands," *Proc. 35th IEEE Int. Conf. Acoustics, Speech, and Signal Processing (ICASSP 2010)*, Dallas, TX, Mar. 14–19, 2010.

[146] J. Oksanen, J. Lundén, and V. Koivunen, "Characterization of spatial diversity in cooperative spectrum sensing," *Proc. 4th Int. Symp. Communications, Control, and Signal Processing (ISCCSP 2010)*, Limassol, Cyprus, Mar. 3–5, 2010.

[147] J. Oksanen, J. Lundén, and V. Koivunen, "Reinforcement learning-based multiband sensing policy for cognitive radios," *Proc. 2nd Int. Workshop Cognitive Information Processing (CIP 2010)*, Elba Island, Tuscany, Italy, June 14–16, 2010.

[148] J. Oksanen, J. Lundén, and V. Koivunen, "Reinforcement learning method for energy efficient cooperative multiband spectrum sensing," *Proc. IEEE Int. Workshop Machine Learning for Signal Processing (MLSP 2010)*, Kittilä, Finland, Aug. 29–Sep. 1, 2010.

[149] J. Oksanen, J. Lundén, and V. Koivunen, "Reinforcement learning based sensing policy optimization for energy efficient cognitive radio networks," *Neurocomputing – Special Issue on Machine Learning for Signal Processing 2010*, **80**, 102–110, Mar. 2012.

[150] T. Oskiper and H. V. Poor, "Online activity detection in a multiuser environment using the matrix CUSUM algorithm," *IEEE Trans. Inf. Theory*, **46**, no. 2, 477–493, Feb. 2002.

[151] E. S. Page, "Continuous inspection schemes," *Biometrika*, **41**, no. 1–2, 100–115, 1954.

[152] D. G. Pandelis and D. Teneketzis, "On the optimality of the Gittins index rule for multi-armed bandits with multiple plays," *Math. Meth. Oper. Res.*, **50**, no. 3, 449–461, 1999.

[153] S. Pandey, D. Chakrabarti, and D. Agarwal, "Multi-armed bandit problems with dependent arms," *Proc. 24th Int. Conf. Machine Learning (ICML 2007)*, Corvallis, OR, USA, June 20–24, 2007.

[154] J.-S. Pang, G. Scutari, and D. P. Palomar, "Design of cognitive radio systems under temperature-interference constraints: a variational inequality approach," *IEEE Trans. Signal Process.*, **58**, no. 6, 3251–3271, June 2010.

[155] E. C. Y. Peh, Y.-C. Liang, Y. L. Guan, and Y. Zeng, "Optimization of cooperative sensing in cognitive radio networks: a sensing-throughput tradeoff view," *IEEE Trans. Veh. Technol.*, **58**, no. 9, 5294–5299, Nov. 2009.

[156] B. Peleg and P. Sudhölter, *Introduction to the Theory of Cooperative Games* (2nd edn.). Springer-Verlag, 2007.

[157] Q. Peng, K. Zeng, J. Wang, and S. Li, "A distributed spectrum sensing scheme based on credibility and evidence theory in cognitive radio context," *Proc. 17th IEEE Int. Symp. Personal, Indoor, and Mobile Radio Communications (PIMRC 2006)*, Helsinki, Finland, Sep. 11–14, 2006, pp. 1–5.

[158] M. Pollak, "A robust changepoint detection method," *Sequential Analysis*, **29**, no. 2, 146–161, Apr. 2010.

[159] M. Pollak and A. G. Tartakovsky, "Optimality properties of the Shiryaev-Roberts procedure," *Statist. Sinica*, **19**, 1729–1739, 2009.

[160] A. S. Polunchenko and A. G. Tartakovsky, "On optimality of the Shiryaev-Roberts procedure for detecting a change in distribution," *Ann. Stat.*, **38**, no. 6, 3445–3457, 2010.

[161] H. V. Poor, *An Introduction to Signal Detection and Estimation* (2nd edn.). Springer-Verlag, 1994.

[162] H. V. Poor and O. Hadjiliadis, *Quickest Detection*, Cambridge University Press, 2009.

[163] Z. Quan, S. Cui, H. V. Poor, and A. H. Sayed, "Collaborative wideband sensing for cognitive radios," *IEEE Signal Process. Mag. – Special Issue on Signal Processing for Cognitive Radio Networks*, **25**, no. 6, 60–73, Nov. 2008.

[164] Z. Quan, S. Cui, and A. H. Sayed, "Optimal linear cooperation for spectrum sensing in cognitive radio networks," *IEEE J. Sel. Topics Signal Process.*, **2**, no. 4, 2431–2436, Apr. 2010.

[165] Z. Quan, S. Cui, A. H. Sayed, and H. V. Poor, "Optimal multiband joint detection for spectrum sensing in cognitive radio networks," *IEEE Trans. Signal Process.*, **57**, no. 3, 1128–1140, Mar. 2009.

[166] C. Rago, P. Willett, and Y. Bar-Shalom, "Censoring sensors: a low-communication-rate scheme for distributed detection," *IEEE Trans. Aerosp. Electron. Syst.*, **32**, no. 2, 554–568, Apr. 1996.

[167] J. Rajasekharan, J. Eriksson, and V. Koivunen, "Cooperative game-theoretic solutions to spectrum sharing in cognitive radios," *Proc. 44th Asilomar Conf. Signals, Systems, and Computers*, Pacific Grove, CA, Nov. 7–10, 2010.

[168] S. W. Roberts, "A comparison of some control chart procedures," *Technometrics*, **8**, no. 3, 411–430, Aug. 1966.

[169] G. A. Rummery and M. Niranjan, "On-line Q-learning using connectionist systems," *Technical Report CUED/F-INFENG/TR 166*, Cambridge University Engineering Department, Cambridge, UK, Sep. 1994.

[170] W. Saad, Z. Han, T. Başar, M. Debbah, and A. Hjørungnes, "Coalition formation games for collaborative spectrum sensing," *IEEE Trans. Veh. Technol.*, **60**, no. 1, 276–297, Jan. 2011.

[171] W. Saad, Z. Han, M. Debbah, A. Hjørungnes, and T. Başar, "Coalitional game theory for communication networks: a tutorial," *IEEE Signal Process. Mag.*, **26**, no. 5, 77–97, Sep. 2009.

[172] M. Sanna and M. Murroni, "Optimization of non-convex multiband cooperative sensing with genetic algorithms," *IEEE J. Sel. Topics Signal Process.*, **5**, no. 1, 87–96, Feb. 2011.

[173] S. Sardellitti, M. Giona, and S. Barbarossa, "Fast distributed average consensus algorithms based on advection-diffusion processes," *IEEE Trans. Signal Process.*, **58**, no. 2, 826–842, Feb. 2010.

[174] G. Scutari and D. P. Palomar, "MIMO cognitive radio: a game theoretical approach," *IEEE Trans. Signal Process.*, **58**, no. 2, 761–780, Feb. 2010.

[175] Y. Shei and Y. T. Su, "A sequential test based cooperative spectrum sensing scheme for cognitive radios," *Proc. 19th Int. Symp. Personal, Indoor, and Mobile Radio Communications (PIMRC 2008)*, Cannes, France, Sep. 15–18, 2008.

[176] S. Shetty, M. Song, C. Xin, and E. K. Park, "A learning-based multiuser opportunistic spectrum access approach in unslotted primary networks," *Proc. IEEE Conf. Computer Communications (INFOCOM 2009)*, Rio de Janeiro, Brazil, Apr. 19–25, 2009.

[177] A. N. Shiryaev, "On optimum methods in quickest detection problems," *Theory Probab. Appl.*, **8**, no. 1, 22–46, Jan. 1963.

[178] A. N. Shiryaev, "Quickest detection problems: fifty years later," *Sequential Anal.*, **29**, no. 4, 345–385, Oct. 2010.

[179] Y. Shoham and K. Leyton-Brown, *Multiagent Systems: Algorithmic, Game-Theoretic, and Logical Foundations*. Cambridge University Press, 2009.

[180] D. Siegmund, *Sequential Analysis*, Springer-Verlag, 1985.

[181] S. P. Singh, T. Jaakkola, M. L. Littman, and C. Szepesvári, "Convergence results for single-step on-policy reinforcement-learning algorithms," *Mach. Learn.*, **39**, 287–308, 2000.

[182] I. M. Sonin, "A generalized Gittins index for a Markov chain and its recursive calculation," *Technical Report*, Department of Mathematics, University of North Carolina at Charlotte, Charlotte, NC, 2005.

[183] R. S. Sutton and A. G. Barto, *Reinforcement Learning: An Introduction*, MIT Press, 1998.

[184] C. Szepesvári, *Algorithms for Reinforcement Learning*, Synthesis Lectures on Artificial Intelligence and Machine Learning, Morgan & Claypool, 2010.

[185] A. Taherpour, Y. Norouzi, M. Nasiri-Kenari, A. Jamshidi, and Z. Zeinalpour-Yazdi, "Asymptotically optimum detection of primary user in cognitive radio networks," *IET Commun.*, **1**, no. 6, 1138–1145, Dec. 2007.

[186] S. Tantaratana and H. V. Poor, "Asymptotic efficiencies of truncated sequential tests," *IEEE Trans. Inf. Theory*, **28**, no. 6, 911–923, Nov. 1982.

[187] G. Taricco, "Optimization of linear cooperative spectrum sensing for cognitive radio networks," *IEEE J. Sel. Topics Signal Process.*, **5**, no. 1, 77–86, Nov. 2009.

[188] A. G. Tartakovsky and G. V. Moustakides, "State-of-the-art in Bayesian changepoint detection," *Sequential Anal.*, **29**, no. 2, 125–145, Apr. 2009.

[189] A. G. Tartakovsky, M. Pollak, and A. S. Polunchenko, "Third-order asymptotic optimality of the generalized Shiryaev-Roberts changepoint detection procedures," *Teor. Veroyatnost. i Primenen.*, **56**, no. 3, 534–565, 2011, and *Theory Probab. Appl.*, to be published.

[190] A. G. Tartakovsky and A. S. Polunchenko, "Quickest changepoint detection in distributed multi-sensor systems under unknown parameters," *Proc. 11th Int. Conf. Information Fusion*, Cologne, Germany, June 30–July 3, 2008.

[191] A. G. Tartakovsky and V. V. Veeravalli, "Change-point detection in multichannel and distributed systems with applications," in N. Mukhopadhyay, S. Datta, and S. Chattopadhyay (eds.), *Applications of Sequential Methodologies*, pp. 331–363, Marcel Dekker, Inc., 2004.

[192] A. G. Tartakovsky and V. V. Veeravalli, "Asymptotically optimal quickest change detection in distributed sensor systems," *Sequential Anal.*, **27**, no. 4, 441–475, 2008.

[193] C. Tekin and M. Liu, "Online learning in opportunistic spectrum access: a restless bandit approach," *Proc. 30th IEEE Int. Conf. Computer Communications*, Shanghai, China, Apr. 10–15, 2011.

[194] S. C. A. Thomopoulos, R. Viswanathan, and D. K. Bougoulias, "Optimal distributed decision fusion," *IEEE Trans. Aerosp. Electron. Syst.*, **25**, no. 5, 761–765, Sep. 1989.

[195] M. Tokic, "Adaptive ϵ-greedy exploration in reinforcement learning based on value differences," in R. Dillmann, J. Beyerer, U. D. Hanebeck, and T. Schultz (eds.), *KI 2010: Advances in Artificial Intelligence*, Lecture Notes in Computer Science, vol. 6359, pp. 203–210, Springer, 2010.

[196] J. N. Tsitsiklis, "Asynchronous stochastic approximation and Q-learning," *Mach. Learn.*, **16**, no. 3, 185–202, Sep. 1994.

[197] J. Unnikrishnan and V. V. Veeravalli, "Algorithms for dynamic spectrum access with learning for cognitive radio," *IEEE Trans. Signal Process.*, **58**, no. 2, 750–760, Feb. 2010.

[198] J. Unnikrishnan, V. V. Veeravalli, and S. Meyn, "Minimax robust quickest change detection," *IEEE Trans. Inf. Theory*, **57**, no. 3, 1604–1614, Mar. 2011.

[199] P. P. Varaiya, J. C. Walrand, and C. Buyukkoc, "Extensions of the multiarmed bandit problem: the discounted case," *IEEE Trans. Autom. Control*, **AC–30**, no. 5, 426–439, May 1985.

[200] P. K. Varshney, *Distributed Detection and Data Fusion*, Springer-Verlag, 1996.

[201] V. V. Veeravalli, "Decentralized quickest change detection," *IEEE Trans. Inf. Theory*, **47**, no. 4, 1657–1665, May 2001.

[202] V. V. Veeravalli, T. Basar, and H. V. Poor, "Decentralized sequential detection with a fusion center performing the sequential test," *IEEE Trans. Inf. Theory*, **39**, no. 2, 433–442, Mar. 1993.

[203] V. V. Veeravalli and C. W. Baum, "Asymptotic efficiency of a sequential multihypothesis test," *IEEE Trans. Inf. Theory*, **41**, no. 6, 1994–1997, Nov. 1995.

[204] J. Vermorel and M. Mohri, "Multi-armed bandit algorithm and empirical evaluation," in J. Gama, R. Camacho, P. Brazdil, A. Jorge, and L. Torgo (eds.), *Machine Learning: ECML 2005*, Lecture Notes in Computer Science, **3720/2005**, 437–448, Springer-Verlag 2005.

[205] W. Vickrey, "Counterspeculation, auctions, and competitive sealed tenders," *J. Financ.*, **16**, no. 1, 8–37, Mar. 1961.

[206] R. Viswanathan and P. K. Varshney, "Distributed detection with multiple sensors: Part I – Fundamentals," *Proc. IEEE*, **85**, no. 1, 54–63, Jan. 1997.

[207] A. Wald, "Sequential tests of statistical hypotheses," *Ann. Math. Stat.*, **16**, no. 2, 117–186, June 1945.

[208] A. Wald and J. Wolfowitz, "Optimum character of the sequential probability ratio test," *Ann. Math. Stat.*, **19**, no. 3, 326–339, 1948.

[209] B. Wang, Y. Wu, and K. J. R. Liu, "Game theory for cognitive radio networks: an overview," *Comput. Netw.*, **54**, no. 14, 2537–2561, Oct. 2010.

[210] C.-C. Wang, S. R. Kulkarni, and H. V. Poor, "Bandit problems with side observations," *IEEE Trans. Autom. Control*, **50**, no. 3, 338–355, Mar. 2005.

[211] C.-C. Wang, S. R. Kulkarni, and H. V. Poor, "Arbitrary side observations in bandit problems," *Adv. Appl. Math. – Special Issue Dedicated to Dr. David P. Robbins*, **34**, no. 4, 903–938, May 2005.

[212] H. Wang, G. Noh, D. Kim, S. Kim, and D. Hong, "Advanced sensing techniques of energy detection in cognitive radios," *J. Commun. Netw.*, **12**, no. 1, 19–29, Feb 2010.

[213] J. Wang, G. Scutari, and D. P. Palomar, "Robust MIMO cognitive radio via game theory," *IEEE Trans. Signal Process.*, **59**, no. 3, 1183–1201, Mar. 2011.

[214] K. Wang and L. Chen, "On optimality of myopic policy for restless multi-armed bandit problem: an axiomatic approach," *IEEE Trans. Signal Process.*, **60**, no. 1, 300–309, Jan. 2012.

[215] C. J. C. H. Watkins, *Learning from Delayed Rewards*, Ph.D. Dissertation, University of Cambridge, Cambridge, UK, 1989.

[216] C. J. C. H. Watkins and P. Dayan, "Q-learning," *Mach. Learn.*, **8**, 279–292, 1992.

[217] R. R. Weber and G. Weiss, "On an index policy for restless bandits," *J. Appl. Prob.*, **27**, no. 3, 637–648, Sep. 1990.

[218] R. R. Weber and G. Weiss, "Addendum to 'On an index policy for restless bandits'," *Adv. Appl. Prob.*, **23**, no. 2, 429–430, June 1991.

[219] P. Whittle, "Arm-acquiring bandits," *Ann. Probab.*, **9**, no. 2, 284–292, Apr. 1981.

[220] P. Whittle, "Restless bandits: activity allocation in a changing world," *J. Appl. Prob.*, **25**, 287–298, 1988.

[221] B. Wild and K. Ramchandran, "Detecting primary receivers for cognitive radio applications," *Proc. 1st IEEE Int. Symp. New Frontiers in Dynamic Spectrum Access Networks (DySPAN 2005)*, Baltimore, MD, USA, Nov. 8–11, 2005, pp. 124–130.

[222] L. A. Wolsey, *Integer Programming*, Wiley, 1998.

[223] C. Wu, K. Chowdhury, M. Di Felice, and W. Meleis, "Spectrum management of cognitive radio using multi-agent reinforcement learning," *Proc. 9th Int. Conf. Autonomous Agents and Multiagent Systems (AAMAS 2010)*, Toronto, Canada, May 10–14, 2010.

[224] Y. Xin, H. Zhang, and S. Rangarajan, "SSCT: a simple sequential spectrum sensing scheme for cognitive radio," *Proc. IEEE Global Communications Conf. (GLOBECOM 2009)*, Honolulu, HI, Nov. 30–Dec. 4, 2009.

[225] K. Yamazaki, *Essays on Sequential Analysis: Multi-Armed Bandit with Availability Constraints and Sequential Change Detection and Identification*, Ph.D. Dissertation, Princeton University, Princeton, NJ, Apr. 2009.

[226] E. Yang and D. Gu, "Multiagent reinforcement learning for multi-robot systems: a survey," *Technical Report CSM-404*, Department of Computer Science, University of Essex, UK, 2004.

[227] K.-L. A. Yau, P. Komisarczuk, and P. D. Teal, "A context-aware and intelligent dynamic channel selection scheme for cognitive radio networks," *Proc. Conf. Cognitive Radio Oriented Wireless Networks and Communications (CROWNCOM 2009)*, Hanover, Germany, June 2009.

[228] K.-L. A. Yau, P. Komisarczuk, and P. D. Teal, "Context awareness and intelligence in distributed cognitive radio networks: a reinforcement learning approach," *Proc. 11th Australian Communications Theory Workshop (AusCTW 2010)*, Canberra, Australia, Feb. 2010.

[229] K.-L. A. Yau, P. Komisarczuk, and P. D. Teal, "Achieving efficient and optimal joint action in distributed cognitive radio networks using payoff propagation," *Proc. IEEE Int. Conf. Communications (ICC 2010)*, Cape Town, South Africa, May 23–27, 2010.
[230] W. Yuan, H. Leung, S. Chen, and W. Cheng, "A distributed sensor selection mechanism for cooperative spectrum sensing," *IEEE Trans. Signal Process.*, **59**, no. 12, 6033–6044, Dec. 2011.
[231] L. Zacharias and R. Sundaresan, "Decentralized sequential change detection using physical layer fusion," *IEEE Trans. Wireless Commun.*, **7**, no. 12, 4999–5008, Dec. 2008.
[232] Y. Zeng, Y. C. Liang, E. C. Y. Peh, and A. T. Hoang, "Cooperative covariance and eigenvalue based detections for robust sensing," *Proc. IEEE Global Communications Conf. (GLOBECOM 2009)*, Honolulu, HI, Nov. 30–Dec. 4, 2009, pp. 1–6.
[233] Q. Zhao, B. Krishnamachari, and K. Liu, "On myopic sensing for multi-channel opportunistic access: structure, optimality, and performance," *IEEE Trans. Wireless Commun.*, **7**, no. 12, 5431–5440, Dec. 2008.
[234] Q. Zhao, L. Tong, A. Swami, and Y. Chen, "Decentralized cognitive MAC for opportunistic spectrum access in ad hoc networks: a POMDP framework," *IEEE J. Sel. Areas Commun.*, **25**, no. 3, 589–600, Apr. 2007.
[235] Q. Zhao and J. Ye, "Quickest detection in multiple on-off processes," *IEEE Trans. Signal Process.*, **58**, no. 12, 5994–6006, Dec. 2010.
[236] K. Zheng, H. Li, S. M. Djouadi, and J. Wang, "Spectrum sensing in low SNR regime via stochastic resonance," *Proc. 44th Annu. Conf. Information Sciences and Systems (CISS 2010)*, Princeton, NJ, Mar. 17–19, 2010.
[237] F. Zhu and S.-W. Seo, "Enhanced robust cooperative spectrum sensing in cognitive radio," *J. Commun. Netw.*, **11**, no. 2, 122–133, Apr. 2009.
[238] Q. Zou, S. Zheng, and A. H. Sayed, "Cooperative sensing via sequential detection," *IEEE Trans. Signal Process.*, **58**, no. 12, 6266–6283, Dec. 2010.

Bibliography

M. Abramowitz and I. A. Stegun (eds.), *Handbook of Mathematical Functions*, Dover Publ. Inc., 1972.

J. Acharya, H. Viswanathan, and S. Venkatesan, "Timing acquisition for non-contiguous OFDM based dynamic spectrum access," *IEEE International Symposium on New Frontiers in Dynamic Spectrum Access Networks, 2008*, Oct. 2008, pp. 1–10.

J. Acharya and R. Yates, "Dynamic spectrum allocation for uplink users with heterogeneous utilities," *IEEE Trans. Wireless Commun.*, **8**, no. 3, 1405–1413, Mar. 2009.

P. Agrawal and N. Patwari, "Correlated link shadow fading in multi-hop wireless networks," *IEEE Trans. Wireless Commun.*, **8**, no. 8, 4024–4036, Aug. 2009.

R. Agrawal, M. V. Hedge, and D. Teneketzis, "Asymptotically efficient adaptive allocation rules for the multiarmed bandit problem with switching cost," *IEEE Trans. Autom. Control*, **33**, no. 10, 899–906, Oct. 1988.

R. Agrawal, M. V. Hedge, and D. Teneketzis, "Multi-armed bandit problems with multiple plays and switching cost," *Stochast. Stochast. Rep.*, **29**, 437–459, 1990.

S. Ahmad and M. Liu, "Multi-channel opportunistic access: a case of restless bandits with multiple plays," *Proc. Allerton Conf. Communication, Control, and Computing*, Monticello, IL, Oct. 2009, pp. 1361–1368.

S. H. Ahmad, M. Liu, T. Javidi, Q. Zhao, and B. Krishnamachari, "Optimality of myopic sensing in multi-channel opportunistic access," *IEEE Trans. Inf. Theory*, **55**, no. 9, 4040–4050, Sep. 2009.

A. Algans, K. I. Pedersen, and P. E. Mogensen, "Experimental analysis of the joint statistical properties of azimuth spread, delay spread, and shadow fading," *IEEE J. Sel. Areas Commun.*, **20**, no. 3, 523–531, Apr. 2002.

K. Amiri, Y. Sun, P. Murphy, C. Hunter, J. R. Cavallaro, and A. Sabharwal, "WARP, a modular testbed for configurable wireless network research at Rice," *Proc. IEEE SWRIF*, 2007.

K. Amiri, Y. Sun, P. Murphy, C. Hunter, J. Cavallaro, and A. Sabharwal, "WARP, a unified wireless network testbed for education and research," in *IEEE International Conference on Microelectronic Systems Education*, June 2007, pp. 53–54.

A. Anandkumar, N. Michael, and A. K. Tang, "Opportunistic spectrum access with multiple users: learning under competition," *Proc. IEEE Conf. Computer Communications (INFOCOM 2010)*, San Diego, CA, USA, Mar. 15–19, 2010.

A. Anandkumar, N. Michael, A. K. Tang, and A. Swami, "Distributed algorithms for learning and cognitive medium access with logarithmic regret," *IEEE J. Sel. Areas in Commun.*, **29**, no. 4, 731–745, Apr. 2011.

V. Anantharam, P. Varaiya, and J. Walrand, "Asymptotically efficient allocation rules for the multiarmed bandit problem with multiple plays – part I: i.i.d. rewards," *IEEE Trans. Autom. Control*, **AC–32**, no. 11, 968–977, Nov. 1987.

V. Anantharam, P. Varaiya, and J. Walrand, "Asymptotically efficient allocation rules for the multiarmed bandit problem with multiple plays – part II: Markovian rewards," *IEEE Trans. Autom. Control*, **AC–32**, no. 11, 977–982, Nov. 1987.

J. Andrews, A. Ghosh, and R. Muhamed, *Fundamentals of WiMAX*, Prentice-Hall, 2007.

V. S. Annapureddy and V. Veeravalli, "Gaussian interference networks: sum capacity in the low interference regime and new outer bounds on the capacity region," *IEEE Trans. Inf. Theory*, **55**, no. 7, 3032–3050, July 2009.

"Annex 12 to working party 5A chairman's report," Radiocommunication Study Groups, ITU, Tech. Rep., 2010.

M. R. Aref, *Information Flow in Relay Networks*, Ph.D Dissertation, Stanford University, Stanford, CA, 1980.

E. Arıkan, "Some complexity results about packet radio networks," *IEEE Trans. Inf. Theory*, **30**, no. 4, 681–685, July 1984.

A. A. Arowojolu, A. M. D. Turkmani, and J. D. Parsons, "Time dispersion measurements in urban microcellular environments," *Proc. IEEE Veh. Technol. Conf.*, pp. 150–154, 1994.

M. Asawa and D. Teneketzis, "Multi-armed bandits with switching penalties," *IEEE Trans. Autom. Control*, **41**, no. 3, 328–348, Mar. 1996.

H. Asplund, A. A. Glazunov, A. F. Molisch, K. I. Pedersen, and M. Steinbauer, "The COST259 directional channel model – Part II: Macrocells," *IEEE Trans. Wireless Commun.*, **5**, no. 12, 3434–3450, Dec. 2006.

J.-Y. Audibert, R. Munos, and C. Szepesvári, "Exploration–exploitation tradeoff using variance estimates in multi-armed bandits," *Theor. Comput. Sci.*, **410**, no. 19, 1876–1902, Apr. 2009.

P. Auer, N. Cesa-Bianchi, and P. Fischer, "Finite-time analysis of the multiarmed bandit problem," *Mach. Learn.*, **47**, no. 2–3, 235–256, 2002.

P. Auer and R. Ortner, "UCB revisited: improved regret bounds for the stochastic multi-armed bandit problems," *Period. Math. Hungar.*, **61**, no. 1–2, 55–65, 2010.

E. Axell and E. G. Larsson, "Optimal and sub-optimal spectrum sensing of OFDM signals in known and unknown noise variance," *IEEE J. Sel. Areas Commun.*, **29**, no. 2, 290–304, Feb. 2011.

C. Bae and W. E. Stark, "End-to-end energy/bandwidth tradeoff in multihop wireless networks," *IEEE Trans. Inf. Theory*, **55**, no. 9, 4051–4066, Sep. 2009.

P. Bahl, R. Chandra, T. Moscibroda, R. Murty, and M. Welsh, "White space networking with WiFi like connectivity," *ACM Sigcomm*, 2009.

V. Bahl, "The Promises and Challenges of the Wireless Frontier – from 600 MHz to 60 GHz," *IEEE DySPAN 2011*, Plenary lecture, May 2011.

T. Banerjee, V. Sharma, V. Kavitha, and A. K. Jayaprakasam, "Generalized analysis of a distributed energy efficient algorithm for change detection," *IEEE Trans. Wireless Commun.*, **10**, no. 1, 91–101, Jan. 2011.

J. Banks and R. Sundaram, "Switching costs and the Gittins index," *Econometrica*, **62**, no. 3, 687–694, May 1994.

M. Basseville and I. V. Nikiforov, *Detection of Abrupt Changes – Theory and Application*, Prentice-Hall, 1993.

D. S. Baum, H. El-Sallabi, T. Jämsä, *et al.*, "Final report on link level and system level channel models," WINNER, IST-2003-507581, D5.4, 2005.

D. S. Baum, D. Gore, R. Nabar, *et al.*, "Measurements and characterization of broadband MIMO fixed wireless channels at 2.5 GHz," *Proc. ICPWC*, Dec. 2000.

BBN Technologies, "Next generation (XG) architecture and protocol development (XAP)," Tech. Rep., Aug. 2005. www.dtic.mil/cgi-bin/GetTRDoc?Location=U2&doc=GetTRDoc.pdf&AD=ADA437096.

BBN Technologies, "The XG Vision–RFC," Tech. Rep. www.ir.bbn.com/~ramanath/pdf/rfc_vision.pdf.

R. Bellman, *Dynamic Programming*, Princeton University Press, 2010.

P. Bello, "Characterization of randomly time-variant linear channels," *IEEE Trans. Commun.*, **11**, no. 4, 360–393, Dec. 1963.

S. Benedetto and E. Biglieri, *Principles of Digital Transmission With Wireless Applications*. Kluwer Academic /Plenum Publishers, 1999.

J. E. Berg, J. Ruprecht, J. P. de Weck, and A. Mattsson, "Specular reflections from high-rise buildings in 900 MHz cellular systems," *Proc. IEEE Veh. Technol. Conf.*, pp. 594–599, 1991.

C. Bergljung and P. Karlsson, "Propagation characteristics for indoor broadband radio access networks in the 5 GHz band," *Proc. IEEE International Symposium on Personal, Indoor and Mobile Radio Communications*, pp. 612–616, 1998.

D. A. Berry and B. Fristedt, *Bandit Problems: Sequential Allocation of Experiments*, Chapman and Hall, 1985.

U. Berthold, F. Fu, M. van der Schaar, and F. K. Jondral, "Detection of spectral resources in cognitive radios using reinforcement learning," *Proc. 3rd IEEE Int. Symp. New Frontiers in Dynamic Spectrum Access Networks (DySPAN 2008)*, Chicago, IL, Oct. 14–17, 2008.

D. P. Bertsekas, *Dynamic Programming and Optimal Control* (3rd edn.), Vols. 1 and 2, Athena Scientific, 2007.

D. Bertsimas and J. Niño-Mora, "Conservations laws, extended polymatroids and multiarmed bandit problems; a polyhedral approach to indexable systems," *Math. Oper. Res.*, **21**, no. 2, 257–306, May 1996.

E. Biglieri, R. Calderbank, A. Constantinides, A. Goldsmith, A. Paulraj, and H. V. Poor, *MIMO Wireless Communications*, Cambridge University Press, 2007.

E. Biglieri, J. Proakis, and S. Shamai (Shitz), "Fading channels: information theoretic and communication aspects," *IEEE Trans. Inf. Theory*, **44**, no. 6, 2619–2692, Oct. 1998.

S. Biswas and R. Morris, "Opportunistic routing in multi-hop wireless networks," *Proc. ACM SIGCOMM*, 2004.

R. S. Blum, S. A. Kassam, and H. V. Poor, "Distributed detection with multiple sensors: Part II – advanced topics," *Proc. IEEE*, **85**, no. 1, 64–79, Jan. 1997.

R. S. Blum and B. M. Sadler, "Energy efficient signal detection in sensor networks using ordered transmissions," *IEEE Trans. Signal Process.*, **56**, no. 7, 3229–3235, July 2008.

H. Bölcskei, R. U. Nabar, O. Oyman, and A. J. Paulraj, "Capacity scaling laws in MIMO relay networks," *IEEE Trans. Wireless Commun.*, **5**, no. 6, 1433–1444, June 2006.

M. Bowling and M. Veloso, "Multiagent learning using a variable learning rate," *Artif. Intell.*, **136**, no. 2, 215–250, Apr. 2002.

R. Branzei, D. Dimitrov, and S. Tijs, *Models in Cooperative Game Theory*, 2nd edition, Springer-Verlag, 2008.

V. Brik, E. Rozner, S. Banerjee, and P. Bahl, "DSAP: a protocol for coordinated spectrum access," *IEEE International Symposium on New Frontiers in Dynamic Spectrum Access Networks, 2005*, Nov. 2005, pp. 611–614.

J. Brito, "The spectrum commons in theory and practice," Stanford Technology Law Review, Discussion Papers, 2007. `stlr.stanford.edu/pdf/brito-commons.pdf`.

R. Brodersen, A. Wolisz, D. Cabric, S. M. Mishra, and D. Willkomm, "CORVUS: a cognitive radio approach for usage of virtual unlicensed spectrum," *White Paper: Berkeley Wireless Research Center*. `bwrc.eecs.berkeley.edu/research/mcma/CR_White_paper_final1.pdf`.

B. E. Brodsky and B. S. Darkhovsky, *Nonparametric Methods in Change-Point Problems*. Kluwer Academic Publishers, 1993.

B. E. Brodsky and B. S. Darkhovsky, "Minimax sequential tests for many composite hypothesis. I," *Theory Probab. Appl.*, **52**, no. 4, 565–579, 2008.

B. E. Brodsky and B. S. Darkhovsky, "Minimax sequential tests for many composite hypothesis. II," *Theory Probab. Appl.*, **53**, no. 1, 1–12, 2009.

I. Broustis, J. Eriksson, S. Krishnamurthy, and M. Faloutsos, "A blueprint for a manageable and affordable wireless testbed: design, pitfalls and lessons learned," *3rd International Conference on Testbeds and Research Infrastructure for the Development of Networks and Communities, TridentCom 2007*, May 2007, pp. 1–6.

T. X. Brown, "An analysis of unlicensed device operation in licensed broadcast service bands," in *Proc. 1st IEEE Int. Symp. New Frontiers in Dynamic Spectrum Access Networks (DySPAN 2005)*, Baltimore, MD, Nov. 8–11, 2005, pp. 11–29.

T. Brown and A. Sethi, "Potential cognitive radio denial-of-service vulnerabilities and protection countermeasures: a multi-dimensional analysis and assessment," *2nd International Conference on Cognitive Radio Oriented Wireless Networks and Communications, CrownCom.*, Aug. 2007, pp. 456–464.

M. Buddhikot, "Understanding dynamic spectrum access: models, taxonomy and challenges," *2nd IEEE International Symposium on New Frontiers in Dynamic Spectrum Access Networks, DySPAN*, Apr. 2007.

M. Buddhikot, P. Kolodzy, S. Miller, K. Ryan, and J. Evans, "DIMSUMNet: new directions in wireless networking using coordinated dynamic spectrum access," *IEEE WoWMoM*, June 2005.

T. N. Bui, V. Krishnamurthy, and H. V. Poor, "Online Bayesian activity detection in DS/CDMA network," *IEEE Trans. Signal Process.*, **53**, no. 1, 371–375, Jan. 2005.

L. Buşoniu, R. Babuška, and B. De Schutter, "A comprehensive survey of multiagent reinforcement learning," *IEEE Trans. Syst., Man, Cybern. C, Appl. Rev.*, **38**, no. 2, 156–172, Mar. 2008.

L. Buşoniu, R. Babuška, B. De Schutter, and D. Ernst, *Reinforcement Learning and Dynamic Programming Using Function Approximators*, CRC Press, 2010.

V. R. Cadambe and S. A. Jafar, "Interference alignment and degrees of freedom of the K-user interference channel," *IEEE Trans. Inf. Theory*, **54**, no. 8, 3425–3441, Aug. 2008.

V. R. Cadambe and S. A. Jafar, "Parallel Gaussian interference channels are not always separable," *IEEE Trans. Inf. Theory*, **55**, no. 9, 3983–3990, Sep. 2009.

G. Calcev, D. Chizhik, B. Goransson, *et al.*, "A Wideband Spatial Channel Model for System-Wide Simulations," *IEEE Trans. Veh. Technol.*, **56**, no. 2, 389–403, Mar. 2007.

Y. Cao and B. Chen, "Interference channel with one cognitive transmitter," *Proc. of the Asilomar Conference on Signals, Systems, and Computers*, Pacific Grove, CA, Oct. 26–29, 2008.

A. B. Carleial, "A case where interference does not reduce capacity," *IEEE Trans. Inf. Theory*, **21**, no. 5, 569–570, Sep. 1975.

A. B. Carleial, "Interference channels," *IEEE Trans. Inf. Theory*, **24**, no. 1, 60–70, Jan. 1978.

D. Cassioli, M. Z. Win, and A. Molisch, "The ultra-wide bandwidth indoor channel: from statistical model to simulations", *IEEE J. Sel. Areas Commun.*, **20**, no. 6, 1247–1257, Aug. 2002.

H. Celebi and H. Arslan, "Adaptive positioning systems for cognitive radios," *Proc. 2nd IEEE Int. Symp. New Frontiers in Dynamic Spectrum Access Networks (DySPAN 2007)*, Dublin, Ireland, Apr. 17–20, 2007, pp. 78–84.

H. Celebi and H. Arslan, "Cognitive positioning systems," *IEEE Trans. Wireless Commun.*, **6**, no. 12, 4475–4483, Dec. 2007.

S. Chaudhari and V. Koivunen, "Effect of quantization and channel errors on collaborative spectrum sensing," *Proc. 43rd Asilomar Conf. Signals, Systems, and Computers*, Pacific Grove, CA, Nov. 1–4, 2009, pp. 528–533.

S. Chaudhari, V. Koivunen, and H. V. Poor, "Autocorrelation-based decentralized sequential detection of OFDM signals in cognitive radios," *IEEE Trans. Signal Process.*, **57**, no. 7, 2690–2700, July 2009.

S. Chaudhari, J. Lundén, and V. Koivunen, "BEP walls for collaborative spectrum sensing," *Proc. 36th IEEE Int. Conf. Acoustics, Speech, and Signal Processing (ICASSP 2011)*, Prague, Czech Republic, May 22–27, 2011.

S. Chaudhari, J. Lundén, and V. Koivunen, "Performance limitations for cooperative spectrum sensing with reporting channel errors," *Proc. 22nd IEEE Int. Symp. Personal, Indoor, and Mobile Radio Communications (PIMRC 2011)*, Toronto, Canada, Sep. 11–14, 2011.

S. Chaudhari, J. Lundén, V. Koivunen, and H. V. Poor, "Cooperative sensing with imperfect reporting channels: Hard decisions or soft decisions?," *IEEE Trans. Signal Process.*, **60**, no. 1, 18–28, Jan. 2012.

B. Chen and H. Wang, "Maximum likelihood estimation of OFDM carrier frequency offset," in *IEEE International Conference on Communications, 2002. ICC 2002*, **1**, May 2002, 49 –53.

B. Chen and P. K. Willett, "On the optimality of the likelihood-ratio test for local sensor decision rules in the presence of nonideal channels," *IEEE Trans. Inf. Theory*, **51**, no. 2, 693–699, Feb. 2005.

H. Chen, B. Chen, and P. K. Varshney, "Further results on the optimality of the likelihood-ratio test for local sensor decision rules in the presence of nonideal channels," *IEEE Trans. Inf. Theory*, **55**, no. 2, 828–832, Feb. 2009.

M. Chen and H. Asplund, "Measurements and models for direction of arrival of radio waves in LOS in urban microcells," *Proc. IEEE International Symposium on Personal, Indoor and Mobile Radio Communications*, pp. 100–104, 2001.

R. Chen, J.-M. Park, and K. Bian, "Robust distributed spectrum sensing in cognitive radio networks," *Proc. 27th IEEE Conf. Computer Communications (INFOCOM 2008)*, Phoenix, AZ, Apr. 13–18, 2008.

R. Chen, J.-M. Park, and J. Reed, "Defense against primary user emulation attacks in cognitive radio networks," *IEEE J. Sel. Areas Commun.*, **26**, no. 1, 25–37, Jan. 2008.

S. H. Chen, J. S. Row, and K. L. Wong, "Reconfigurable square-ring patch antenna with pattern diversity," *IEEE Trans. Antennas and Propag.*, **55**, no. 2, 472–475, Feb. 2007.

Y. Chen, "Optimum number of secondary users in collaborative spectrum sensing considering resource usage efficiency," *IEEE Commun. Lett.*, **12**, no. 12, 877–879, Dec. 2008.

Y. Chen, Q. Zhao, and A. Swami, "Joint design and separation principle for opportunistic spectrum access in the presence of sensing errors," *IEEE Trans. Inf. Theory*, **54**, no. 5, 2053–2071, May 2008.

Z. Chen, Z. Hu, and R. C. Qui, "Quickest spectrum detection using hidden Markov model for cognitive radio," *Proc. IEEE Military Communications Conf. (MILCOM 2009)*, Boston, MA, Oct. 18–21, 2009.

M. Chiani and A. Giorgetti, "Coexistence between UWB and narrow-band wireless communication systems," *Proc. IEEE*, **97**, no. 2, 231–254, Feb. 2009.

K. W. Choi, W. S. Jeon, and D. G. Jeong, "Sequential detection of cyclostationary signal for cognitive radio systems," *IEEE Trans. Wireless Commun.*, **8**, no. 9, 4480–4485, Sep. 2009.

C.-C. Chong, C.-M. Tan, D. I. Laurenson, S. McLaughlin, M. A. Beach, and A. R. Nix, "A new statistical wideband spatio-temporal channel model for 5-GHz band WLAN systems," *IEEE J. Sel. Areas Commun.*, **21**, no. 2, 139–150, Feb. 2003.

Cisco, "Cisco visual networking index: Global mobile data traffic forecast update, 2010–2015," Tech. Rep., 2011. www.cisco.com/en/US/solutions/collateral/ns341/ns525/ns537/ns705/ns827/white_paper_c11-520862.pdf.

T. C. Clancy, "Achievable capacity under the interference temperature model," *Proc. of the IEEE International Conference on Computer Communications (INFOCOM)*, Anchorage, AK, May 6–12 2007, pp. 794–802.

T. C. Clancy, "Formalizing the interference temperature model," *J. Wireless Commun. Mobile Comput.*, **7**, no. 9, 1077-1086, Nov. 2007.

T. C. Clancy, "On the use of interference temperature for dynamic spectrum access," *Ann. Telecomm.*, **64**, no. 7-8, 573–592, Aug. 2009.

T. C. Clancy and D. Walker, "Spectrum shaping for interference management in cognitive radio networks," *SDR Forum Technical Conference*, Orlando, FL, Nov. 13–17, 2006.

R. H. Clarke, "A statistical theory of mobile radio reception," *Bell Syst. Tech. J.*, **47**, no. 6, 957–1000, 1968.

"Cognitive radio network testbed (CORNET)," cornet.wireless.vt.edu/.

"Cognitive Radio Research at Nokia Research Center," www.research.nokia.com/cognitive_radio.

R. H. Coase, "The Federal Communications Commission," *Journal of Law and Economics*, **2**, 1–40, Oct. 1959.

C. Cordeiro and K. Challapali, "C-MAC: a cognitive MAC protocol for multi-channel wireless networks," *IEEE International Symposium on New Frontiers in Dynamic Spectrum Access Networks*, Apr. 2007, pp. 147–157.

C. B. Cormio and K. R. A. Chodhury, "A survey on MAC protocols for cognitive radio networks," *Elsevier J. Ad Hoc Netw.*, **7**, no. 7, 1315–1329, July 2009.

M. H. M. Costa, "Writing on dirty paper," *IEEE Trans. Inf. Theory*, **29**, no. 3, 439–441, May 1983.

M. H. M. Costa and A. El Gamal, "The capacity region of the discrete memoryless interference channel with strong interference," *IEEE Trans. Inf. Theory*, **33**, no. 5, 710–711, Sep. 1987.

T. Cover, "Broadcast channels," *IEEE Trans. Inf. Theory*, **18**, no. 1, 2–14, Jan. 1972.

T. Cover and A. El Gamal, "Capacity theorems for the relay channel," *IEEE Trans. Inf. Theory*, **25**, no. 5, 572–584, Sep. 1979.

T. Cover and J. Thomas, *Elements of Information Theory*, 2nd edition, J. Wiley & Sons, 2006.

D. C. Cox, "910 MHz urban mobile radio propagation: multipath characteristics in New York City," *IEEE Trans. Commun.*, **21**, no. 11, 1188–1194, Nov. 1973.

R. J. M. Cramer, R. A. Scholtz, and M. M. Z. Win, "Evaluation of an ultra-wide-band propagation channel," *IEEE Trans. Antennas Propag.*, **50**, no. 5, 561–570, May 2002.

I. Cuinas and M. G. Sanchez, "Measuring, modeling, and characterizing of indoor radio channel at 5.8 GHz," *IEEE Trans. Veh. Technol.*, **50**, no. 2, 526–535, Mar. 2001.

D. Čabrić, "Addressing feasibility of cognitive radios," *IEEE Signal Processing Mag.*, **25**, no. 6, 85–93, Nov. 2008.

D. Čabrić, S. M. Mishra, and R. W. Brodersen, "Implementation issues in spectrum sensing for cognitive radios," in *Proc. 38th Asilomar Conference on Signals, Systems and Computers*, 2004.

A. Damnjanović, J. Montojo, Y. Wei, *et al.*, "A survey on 3GPP heterogeneous networks," *IEEE Trans. Commun.*, **18**, no. 3, 10–21, June 2011.

A. V. Dandawaté and G. B. Giannakis, "Statistical tests for presence of cyclostationarity," *IEEE Trans. Signal Process.*, **42**, no. 9, 2355–2369, Sep. 1994.

S. Dayanik, C. Goulding, and H. V. Poor, "Bayesian sequential change diagnosis," *Math. Oper. Res.*, **33**, no. 2, 475–496, May 2008.

S. Dayanik, W. Powell, and K. Yamazaki, "Index policies for discounted bandit problems with availability constraints," *Adv. Appl. Prob.*, **40**, no. 2, 377–400, June 2008.

"Delay dispersion in UWB residential/office environments," *IEEE 802.15.3a*.

"Delay dispersion in UWB channels (office, residential, outdoor, industrial)," *IEEE 802.15.4a*.

D. Devasirvatham, "Time delay spread and signal level measurements of 850 MHz radio waves in building environments," *IEEE Trans. Antennas Propag.*, **34**, no. 11, 1300–1305, Nov. 1986.

D. Devasirvatham, "A comparison of time delay spread and signal level measurements within two dissimilar office buildings," *IEEE Trans. Antennas Propag.*, **35**, no. 3, 319–324, Mar. 1987.

N. Devroye, P. Mitran, and V. Tarokh, "Achievable rates in cognitive radio channels," *IEEE Trans. Inf. Theory*, **52**, no. 5, 1813–1827, May 2006.

N. Devroye, M. Vu, and V. Tarokh, "Cognitive radio networks," *IEEE Signal Processing Mag.*, **25**, no. 6, 12–23, Nov. 2008.

M. Dianati, X. Ling, K. Naik, and X. Shen, "A node-cooperative ARQ scheme for wireless ad hoc networks," *IEEE Trans. Veh. Technol.*, **55**, no. 3, 1032–1044, May 2006.

M. Di Felice, K. R. Chowdhury, W. Meleis, and L. Bononi, "To sense or to transmit: a learning-based spectrum management scheme for cognitive radio mesh networks," *Proc. 5th IEEE Workshop Wireless Mesh Networks (WIMESH 2010)*, Boston, MA, June 21, 2010.

F. F. Digham, M.-S. Alouini, and M. K. Simon, "On the energy detection of unknown signals over fading channels," *Proc. IEEE Int. Conf. Communications (ICC 2003)*, Anchorage, AK, May 11–15, 2003, pp. 3575–3579.

F. F. Digham, M.-S. Alouini, and M. K. Simon, "On the energy detection of unknown signals over fading channels," *IEEE Trans. Commun.*, **55**, no. 1, 21–24, Jan. 2007.

C. Doerr, M. Neufeld, J. Fifield, T. Weingart, D. Sicker, and D. Grunwald, "Multimac – an adaptive MAC framework for dynamic radio networking," *IEEE International Symposium on New Frontiers in Dynamic Spectrum Access Networks, 2005*, Nov. 2005, pp. 548–555.

M. Dresher, "Moment spaces and inequalities," *Duke Math. J.*, **20**, 261–271, June 1953.

P. E. Driessen, "Prediction of multipath delay profiles in mountainous terrain," *IEEE J. Sel. Areas Commun.*, **18**, no. 3, 336–346, Mar. 2000.

G. D. Durgin, V. Kukshya, and T. S. Rappaport, "Wideband measurements of angle and delay dispersion for outdoor and indoor peer-to-peer radio channels at 1920 MHz," *IEEE Trans. Antennas Propag.*, **51**, no. 5, 936–944, May 2003.

M. Effros, A. Goldsmith, and Y. Liang, "Generalizing capacity: new definitions and capacity theorems for composite channels," *IEEE Trans. Inf. Theory*, **56**, no. 7, July 2010.

A. El Gamal, "On information flow in relay networks," *Proc. of the IEEE National Telecommunications Conference*, **2**, Nov. 1981, D4.1.1–D4.1.4.

A. El Gamal and Y.-H. Kim, *Network Information Theory*, Cambridge University Press, 2012.

A. El Gamal, J. Mammen, B. Prabhakar, and D. Shah, "Optimal throughput-delay scaling in wireless networks – part I: the fluid model," *IEEE Trans. Inf. Theory*, **52**, no. 6, 2568–2592, June 2006.

H. El Gamal, G. Caire, and M. O. Damen, "The MIMO ARQ channel: diversity-multiplexing-delay tradeoff," *IEEE Trans. Inf. Theory*, **52**, no. 8, 3601–3621, Aug. 2006.

"End-to-end efficiency project," ict-e3.eu/project/overview/overview.html.

V. Erceg, K. V. S. Hari, M. S. Smith, *et al.*, "Channel models for fixed wireless applications," *IEEE 802.16d-03/34*.

V. Erceg, L. Schumacher, P. Kyritsi, *et al.*, "TGn channel models," *IEEE 802.11-03/940r4*, May 2004.

V. Erceg, L. J. Greenstein, S. Y. Tjandra, *et al.*, "An empirically based path loss model for wireless channels in suburban environments," *IEEE J. Sel. Areas Commun.*, **17**, no. 7, 1205–1211, Jul. 1999.

V. Erceg, D. G. Michelson, S. S. Ghassemzadeh, *et al.*, "A model for the multipath delay profile of fixed wireless channels," *IEEE J. Sel. Areas Commun.*, **17**, no. 3, 399-410, Mar. 1999.

R. Etkin, D. N. C. Tse, and H. Wang, "Gaussian interference channel capacity to within one bit," *IEEE Trans. Inf. Theory.*, **54**, no. 12, 5534–5562, May 2008.

European Telecommunication Standard, "Digital Audio Broadcasting (DAB) to mobile, portable and fixed receivers," ETSI, Tech. Rep., May 1997.

M. Failli (ed.), Final report of COST 207, "Digital land mobile radio communications," *Commission of the European Communities*, 1989.

S. Fazeli-Dehkordy, K. N. Plataniotis, and S. Pasupathy, "Wide-band collaborative spectrum search strategy for cognitive radio networks," *IEEE Trans. Signal Process.*, **59**, no. 8, 3903–3914, Aug. 2011.

Federal Communications Commission, *Establishment of an Interference Temperature Metric to Quantify and Manage Interference and to Expand Available Unlicensed*

Operation in Certain Fixed, Mobile and Satellite Frequency Bands. ET Docket No. 03-237, Nov. 2003.

Federal Communications Commission, "Notice of proposed rule making," FCC, ET Docket no. 04-113, May 2004. www.naic.edu/~phil/rfi/fccactions/FCC-04-113A1.pdf.

Federal Communications Commission, "Second memorandum opinion and order," FCC, Tech. Rep., Sep. 2010. transition.fcc.gov/Daily_Releases/Daily_Business/2010/db0923/FCC-10-174A1.pdf

Federal Communications Commission Spectrum Policy Task Force, Report of the Spectrum Efficiency Working Group, *Technical Report 02-135*, (Nov. 2002). www.fcc.gov/sptf/files/SEWGFinalReport_1.pdf.

G. Fellouris and G. V. Moustakides, "Decentralized sequential hypothesis testing using asynchronous communication," *IEEE Trans. Inf. Theory*, **57**, no. 1, 534–548, Jan. 2011.

M. J. Feuerstein, K. L. Blackard, T. S. Rappaport, S. Y. Seidel, and H. H. Xia, "Path loss, delay spread, and outage models as functions of antenna height for microcellular system design," *IEEE Trans. Veh. Technol.*, **43**, no. 3, 487–498, Aug. 1994.

S. Filippi, O. Cappé, F. Clérot, and E. Moulines, "A near optimal policy for channel allocation in cognitive radio," *Recent Advances in Reinforcement Learning: 8th European Workshop (EWRL 2008)*, Villeneuve d'Ascq, France, June 30–July 4, 2008, pp. 69–81.

S. Filippi, O. Cappé, and A. Garivier, "Optimally sensing a single channel without prior information: the tiling algorithm and regret bounds," *IEEE J. Sel. Topics Signal Process.*, **5**, no. 1, 68–76, Feb. 2011.

F. Fitzek and M. Katz, eds., *Cognitive Wireless Networks: Concepts, Methodologies and Visions inspiring the age of Enlightenment of Wireless Communications*, Springer, 2007.

B. H. Fleury, "An uncertainty relation for WSS processes and its application to WSSUS systems," *IEEE Trans. Commun.*, **44**, no. 12, 1632–1634, Dec. 1996.

B. H. Fleury, "First- and second-order characterization of direction dispersion and space selectivity in the radio channel," *IEEE Trans. Information Theory*, **46**, no. 6, 2027–2044, Sep. 2000.

H. T. Friis, "A note on a simple transmission formula," *Proc. IRE*, **34**, no. 5, 254, May 1946.

J. Fuhl, A. F. Molisch, and E. Bonek, "Unified channel model for mobile radio systems with smart antennas," *IEE Proc. Radar, Sonar and Navigation*, **145**, no. 1, 32–41, Feb. 1998.

Y. Gai, B. Krishnamachari, and R. Jain, "Learning multiuser channel allocations in cognitive radio networks: a combinatorial multi-armed bandit formulation," *Proc. IEEE Int. Symp. Dynamic Spectrum Access Networks (DySPAN 2010)*, Singapore, Apr. 6–9, 2010.

R. G. Gallager, *Information Theory and Reliable Communication*, J. Wiley & Sons, 1968.

F. Gao, W. Yuan, W. Liu, W. Cheng, and S. Wang, "A robust and efficient cooperative spectrum sensing scheme in cognitive radio networks," *Proc. IEEE Int. Conf. Communications (ICC 2010)*, Cape Town, South Africa, May 23–27, 2010.

W. A. Gardner, "Signal interception: a unifying theoretical framework for feature detection," *IEEE Trans. Commun.*, **36**, no. 8, 897–906, Aug. 1988.

W. A. Gardner, "Exploitation of spectral redundancy in cyclostationary signals," *IEEE Signal Process. Mag.*, **8**, no. 2, 14–36, Apr. 1991.

W. A. Gardner (ed.), *Cyclostationarity in Communications and Signal Processing*. IEEE Press, 1994.

M. Gastpar, "On capacity under receive and spatial spectrum-sharing constraints," *IEEE Trans. Inf. Theory*, **53**, no. 2, 471–487, Feb. 2007.

S. I. Gelfand and M. S. Pinsker, "Coding for channel with random parameters," *Probl. Peredachi Informatsii*, **9**, no. 1, 19–31, Jan. 1980.

C. Gerami, *Design Methodology for Backhaul and Distribution Networks using TV White Spaces*, M.S. Dissertation, Rutgers University, May 2011.

C. Gerami, N. Mandayam, and L. Greenstein, "Backhauling in TV white spaces," *2010 IEEE Global Telecommunications Conference*, Dec. 2010, pp. 1–6.

S. Gezici, H. Celebi, H. V. Poor, and H. Arslan, "Fundamental limits on time delay estimation in dispersed spectrum cognitive radio systems," *IEEE Trans. Wireless Commun.*, **8**, no. 1, 78–83, Jan. 2009.

A. Ghasemi and E. Sousa, "Fundamental limits of spectrum-sharing in fading environments," *IEEE Trans. Wireless Commun.*, **6**, no. 2, 649–658, Feb. 2007.

A. Ghasemi and E. S. Sousa, "Spectrum sensing in cognitive radio networks: the cooperation-processing tradeoff," *Wirel. Commun. Mob. Comput.*, **7**, no. 9, 1049–1060, Sept. 2007.

A. Ghasemi and E. S. Sousa, "Interference aggregation in spectrum-sensing cognitive wireless networks," *IEEE J. Sel. Topics Signal Process.*, **2**, no. 1, 41–56, Feb. 2008.

A. Ghasemi and E. S. Sousa, "Spectrum sensing in cognitive radio networks: Requirements, Challenges and design trade-offs," *IEEE Commun. Mag.*, **46**, no. 4, 32–39, Apr. 2008.

S. S. Ghassemzadeh, L. J. Greenstein, T. Sveinsson, and V. Tarokh, "UWB delay profile models for residential and commercial indoor environments," *IEEE Trans. Veh. Technol.*, **54**, no. 4, 1235-1244, July 2005.

S. S. Ghassemzadeh, R. Jana, C. W. Rice, W. Turin, and V. Tarokh, "Measurement and modeling of an ultra-wide bandwidth indoor channel," *IEEE Trans. Commun.*, **52**, no. 10, 1786–1796, Oct. 2004.

B. K. Ghosh, *Sequential Tests of Statistical Hypothesis*, Addison-Wesley, 1970.

R. Gibbons, *A Primer in Game Theory*, Pearson Education Limited, 1992.

J. C. Gittins, "Bandit processes and dynamic allocation indices," *J. Roy. Statist. Soc. Ser. B*, **41**, no. 2, 148–177, 1979.

J. C. Gittins and D. M. Jones, "A dynamic allocation index for the sequential design of experiments," in J. Gani (ed.) *Progress in Statistics*, North Holland, pp. 241–266, 1974.

K. D. Glazebrook, D. Ruiz-Hernandez, and C. Kirkbride, "Some indexable families of restless bandit problems," *Adv. Appl. Prob.*, **38**, no. 3, 643–672, 2006.

A. A. Glazunov, H. Asplund, and J. E. Berg, "Statistical analysis of measured short-term impulse response functions of 1.88 GHz radio channels in Stockholm with corresponding channel model," *Proc. IEEE Veh. Technol. Conf.*, pp. 107–111, 1999.

A. Goldsmith, *Wireless Communications*, Cambridge University Press, 2005.

A. Goldsmith, M. Effros, R. Koetter, M. Médard, A. Ozdaglar, and L. Zheng, "Beyond Shannon: the quest for fundamental performance limits of wireless ad hoc networks," *IEEE Commun. Mag.*, **49**, no. 5, 195–205, May 2011.

A. J. Goldsmith and L. J. Greenstein, "A measurement-based model for predicting coverage areas of urban microcells", *IEEE J. Sel. Areas Commun.*, **11**, no. 7, 1013-1023, Sept. 1993.

A. J. Goldsmith, S. A. Jafar, N. Jindal, and S. Vishwanath, "Capacity limits of MIMO channels," *IEEE J. Sel. Areas Commun.*, **21**, no. 5, 684–702, June 2003.

A. Goldsmith, S. Jafar, I. Marić, and S. Srinivasa, "Breaking spectrum gridlock with cognitive radios: an information theoretic perspective," *Proc. IEEE*, **97**, no. 5, 894–914, May 2009.

F. Graziosi and F. Santucci, "A general correlation model for shadow fading in mobile radio systems," *IEEE Commun. Letters*, **6**, no. 3, 102–104, 2002.

L. J. Greenstein and V. Erceg, "Gain reductions due to scatter on wireless paths with directional antennas," *IEEE Commun. Letters*, **3**, no. 6, 169–171, Jun. 1999.

L. J. Greenstein, V. Erceg, Y. S. Yeh, and M. V. Clark, "A new path-gain/delay-spread propagation model for digital cellular channels," *IEEE Trans. Veh. Technol.*, **46**, no. 2, 477–485, May 1997.

L. J. Greenstein, S. S. Ghassemzadeh, V. Erceg, and D. G. Michelson, "Ricean K-factors in narrowband fixed wireless channels: theory, experiments and statistical models," *IEEE Trans. Veh. Technol.*, **58**, no. 9, 4000–4012, Oct. 2009.

L. J. Greenstein, S. S. Ghassemzadeh, S. C. Hong, and V. Tarokh, "Comparison study of UWB indoor channel models," *IEEE Trans. Wireless Commun.*, **6**, no. 1, 128-135, Jan. 2007.

M. Grossglauser and D. N. C. Tse, "Mobility increases the capacity of ad hoc wireless networks," *IEEE/ACM Trans. Netw.*, **10**, no. 4, 477–486, Aug. 2002.

P. Grover and A. Sahai, "What is needed to exploit knowledge of primary transmissions?," *Proc. of the IEEE International Symposium on New Frontiers in Dynamic Spectrum Access Networks (DySPAN)*, Dublin, Ireland, Apr. 17–20, 2007, pp. 462–471.

M. Gudmundson, "Correlation model for shadow fading in mobile radio systems," *Electronics Letters*, **27**, no. 23, 2145–2146, 7 Nov. 1991.

M. Guerriero, V. Pozdnyakov, J. Glaz, and P. Willett, "A repeated significance test with applications to sequential detection in sensor networks," *IEEE Trans. Signal Process.*, **58**, no. 7, 3426–3435, July 2010.

Guidelines for Evaluation of Radio Transmission Technologies for IMT-2000, Recommendation ITU-R M.1225, 1997.

P. Gupta and P. R. Kumar, "The capacity of wireless networks," *IEEE Trans. Inf. Theory*, **42**, no. 2, 388–404, Mar. 2000.

P. Gupta and P. R. Kumar, "Towards an information theory of large networks: An achievable rate region," *IEEE Trans. Inf. Theory*, **49**, no. 8, 1877–1894, Aug. 2003.

O. Hadjiliadis, H. Zhang, and H. V. Poor, "One shot schemes for decentralized quickest change detection," *IEEE Trans. Inf. Theory*, **55**, no. 7, 3346–3359, July 2009.

A. Haghparast, T. Abrudan, and V. Koivunen, "OFDM ranging in multipath channels using time reversal method," *Proc. 10th IEEE Int. Workshop Signal Processing Advances for Wireless Communications (SPAWC 2009)*, Perugia, Italy, June 21–24, 2009, pp. 568–572.

P. S. Hall, P. Gardner, J. Kelly, E. Ebrahimi, M. R. Hamid, and F. Ghanem, "Antenna challenges in cognitive radio," *Proc. ISAP 08*, Oct. 2008, pp. 141–144.

D. Hampicke, A. Richter, A. Schneider, G. Sommerkorn, R. S. Thoma, and U. Trautwein, "Characterization of the directional mobile radio channel in industrial scenarios, based on wideband propagation measurements," *Proc. Veh. Technol. Conf.*, pp. 2258–2262, 1999.

B. Hamdaoui, P. Venkatraman, and M. Guizani, "Opportunistic exploitation of bandwidth resources through reinforcement learning," *Proc. IEEE Global Communications Conf. (GLOBECOM 2009)*, Honolulu, HI, Nov. 30–Dec. 4, 2009.

T. Han and K. Kobayashi, "A new achievable rate region for the interference channel," *IEEE Trans. Inf. Theory*, **27**, no. 1, 49–60, Jan. 1981.

Z. Han, R. Zheng, and H. V. Poor, "Repeated auctions with Bayesian nonparametric learning for spectrum access in cognitive radio networks," *IEEE Trans. Wireless Commun.*, **10**, no. 2, 890–900, Mar. 2011.

L. Hanzo and T. Keller, *OFDM and MC-CDMA: a Primer*, IEEE Press, J. Wiley & Sons, 2006.

G. Hardin, "Tragedy of the commons," *Science*, pp. 1243–1248, Dec. 1968.

K. V. S. Hari and C. Bushue, "Interim channel models for G2 MMDS fixed wireless applications," *IEEE 802.16.3c-00/49r2*.

F. Harryson, J. Medbo, A. F. Molisch, *et al.*, "Efficient experimental evaluation of MIMO and set with user influence," *IEEE Trans. Wireless Commun.*, **9**, no. 2, 853–863, Feb. 2010.

S. Hart and A. Mas-Colell, "A reinforcement procedure leading to correlated equilibrium," in G. Debreu, W. Neuefeind, and W. Trockel (eds.), *Economics Essays*, pp. 181–200, Springer, 2001.

H. Hashemi and D. Tholl, "Statistical modeling and simulation of the RMS delay spread of indoor radio propagation channels," *IEEE Trans. Veh. Technol.*, **43**, no. 1, 110–120, Feb. 1994.

M. Hata, "Empirical formula for propagation loss in land mobile radio services," *IEEE Trans. Veh. Technol.*, **29**, no. 3, 317–325, Aug. 1980.

M. H. Hayes, *Statistical Digital Signal Processing and Modeling*, J. Wiley & Sons, 1996.

S. Haykin, "Cognitive radio: brain-empowered wireless communications," *IEEE J. Sel. Areas Commun.*, **23**, no. 2, 201–220, Feb. 2005.

S. Haykin, D. J. Thomson, and J. H. Reed, "Spectrum sensing for cognitive radio," *Proc. IEEE*, **97**, no. 5, 849–877, May 2009.

S. P. Herath, N. Rajatheva, and C. Tellambura, "Energy detection of unknown signals in fading and diversity reception," *IEEE Trans. Commun.*, **59**, no. 9, 2443–2453, Sep. 2011.

W. Hirt and J. L. Massey, "Capacity of the discrete-time Gaussian channel with inter-symbol interference," *IEEE Trans. Inf. Theory*, **34**, no. 3, 380–388, May 1998.

M. Hong and A. Garcia, "Equilibrium pricing of interference in cognitive radio networks," *IEEE Trans. Signal Process.*, **59**, no. 12, 6058–6072, Dec. 2011.

B. Horine and D. Turgut, "Link rendezvous protocol for cognitive radio networks," *IEEE International Symposium on New Frontiers in Dynamic Spectrum Access Networks, 2007*, Apr. 2007, pp. 444–447.

S. H. Hsu and K. Chang, "A novel reconfigurable microstrip antenna with switchable circular polarization," *IEEE Antennas and Wireless Propag. Lett.*, **6**, 160–162, 2007.

T.-C. Hsu, T.-Y. Wang, and Y.-W. P. Hong, "Collaborative change detection for efficient spectrum sensing in cognitive radio networks," *Proc. 71st IEEE Vehicular Technology Conf. (VTC 2010-Spring)*, Taipei, Taiwan, May 16–19, 2010.

J. Huang, R. A. Berry, and M. L. Honig, "Spectrum sharing with distributed interference compensation," *Proc. of the IEEE International Symposium on New Frontiers in Dynamic Spectrum Access Networks (DySPAN)*, Baltimore, MD, Nov. 8–11, 2005, pp. 88–93.

Huawei Technolgies and UESTC, "Sensing schemes for DVB-T," *IEEE Std. 802.22-06/0127r1*, July 2006.

G. H. Huff, J. Feng, S. Zhang, and J. T. Bernhard, "A novel radiation pattern and frequency reconfigurable single turn square spiral microstrip antenna," *IEEE Trans. Microw. Wireless Compon. Lett.*, **13**, no. 2, 57–59, Feb. 2003.

A. M. Hussain, "Multisensor distributed sequential systems," *IEEE Trans. Aerosp. Electron. Syst.*, **30**, no. 3, 698–708, July 1994.

IEEE 802.11, "Wireless LAN medium access control (MAC) and physical layer (PHY) specifications," Tech. Rep., June 2007.

IEEE 802.16.j standardization group, "Multi-hop relay system evaluation methodology (channel model and performance metric)," *IEEE 802.16j-06/013r3*, Feb. 2, 2007.

Ö. Ileri and N. Mandayam, "Dynamic spectrum access models: toward an engineering perspective in the spectrum debate," *IEEE Commun. Magazine*, **46**, no. 1, 153–160, Jan. 2008.

Ö. Ileri, D. Samardzija, T. Sizer, and N. Mandayam, "Demand responsive pricing and competitive spectrum allocation via a spectrum policy server," *Proc. IEEE DySPAN*, Nov. 2005, Baltimore, MD.

T. Jaakkola, M. I. Jordan, and S. P. Singh, "On the convergence of stochastic iterative dynamic programming algorithms," *Neural Comput.*, **6**, 1185–1201, 1994.

S. A. Jafar and S. Shamai (Shitz), "Degrees of freedom region for the MIMO X channel," *IEEE Trans. Inf. Theory*, **54**, no. 1, 151–170, Jan. 2008.

S. A. Jafar and S. Srinivasa, "Capacity limits of cognitive radio with distributed and dynamic spectral activity," *IEEE J. Sel. Areas Commun.*, **25**, no. 3, 529–537, Apr. 2007.

W. C. Jakes, *Microwave Mobile Communications*, J. Wiley & Sons, 1974 (Reprinted by IEEE Press, 1994).

A. K. Jayaprakasam and V. Sharma, "Sequential detection based cooperative spectrum sensing algorithms in cognitive radio," *Proc. 1st IEEE UK-India Int. Workshop Cognitive Wireless Systems (UKIWCWS 2009)*, IIT Delhi, India, Dec. 11–12, 2009.

A. K. Jayaprakasam, V. Sharma, C. R. Murthy, and P. Narayanan, "Cyclic prefix based cooperative sequential spectrum sensing algorithms for OFDM," *Proc. IEEE Int. Conf. Communication (ICC 2010)*, Cape Town, South Africa, May 2010.

S.-W. Jeon, N. Devroye, M. Vu, S.-Y. Chung, and V. Tarokh, "Cognitive networks achieve throughput scaling of a homogeneous network," *IEEE Trans. Inf. Theory*, **57**, no. 8, 5103–5115, Aug. 2011.

H. Jiang, L. Lai, R. Fan, and H. V. Poor, "Optimal selection of channel sensing order in cognitive radio," *IEEE Trans. Wireless Commun.*, **8**, no. 1, 297–307, Jan. 2009.

J. Jiang, I. Marić, A. Goldsmith, S. Shamai (Shitz), and S. Cui, "On the capacity of a class of cognitive Z-interference channels," *Proc. of the IEEE International Conference on Communications (ICC)*, Kyoto, Japan, June 5–9, 2011.

J. Jiang, Y. Xin, and H. Garg, "Interference channels with common information," *IEEE Trans. Inf. Theory*, **54**, no. 1, 171–187, Jan. 2008.

X. Jing and D. Raychaudhuri, "Global control plane architecture for cognitive radio networks," *IEEE International Conference on Communications, 2007. ICC '07*, June 2007, pp. 6466–6470.

A. Jovičić and P. Viswanath, "Cognitive radio: an information-theoretic perspective," *IEEE Trans. Inf. Theory*, **55**, no. 9, 3945–3958, Sep. 2009.

T. Jun, "A survey on the bandit problem with switching costs," *De Economist*, **152**, no. 4, 513–541, 2004.

L. P. Kaebling, M. L. Littman, and A. W. Moore, "Reinforcement learning: a survey," *J. Artif. Intell. Res.*, **4**, no. 1, 237–285, 1996.

P. Kaewprapha, R. Wu, B. C. Ng, and T. J. Li, "Cooperative spectrum sensing for cognitive radios: bounds and algorithms," *Proc. IEEE Wireless Communications and Networking Conf. (WCNC 2010)*, Sydney, Australia, Apr. 18–21, 2010, pp. 1–6.

A. Kanatas, N. Moraitis, G. Pantos, and P. Constantinou, "Wideband characterization of microcellular suburban mobile radio channels at 1.89 GHz," *Proc. IEEE Veh. Technol. Conf.*, pp. 1060–1064, 2002.

S. Kandeepan, S. Reisenfeld, T. C. Aysal, D. Lowe, and R. Piesiewicz, "Bayesian tracking in cooperative localization for cognitive radio networks," *Proc. 69th IEEE Vehicular Technology Conf. (VTC 2009-Spring)*, Barcelona, Spain, Apr. 26–29, 2009.

J. Karedal, N. Czink, A. Paier, F. Tufvesson, and A. F. Molisch, "Path loss modeling for vehicle-to-vehicle communications," *IEEE Trans. Veh. Technol.*, **60**, no. 1, pp. 323–328, Jan. 2011.

P. Karlsson, C. Bergljung, E. Thomsen, and H. Borjeson, "Wideband measurement and analysis of penetration loss in the 5 GHz band," *Proc. IEEE Veh. Technol. Conf.*, pp. 2323–2328, 1999.

A. Kashyap, "Comments on the "On the optimality of the likelihood-ratio test for local sensor decision rules in the presence of nonideal channels"," *IEEE Trans. Inf. Theory*, **52**, no. 3, 1274–1275, Mar. 2006.

A.-K. Katta and J. Sethuraman, "A note on bandits with a twist," *SIAM J. Discrete Math.*, **18**, no. 1, 110–113, 2004.

R. Kattenbach, "Statistical modeling of small-scale fading in directional radio channels," *IEEE J. Sel. Areas Commun.*, **20**, no. 3, 584–592, Apr. 2002.

S. M. Kay, *Fundamentals of Statistical Signal Processing: Detection Theory*, Prentice-Hall PTR, 1998.

J. R. Kelly, E. Ebrahimi, P. S. Hall, P. Gardner, and F. Ghanem, "Combined wideband and narrowband antennas for cognitive radio applications," *IET Seminar on Cognitive Radio and Software Defined Radios: Technologies and Techniques*, pp. 1–4, Sept. 2008.

J. F. Kepler, T. P. Krauss, and S. Mukthavaram, "Delay spread measurements on a wideband MIMO channel at 3.7 GHz," *Proc. IEEE Veh. Technol. Conf.*, pp. 2498–2502, 2002.

H. Kim and K. G. Shin, "Efficient discovery of spectrum opportunities with MAC-layer sensing in cognitive radio networks," *IEEE Trans. Mobile Comput.*, **7**, no. 5, 533–545, May 2008.

S.-J. Kim and G. B. Giannakis, "Rate-optimal and reduced-complexity sequential sensing algorithms for cognitive OFDM radios," *EURASIP J. Adv. Signal Process. – Special Issue on Dynamic Spectrum Access for Wireless Networking*, **2009**, Article ID 421540, 2009.

S. Kim, H. Jeon, and J. Ma, "Robust localization with unknown transmission power for cognitive radio," *Proc. IEEE Military Communications Conf. (MILCOM 2007)*, Orlando, FL, Oct. 29–31, 2007.

S.-J. Kim and G. B. Giannakis, "Sequential and cooperative sensing for multi-channel cognitive radios," *IEEE Trans. Signal Process.*, **58**, no. 8, 4239–4253, Aug. 2010.

J. Kivinen, X. Zhao, and P. Vainikainen, "Empirical characterization of wideband indoor radio channel at 5.3 GHz," *IEEE Trans. Antennas Propag.*, **49**, no. 8, 1192–1203, Aug. 2001.

F. Kocak, H. Celebi, S. Gezici, K. A. Qarake, H. Arslan, and H. V. Poor, "Time delay estimation in dispersed spectrum cognitive radio systems," *EURASIP J. Adv. Signal Process. – Special Issue on Advanced Signal Processing for Cognitive Radio Networks*, **2010**, Article ID 675959, 10 pages, 2010.

R. Koetter, M. Effros, and M. Médard, "On a theory of network equivalence," *Proc. of the IEEE Information Theory Workshop (ITW)*, Volos, Greece, June 10–12 2009. arxiv.org/abs/1007.1033

J. R. Kok and N. Vlassis, "Collaborative multiagent reinforcement learning by payoff propagation," *J. Mach. Learn. Res.*, **7**, 1789–1828, Dec. 2006.

M. Komulainen, M. Berg, H. Jantunen, E. T. Salonen, and C. Free, "A frequency tuning method for a planar inverted-F antenna," *IEEE Trans. Antennas and Propag.*, **56**, no. 4, 944–950, Apr. 2008.

V. I. Kostylev, "Energy detection of a signal with random amplitude," in *Proc. IEEE Int. Conf. Commun. (ICC 2002)*, **3**, 1606–1610, Apr. 28–May 2, 2002.

S. Kozono and A. Taguchi, "Mobile propagation loss and delay spread characteristics with a low base station antenna on an urban road," *IEEE Trans. Veh. Technol.*, **42**, no. 1, 103–109, Feb. 1993.

G. Kramer, I. Marić, and R. D. Yates, "Cooperative communications," *NOW J. Found. Trends Netw.*, **1**, no. 3-4, 271–425, 2006.

G. Kramer and S. Savari, "Capacity bounds for relay networks," *Proc. of the UCSD Workshop on Information Theory and Applications*, La Jolla, CA, Feb. 6–10, 2006.

G. Kramer and S. Savari, "Edge-cut bounds on network coding rates," *J. Netw. Syst. Management*, **14**, no. 1, 49–67, Mar. 2006.

M. G. Kreĭn and A. A. Nudel'man, *The Markov Moment Problem and Extremal Problems*. American Mathematical Society, 1977.

A. Kuchar, J. P. Rossi, and E. Bonek, "Directional macro-cell channel characterization from urban measurements," *IEEE Trans. Antennas Propag.*, **48**, no. 2, 137–146, Feb. 2000.

N. Kundargi and A. Tewfik, "A performance study of novel sequential energy detection methods for spectrum sensing," *Proc. IEEE Int. Conf. Acoustics, Speech, and Signal Processing (ICASSP 2010)*, Dallas, TX, Mar. 14–19, 2010.

H. Kushwaha, Y. Xing, R. Chandramouli, and H. Heffes, "Reliable multimedia transmission over cognitive radio networks using fountain codes," *Proc. IEEE*, **96**, no. 1, 155–165, Jan. 2008.

H. Kushwaha, Y. Xing, R. Chandramouli, and K. P. Subbalakshmi, "Erasure tolerant coding for cognitive radios," in Q. H. Mahmoud (ed.), *Cognitive Networks – Towards Self-Aware Networks*. J. Wiley & Sons, 2007.

H. Kushwaha, R. Chandramouli, and K. P. Subbalakshmi, "Cognitive networks: towards self-aware networks," in Q. H. Mahmoud (ed.), *Cognitive Networks – Towards Self-Aware Networks*. J. Wiley & Sons, 2007.

U. Ladebusch and C. Liss, "Terrestrial DVB (DVB-T): a broadcast technology for stationary portable and mobile use," *Proc. IEEE*, **94**, no. 1, pp. 183–193, Jan. 2006.

L. Lai, H. El Gamal, H. Jiang, and H. V. Poor, "Cognitive medium access: Exploration, exploitation, and competition," *IEEE Trans. Mobile Comput.*, **10**, no. 2, 239–253, Feb. 2011.

L. Lai, Y. Fan, and H. V. Poor, "Quickest detection in cognitive radio: A sequential change detection framework," *Proc. IEEE Global Communications Conf. (GLOBECOM 2008)*, New Orleans, LA, Nov. 30–Dec. 4, 2008.

L. Lai, H. Jiang, and H. V. Poor, "Medium access in cognitive radio networks: A competitive multi-armed bandit framework," *Proc. 42nd Asilomar Conf. Signals, Systems, and Computers*, Pacific Grove, CA, Oct. 26–29, 2008, pp. 98–102.

T. L. Lai, "Asymptotic optimality of invariant sequential probability ratio tests," *Ann. Stat.*, **9**, no. 2, 318–333, 1981.

T. L. Lai, "Nearly optimal sequential tests of composite hypotheses," *Ann. Stat.*, **16**, no. 2, 856–886, 1988.

T. L. Lai, "On optimal stopping problems in sequential hypothesis testing," *Statist. Sinica*, **7**, 33–51, 1997.

T. L. Lai, "Sequential analysis: some classical problems and new challenges," *Statist. Sinica*, **11**, no. 2, 303–408, Apr. 2001.

T. L. Lai and H. Robbins, "Asymptotically efficient adaptive allocation rules," *Adv. Appl. Math.*, **6**, no. 1, 4–22, Mar. 1985.

T. L. Lai and H. Xing, "Sequential change-point detection when the pre- and post-change parameters are unknown," *Technical Report no. 2009-5*, Department of Statistics, Stanford University, Stanford, CA, Apr. 2009.

T. L. Lai and L. Zhang, "A modification of Schwarz's sequential likelihood ratio tests in multivariate sequential analysis," *Sequential Anal.: Design Methods Appl.*, **13**, no. 2, 79–96, 1994.

J. Langford and T. Zhang, "The epoch-greedy algorithm for contextual multi-armed bandits," in J.C. Platt, D. Koller, Y. Singer, and S. Roweis (eds.), *Advances in Neural Information Processing Systems 20*, pp. 817–824, MIT Press, 2008.

M. Larsson, "Spatio-temporal channel measurements at 1800 MHz for adaptive antennas," *Proc. IEEE Veh. Technol. Conf.*, pp. 376–380, 1999.

P. Laspougeas, P. Pajusco, and J. C. Bic, "Radio propagation in urban small cells environment at 2 GHz: experimental spatio-temporal characterization and spatial wideband channel model," *Proc. IEEE Veh. Technol. Conf.*, pp. 885–892, 2000.

G. S. Lauer and N. R. Sandell Jr., "Distributed detection of known signal in correlated noise," *Report ALPHATECH*, Burlington, MA, Mar. 1982.

J. Laurila, K. Kalliola, M. Toeltsch, K. Hugl, P. Vainikainen, and E. Bonek, "Wideband 3D characterization of mobile radio channels in urban environment," *IEEE Trans. Antennas Propag.*, **50**, no. 2, 233–243, Feb. 2002.

W.-Y. Lee and I. F. Akyildiz, "Optimal spectrum sensing framework for cognitive radio networks," *IEEE Trans. Wireless Commun.*, **7**, no. 10, 3845–3857, Oct. 2008.

Z. Lei and F. Chin, "OFDM signal sensing for cognitive radios," *19th International Symp. on Personal, Indoor, and Mobile Radio Communications (PIMRC'08)*, Cannes, France, Sep. 15–18, 2008.

K. B. Letaief and W. Zhang, "Cooperative communications for cognitive radio networks," *Proc. IEEE*, **97**, no. 5, 878–893, May 2009.

K. Leyton-Brown and Y. Shoham, *Essentials of Game Theory: A Concise Multidisciplinary Introduction*, Morgan & Claypool, 2008.

H. Li, "Restless watchdog: selective quickest spectrum sensing in multichannel cognitive radio systems," *EURASIP J. Adv. Signal Process. – Special Issue on Dynamic Spectrum Access for Wireless Networking*, **2009**, Article ID 417457, Aug. 2009.

H. Li, "Learning the spectrum via collaborative filtering in cognitive radio networks," *Proc. IEEE Int. Symp. New Frontiers in Dynamic Spectrum Access Networks (DySPAN 2010)*, Singapore, Apr. 6–9, 2010.

H. Li, "Multiagent Q-learning for Aloha-like spectrum access in cognitive radio systems," *EURASIP J. Wireless Commun. Netw.*, **2010**, Article ID 876216, 2010.

H. Li, H. Dai, and C. Li, "Collaborative quickest spectrum sensing via random broadcast in cognitive radio systems," *IEEE Trans. Wireless Commun.*, **9**, no. 7, 2338–2348, July 2010.

H. Li, C. Li, and H. Dai, "Collaborative quickest detection in adhoc networks with delay constraint – part I: Two-node network," *Proc. Conf. Information Sciences and Systems (CISS 2008)*, Princeton, NJ, Mar. 19–21, 2008, pp. 594–599.

H. Li, C. Li, and H. Dai, "Collaborative quickest detection in adhoc networks with delay constraint – Part II: Multi-node network," *Proc. Conf. Information Sciences and Systems (CISS 2008)*, Princeton, NJ, Mar. 19–21, 2008, pp. 600–605.

T. Li, N. Mandayam, and A. Reznik, "A framework for resource allocation in a cognitive digital home," *2010 IEEE Global Telecommunications Conference*, Dec. 2010, pp. 1–5.

T. Li, N. Mandayam, and A. Reznik, "Distributed algorithms for joint channel and RAT allocation in a cognitive digital home," *International Symposium on Modeling and Optimization in Mobile, Ad Hoc and Wireless Networks (WiOpt)*, May 2011, pp. 213–219.

Z. Li, L. Liu, and C. Zhou, "Fast detection method in cooperative cognitive radio networks," *Int. J. Digit. Multimedia Broadcast.*, **2010**, Article ID 160462, 2010.

Y. Liang, A. Somekh-Baruch, V. Poor, S. Shamai (Shitz), and S. Verdú, "Cognitive interference channels with and without secrecy," *IEEE Trans. Inf. Theory*, **55**, no. 2, 604–619, Feb. 2009.

Y.-C. Liang, Y. Zeng, E. Peh, and A. T. Hoang, "Sensing-throughput tradeoff for cognitive radio networks," *IEEE Trans. Wireless Commun.*, **7**, no. 4, 1326–1337, Apr. 2008.

M. Lienard and P. Degauque, "Natural wave propagation in mine environments," *IEEE Trans. Antennas Propag.*, **48**, no. 9, 1326–1339, Sep. 2000.

M. L. Littman, "Markov games as a framework for multi-agent reinforcement learning," *Proc. 11th Int. Conf. Machine Learning (ICML 1994)*, San Francisco, CA, 1994, pp. 157–163.

H. Liu, K. Liu, and Q. Zhao, "Learning in a changing world: non-Bayesian restless multi-armed bandit," *Technical Report TR-10-03*, University of California, Davis, CA, 2010.

K. Liu and Q. Zhao, "Indexability of restless bandit problems and optimality of Whittle index for dynamic multichannel access," *IEEE Trans. Inf. Theory*, **56**, no. 11, 5547–5567, Nov. 2010.

K. Liu and Q. Zhao, "Distributed learning in multi-armed bandit with multiple players," *IEEE Trans. Signal Process.*, **58**, no. 11, 5667–5681, Nov. 2010.

K. Liu, Q. Zhao, and B. Krishnamachari, "Dynamic multichannel access with imperfect channel state detection," *IEEE Trans. Signal Process.*, **58**, no. 5, 2795–2808, May 2010.

K. J. R. Liu and B. Wang, *Cognitive Radio Networking and Security*, Cambridge University Press, 2011.

N. Liu, I. Marić, A. Goldsmith, and S. Shamai (Shitz), "Bounds and capacity results for the cognitive Z-interference channel," *Proc. of the IEEE International Symposium on Information Theory*, Seoul, Korea, June 28–July 3, 2009.

G. Lorden, "2-SPRT's and the modified Kiefer-Weiss problem of minimizing an expected sample size," *Ann. Stat.*, **4**, no. 2, 281–291, 1976.

Y. Lu, H. He, J. Wang, and S. Li, "Energy-efficient dynamic spectrum access using no-regret learning," *Proc. 7th Int. Conf. Information, Communications, and Signal Processing (ICICS 2009)*, Macau, China, Dec. 8–10, 2009.

J. Lundén, *Spectrum Sensing for Cognitive Radio and Radar Systems*, D.Sc. Dissertation, Helsinki University of Technology, Espoo, Finland, 2009.

J. Lundén, S. A. Kassam, and V. Koivunen, "Robust nonparametric cyclic correlation-based spectrum sensing for cognitive radio," *IEEE Trans. Signal Process.*, **58**, no. 1, 38–52, Jan. 2010.

J. Lundén, V. Koivunen, A. Huttunen, and H. V. Poor, "Collaborative cyclostationary spectrum sensing for cognitive radio systems," *IEEE Trans. Signal Process.*, **57**, no. 11, 4182–4195, Nov. 2009.

J. Lundén, V. Koivunen, S. R. Kulkarni, and H. V. Poor, "Reinforcement learning based distributed multiagent sensing policy for cognitive radio networks," *Proc. IEEE Int. Symp. New Frontiers in Dynamic Spectrum Access Networks (DySPAN 2011)*, Aachen, Germany, May 3–6, 2011.

J. Lundén, V. Koivunen, S. R. Kulkarni, and H. V. Poor, "Exploiting spatial diversity in multiagent reinforcement learning based spectrum sensing," *Proc. 4th IEEE Int. Workshop Computational Advances in Multi-Sensor Adaptive Processing (CAMSAP 2011)*, San Juan, Puerto Rico, Dec. 13–16, 2011.

L. Luo and S. Roy, "Analysis of search schemes in cognitive radio," *Proc. 2nd IEEE Workshop Networking Technologies for Software Defined Radio Networks*, San Diego, CA, USA, June 18–21, 2007, pp. 17–24.

J. Ma, G. Y. Li, and B. H. Juang, "Signal processing in cognitive radio," *Proc. IEEE*, **97**, no. 5, pp. 805–823, May 2009.

J. Ma, G. Zhao, and Y. Li, "Soft combination and detection for cooperative spectrum sensing in cognitive radio networks," *IEEE Trans. Wireless Commun.*, **7**, no. 11, 4502–4507, Nov. 2008.

Z. Ma, W. Chen, K. B. Letaief, and Z. Cao, "A semi range-based iterative localization algorithm for cognitive radio networks," *IEEE Trans. Veh. Technol.*, **59**, no. 2, 704–717, Feb. 2010.

J. T. MacDonald and D. R. Ucci, "Interference temperature limits of IEEE 802.11 protocol radio channels," *IEEE Electro/Information Technology Conference (EIT'07)*, Chicago, IL, May 17–20, 2007.

A. B. MacKenzie and L. A. DaSilva, *Game Theory for Wireless Engineers*, Morgan & Claypool, 2006.

J. MacLellan, S. Lam, and X. Lee, "Residential indoor RF channel characterization," *Proc. IEEE Veh. Technol. Conf.*, pp. 210–213, 1993.

M. A. Maddah Ali, S. A. Motahari, and A. K. Khandani, "Communication over MIMO X channels: interference alignment, decomposition, and performance analysis," *IEEE Trans. Inf. Theory*, **54**, no. 8, 3457–3470, Aug. 2008.

A. Mahajan and D. Teneketzis, "Multi-armed bandit problems," in: A. O. Hero III, D. Castañón, D. Cochran, and K. Kastella (eds.), *Foundations and Applications of Sensor Management*, pp. 121–151, Springer, 2008.

I. Marić, A. Goldsmith, G. Kramer, and S. Shamai (Shitz), "On the capacity of interference channels with a partially-cognitive transmitter," *Proc. of the IEEE International Symposium on Information Theory*, Nice, France, June 24–29, 2007.

I. Marić, R. D. Yates, and G. Kramer, "Strong interference channel with unidirectional cooperation," *Proc. of the UCSD Workshop on Information Theory and Applications*, La Jolla, CA, Feb. 6–10, 2006.

I. Marić, R. D. Yates, and G. Kramer, "Capacity of interference channels with partial transmitter cooperation," *IEEE Trans. Inf. Theory*, **53**, no. 10, 3536–3548, Oct. 2007.

P. Marshall, *Quantitative Analysis of Cognitive Radio and Network Performance*. Artech House, 2010.

U. Martin, "Spatio-temporal radio channel characteristics in urban macrocells," *IEE Proc. Radar, Sonar and Navigation*, **145**, 42–49, Feb. 1998.

K. Marton, "A coding theorem for the discrete memoryless broadcast channel," *IEEE Trans. Inf. Theory*, **25**, no. 3, 306–311, May 1979.

D. Matolak and J. Frolik, "Worse-than-Rayleigh fading: experimental results and theoretical models," *IEEE Commun. Mag.*, **49**, issue 4, 140–146, 2011.

J. W. Matthews, "Sharp error bounds for intersymbol interference," *IEEE Trans. Inf. Theory*, **IT-19**, no. 4, 440–447, July 1973.

M. McHenry, E. Livsics, T. Nguyen, and N. Majumdar, "Xg dynamic spectrum access field test results," *IEEE Commun. Magazine*, **45**, no. 6, 51–57, June 2007.

M. McHenry, D. McCloskey, and G. Lane-Roberts, "New York City Spectrum Occupancy Measurements Sep. 2004," Tech. Rep., Dec. 2004.

C. F. Mecklenbräuker, A. F. Molisch, J. Karedal, *et al.*, "Vehicle channel characterization and its implications for wireless system design and performance," *Proc. IEEE*, **99**, no. 7, 1189–1212, July 2011.

J. Medbo, H. Hallenberg, and J. E. Berg, "Propagation characteristics at 5 GHz in typical radio-LAN scenarios," *Proc. IEEE Veh. Technol. Conf.*, pp. 185–189, 1999.

J. Medbo and P. Schramm, "Channel models for HIPERLAN/2," *ETSI/BRAN document no. 3ERI085B*.

Y. Mei, "Sequential change-point detection when unknown parameters are present in the pre-change distribution," *Ann. Stat.*, **34**, no. 1, 92–122, 2006.

Y. Mei, "Asymptotic optimality theory for decentralized sequential hypothesis testing in sensor networks," *IEEE Trans. Inf. Theory*, **54**, no. 5, 2072–2089, 2008.

Y. Mei, "Is average run length to false alarm always an informative criterion?," *Sequential Anal.*, **27**, no. 4, 354–376, Oct. 2008.

Z. Miljanic, I. Seskar, K. Le, and D. Raychaudhuri, "The WINLAB network centric cognitive radio hardware platform – WiNC2R," *Mobile Networks and Applications*, **13**, no. 5, October 2008.

G. Minden, J. Evans, L. Searl, *et al.*, "Cognitive radios for dynamic spectrum access – an agile radio for wireless innovation," *IEEE Commun. Magazine*, **45**, no. 5, 113–121, May 2007.

J. Mitola III, *Cognitive Radio: Model-Based Competence for Software Radio*, Licentiate Thesis, The Royal Institute of Technology. Stockholm, Sweden, Aug. 1999.

J. Mitola and G. Q. Maguire, "Cognitive radio: making software radios more personal," *IEEE Pers. Commun.*, **6**, no. 4, 13–18, Aug. 1999.

U. Mitra and H. V. Poor, "Activity detection in a multi-user environment," *Wireless Pers. Commun. – Special Issue on Signal Separation and Cancellation for Personal, Indoor and Mobile Radio Communications*, **3**, Nos. 1–2, 149–174, Jan. 1996.

W. Mohr, "Wideband propagation measurements of mobile radio channels in mountainous areas in the 1800 MHz frequency range," *Proc. IEEE Veh. Technol. Conf.*, pp. 49–52, 1993.

A. F. Molisch, *Wireless Communications* (2nd edn.). J. Wiley & Sons, 2011.

A. F. Molisch, H. Asplund, R. Heddergott, M. Steinbauer, and T. Zwick, "The COST259 directional channel model part I: overview and methodology," *IEEE Trans. Wireless Commun.*, **5**, no. 12, 3421–3433, Dec. 2006.

A. F. Molisch, K. Balakrishnan, C. C. Chong, *et al.*, "A comprehensive model for ultrawideband propagation channels," *IEEE Trans. Antennas Propag.*, **54**, no. 11, 3151–3166, Nov. 2006.

A. F. Molisch, J. R. Foerster, and M. Pendergrass, "Channel models for ultra-wideband personal area networks," *IEEE Wireless Commun. Mag.*, **10**, no. 6, 14–21, Dec. 2003.

A. F. Molisch, L. J. Greenstein, and M. Shafi, "Propagation issues for cognitive radio," *Proc. IEEE*, **97**, no. 5, 787–804, May 2009.

A. F. Molisch and H. Hofstetter, "The COST 273 MIMO channel model," in L. Correia (ed.), *Mobile Broadband Multimedia Networks*, Academic Press, 2006.

A. F. Molisch and F. Tufvesson, "Multipath propagation models for broadband wireless systems," in M. Ibnkahla (ed.), *Handbook of Signal Processing for Wireless Commmunications*, CRC Press, 2004.

A. F. Molisch, F. Tufvesson, J. Karedal, and C. Mecklenbraueker, "A survey on vehicle-to-vehicle propagation channels," *IEEE Wireless Comm.* **16**, issue 6, 12–22, 2009.

A. S. Motahari and A. K. Khandani, "Capacity bounds for the Gaussian interference channel," *Proc. of the IEEE International Symposium on Information Theory*, Toronto, Canada, July 6–11, 2008.

A. Motamedi and A. Bahai, "Optimal channel selection for spectrum-agile low-power wireless packet switched networks in unlicensed band," *EURASIP J. Wireless Commun. Netw.*, **2008**, Article ID 896420, 2008.

G. V. Moustakides, "Optimal stopping times for detecting changes in distributions," *Ann. Stat.*, **14**, no. 4, 1379–1387, 1986.

G. V. Moustakides, A. S. Polunchenko, and A. G. Tartakovsky, "A numerical approach to performance analysis of quickest change-point detection procedures," *Statist. Sinica*, **21**, no. 2, 571–596, 2011.

L. Musavian and S. Aissa, "Capacity and power allocation for spectrum-sharing communications in fading channels," *IEEE Trans. Wireless Commun.*, **8**, no. 1, 148–156, Jan. 2009.

M. Nakagami, "The M-distribution: a general formula of intensity of rapid fading," in W. C. Hoffman (ed.), *Statistical Methods in Radio Wave Propagation*. Pergamon Press, 1960.

National Telecommunications and Information Administration (NTIA), "FCC frequency allocation chart," Tech. Rep., 2003. www.ntia.doc.gov/files/ntia/publications/2003-allochrt.pdf.

B. Nazer and M. Gastpar, "Computation over multiple-access channels," *IEEE Trans. Inf. Theory*, **53**, no. 10, Oct. 2007.

B. Nazer, M. Gastpar, S. Jafar, and S. Vishwanath, "Ergodic interference alignment," *Proc. IEEE International Symposium on Information Theory*, Seoul, S. Korea, June 28–July 3, 2009, pp. 1769–1773.

T. Newman, S. Hasan, D. Depoy, T. Bose, and J. Reed, "Designing and deploying a building-wide cognitive radio network testbed," *IEEE Commun. Magazine*, **48**, no. 9, pp. 106–112, Sept. 2010.

S. Nikolaou, R. Bairavasubramanian, C. Lugo Jr., et al., "Pattern and frequency reconfigurable annular slot antenna using pin diodes," *IEEE Trans. Antennas and Propag.*, **54**, no. 2, 439–448, Feb. 2006.

J. Niño-Mora, "Restless bandits, partial conservation laws and indexability," *Adv. Appl. Prob.*, **33**, no. 1, 76–98, 2001.

J. Niño-Mora, "Dynamic priority allocation via restless bandit marginal productivity indices," *TOP*, **15**, no. 2, 161–198, 2007.

J. Niño-Mora, "A $(2/3)n^3$ fast-pivoting algorithm for the Gittins index and optimal stopping of a Markov chain," *INFORMS J. Comput.*, **19**, no. 4, 596–606, 2007.

J. Niño-Mora, "An index policy for dynamic fading-channel allocation to heterogeneous mobile users with partial observations," *Proc. 4th Euro-NGI Conf. Next Generation Internet Networks*, Kraków, Poland, Apr. 28–30, 2008.

J. Niño-Mora, "A restless marginal productivity index for opportunistic spectrum access with sensing errors," in R. Núñez-Queija and J. Resing (eds.), *Network Control and Optimization, 3rd Euro-NF Conf., NET-COOP 2009 Eindhoven, The Netherlands, Nov. 23-25, 2009 Proceedings*, Lecture Notes in Computer Science, vol. 5894, Springer, 2009.

J. Niño-Mora and S. S. Villar, "Multitarget tracking via restless bandit marginal productivity indices and Kalman filter in discrete time," *Proc. Joint 48th IEEE Conf.*

Decision and Control and 28th Chinese Control Conf., Shanghai, China, Dec. 16–18, 2009.

Office of Engineering and Technology in FCC, "Understanding the FCC regulations for low-power, non-licensed transmitters," Tech. Rep., 1993. `transition.fcc.gov/Bureaus/Engineering_Technology/Documents/bulletins/oet63/oet63rev.pdf`.

D.-C. Oh, H.-C. Lee, and Y.-H. Lee, "Linear hard decision combining for cooperative spectrum sensing in cognitive radio systems," *Proc. 72nd IEEE Vehicular Technology Conf. (VTC 2010-Fall)*, Ottawa, ON, Canada, Sep. 6–9, 2010, pp. 1–5.

J. Oksanen, V. Koivunen, J. Lundén, and A. Huttunen, "Diversity-based spectrum sensing policy for detecting primary signals over multiple frequency bands," *Proc. 35th IEEE Int. Conf. Acoustics, Speech, and Signal Processing (ICASSP 2010)*, Dallas, TX, Mar. 14–19, 2010.

J. Oksanen, J. Lundén, and V. Koivunen, "Characterization of spatial diversity in cooperative spectrum sensing," *Proc. 4th Int. Symp. Communications, Control, and Signal Processing (ISCCSP 2010)*, Limassol, Cyprus, Mar. 3–5, 2010.

J. Oksanen, J. Lundén, and V. Koivunen, "Reinforcement learning-based multiband sensing policy for cognitive radios," *Proc. 2nd Int. Workshop Cognitive Information Processing (CIP 2010)*, Elba Island, Tuscany, Italy, June 14–16, 2010.

J. Oksanen, J. Lundén, and V. Koivunen, "Reinforcement learning method for energy efficient cooperative multiband spectrum sensing," *Proc. IEEE Int. Workshop Machine Learning for Signal Processing (MLSP 2010)*, Kittilä, Finland, Aug. 29–Sep. 1, 2010.

J. Oksanen, J. Lundén, and V. Koivunen, "Reinforcement learning based sensing policy optimization for energy efficient cognitive radio networks," *Neurocomputing – Special Issue on Machine Learning for Signal Processing 2010*, **80**, 102–110, Mar. 2012.

Y. Okumura, E. Ohmori, T. Kawano, and K. Fukuda, "Field strength and its variability in UHF and VHF land-mobile radio service," *Rev. Elec. Commun. Lab.*, **16**, no. 9, 1968.

F. W. J. Olvert, D. W. Lozier, R. F. Boisvert, and C. W. Clark, *NIST Handbook of Mathematical Functions*. NIST and Cambridge University Press, 2010.

I. Oppermann, J. Talvitie, and D. Hunter, "Wide-band wireless local loop channel for urban and sub-urban environments at 2 GHz," *Proc. IEEE Int. Conf. Commun.*, pp. 61–65, 1997.

"ORBIT: open-access research testbed for next-generation wireless networks," `www.orbit-lab.org/`.

T. Oskiper and H. V. Poor, "Online activity detection in a multiuser environment using the matrix CUSUM algorithm," *IEEE Trans. Inf. Theory*, **46**, no. 2, 477–493, Feb. 2002.

A. Ozgur and O. Leveque, "Throughput–delay tradeoff for hierarchical cooperation in ad hoc wireless networks," *IEEE Trans. Inf. Theory*, **56**, no. 3, 1369–1377, Mar. 2010.

A. Ozgur, O. Leveque, and D. N. C. Tse. "Hierarchical cooperation achieves optimal capacity scaling in ad hoc networks," *IEEE Trans. Inf. Theory*, **53**, no. 10, 3549–3572, Oct. 2007.

A. Ozgur and D. N. C. Tse, "Achieving linear scaling with interference alignment," *Proc. of the IEEE International Symposium on Information Theory*, Seoul, Korea, June 28–July 3, 2009, pp. 1754–1758.

E. S. Page, "Continuous inspection schemes," *Biometrika*, **41**, no. 1–2, 100–115, 1954.

P. Pajusco, "Experimental characterization of DOA at the base station in rural and urban area," *Proc. IEEE Veh. Technol. Conf.*, pp. 993–997, 1998.

H. Pan, J. T. Bernhard, and V. K. Nair, "Reconfigurable single-armed square spiral microstrip antenna design," *IEEE International Workshop on Antenna Technology Small Antennas and Novel Metamaterials*, **6–8**, 2006, pp. 180–183.

H. K. Pan, J. Tsai, J. Martinez, S. Golden, V. K. Nair, and J. T. Bernhard, "Reconfigurable antenna implementation in multi-radio platform," *IEEE Antennas and Propagation Society International Symposium, 2008*, July 2008, pp. 1–4.

D. G. Pandelis and D. Teneketzis, "On the optimality of the Gittins index rule for multi-armed bandits with multiple plays," *Math. Meth. Oper. Res.*, **50**, no. 3, 449–461, 1999.

S. Pandey, D. Chakrabarti, and D. Agarwal, "Multi-armed bandit problems with dependent arms," *Proc. 24th Int. Conf. Machine Learning (ICML 2007)*, Corvallis, OR, USA, June 20–24, 2007.

J.-S. Pang, G. Scutari, and D. P. Palomar, "Design of cognitive radio systems under temperature-interference constraints: a variational inequality approach," *IEEE Trans. Signal Process.*, **58**, no. 6, 3251–3271, June 2010.

K. I. Pedersen, P. E. Mogensen, and B. H. Fleury, "Power azimuth spectrum in outdoor environments," *Electronics Letters*, **33**, 1583–1584, Aug. 1997.

K. I. Pedersen, P. E. Mogensen, B. H. Fleury, F. Frederiksen, K. Olesen, and S. L. Larsen, "Analysis of time, azimuth, and doppler dispersion in outdoor radio channels," *Proc. ACTS Mobile Communications Summit*, Aalborg, Denmark, pp. 308–313, 1997.

E. C. Y. Peh, Y.-C. Liang, Y. L. Guan, and Y. Zeng, "Optimization of cooperative sensing in cognitive radio networks: a sensing-throughput tradeoff view," *IEEE Trans. Veh. Technol.*, **58**, no. 9, 5294–5299, Nov. 2009.

Y. Pei, A. T. Hoang, and Y.-C. Liang, "Sensing-throughput tradeoff in cognitive radio networks: how frequently should spectrum sensing be carried out?," *IEEE 18th International Symposium on Personal, Indoor and Mobile Radio Communications (PIMRC 2007)*, Athens, Greece, Sep. 3–7, 2007.

B. Peleg and P. Sudhölter, *Introduction to the Theory of Cooperative Games* (2nd edn.). Springer-Verlag, 2007.

Q. Peng, K. Zeng, J. Wang, and S. Li, "A distributed spectrum sensing scheme based on credibility and evidence theory in cognitive radio context," *Proc. 17th IEEE Int. Symp. Personal, Indoor, and Mobile Radio Communications (PIMRC 2006)*, Helsinki, Finland, Sep. 11–14, 2006, pp. 1–5.

C. E. Perkins, *Ad Hoc Networking*, Addison-Wesley, 2001.

M. Pollak, "A robust changepoint detection method," *Sequential Analysis*, **29**, no. 2, 146–161, Apr. 2010.

M. Pollak and A. G. Tartakovsky, "Optimality properties of the Shiryaev-Roberts procedure," *Statist. Sinica*, **19**, 1729–1739, 2009.

A. S. Polunchenko and A. G. Tartakovsky, "On optimality of the Shiryaev-Roberts procedure for detecting a change in distribution," *Ann. Stat.*, **38**, no. 6, 3445–3457, 2010.

H. V. Poor, *An Introduction to Signal Detection and Estimation* (2nd edn.). Springer-Verlag, 1994.

H. V. Poor and O. Hadjiliadis, *Quickest Detection*, Cambridge University Press, 2009.

C. Prehofer and C. Bettstetter, "Self-organization in communication networks: principles and design paradigms," *IEEE Commun. Magazine*, **43**, no. 7, 78–85, July 2005.

"Quality of service and mobility driven cognitive radio systems," www.ict-qosmos.eu/home.html.

Z. Quan, S. Cui, H. V. Poor, and A. H. Sayed, "Collaborative wideband sensing for cognitive radios," *IEEE Signal Process. Mag. – Special Issue on Signal Processing for Cognitive Radio Networks*, **25**, no. 6, 60–73, Nov. 2008.

Z. Quan, S. Cui, and A. H. Sayed, "Optimal linear cooperation for spectrum sensing in cognitive radio networks," *IEEE J. Sel. Topics Signal Process.*, **2**, no. 4, 2431–2436, Apr. 2010.

Z. Quan, S. Cui, A. H. Sayed, and H. V. Poor, "Optimal multiband joint detection for spectrum sensing in cognitive radio networks," *IEEE Trans. Signal Process.*, **57**, no. 3, 1128–1140, Mar. 2009.

C. Rago, P. Willett, and Y. Bar-Shalom, "Censoring sensors: a low-communication-rate scheme for distributed detection," *IEEE Trans. Aerosp. Electron. Syst.*, **32**, no. 2, 554–568, Apr. 1996.

A. Rahman and P. Gburzynski, "Hidden problems with the hidden node problem," *23rd Biennial Symposium on Communications*, pp. 270-273, May 29–June 1, 2006.

J. Rajasekharan, J. Eriksson, and V. Koivunen, "Cooperative game-theoretic solutions to spectrum sharing in cognitive radios," *Proc. 44th Asilomar Conf. Signals, Systems, and Computers*, Pacific Grove, CA, Nov. 7–10, 2010.

R. Rajbanshi, Q. Chen, A. M. Wyglinski, G. J. Minden, and J. B. Evans, "Quantitative comparison of agile modulation techniques for cognitive radio transceivers," *IEEE Consumer Communications and Networking Conference, 2007*, Jan. 2007, pp. 1144–1148.

R. Rajbanshi, A. M. Wyglinski, and G. J. Minden, "An efficient implementation of NC-OFDM transceivers for cognitive radios," *International Conference on Cognitive Radio Oriented Wireless Networks and Communications*, June 2006, pp. 1–5.

R. Rajbanshi, A. M. Wyglinski, and G. J. Minden, "Peak-to-average power ratio analysis for NC-OFDM transmissions," *IEEE Vehicular Technology Conference*, Sept. 2007, pp. 1351–1355.

C. Raman, *Scheduling and Relaying in Interference Limited Wireless Networks*, Ph.D. Dissertation, Rutgers University, 2010.

C. Raman, R. Yates, and N. Mandayam, "Scheduling variable rate links via a spectrum server," *Proc. IEEE DySPAN*, 2005, Baltimore, MD.

T. S. Rappaport, "Characterization of UHF multipath radio channels in factory buildings," *IEEE Trans. Antennas Propag.*, **37**, no. 8, 1058–1069, Aug. 1989.

T. S. Rappaport, *Wireless Communications: Principles and Practice* (2nd edn.). Prentice Hall, 2001.

T. S. Rappaport, S. Y. Seidel, and R. Singh, "900-MHz multipath propagation measurements for US digital cellular radiotelephone," *IEEE Trans. Veh. Technol.*, **39**, no. 2, 132–139, May 1990.

D. Raychaudhuri, N. Mandayam, J. Evans, B. Ewy, S. Seshan, and P. Steenkiste, "CogNet – an architectural foundation for experimental cognitive radio networks within the future internet," *Proc. MobiArch*, 2006.

Recommendation ITU P 1546-3, "Method for point-to-area predictions for terrestrial services in the frequency range 30 MHz to 3000 MHz," 2007.

"Rice University WARP," `warp.rice.edu/`.

S. Rini, D. Tuninetti, and N. Devroye, "New inner and outer bounds for the memoryless cognitive interference channel and some capacity results," *IEEE Trans. Inf. Theory*, **57**, no. 7, 4087–4109, July 2011.

S. W. Roberts, "A comparison of some control chart procedures," *Technometrics*, **8**, no. 3, 411–430, Aug. 1966.

V. Rodoplu and T. H. Meng, "Bits-per-joule capacity of energy-limited wireless networks," *IEEE Trans. Wireless Commun.*, **6**, no. 3, 857–865, Mar. 2007.

G. A. Rummery and M. Niranjan, "On-line Q-learning using connectionist systems," *Technical Report CUED/F-INFENG/TR 166*, Cambridge University Engineering Department, Cambridge, UK, Sep. 1994.

W. Saad, Z. Han, T. Başar, M. Debbah, and A. Hjørungnes, "Coalition formation games for collaborative spectrum sensing," *IEEE Trans. Veh. Technol.*, **60**, no. 1, 276–297, Jan. 2011.

W. Saad, Z. Han, M. Debbah, A. Hjørungnes, and T. Başar, "Coalitional game theory for communication networks: a tutorial," *IEEE Signal Process. Mag.*, **26**, no. 5, 77–97, Sep. 2009.

J. Sachs, I. Marić, and A. J. Goldsmith, "Cognitive cellular systems within the TV spectrum," *Proc. of the IEEE International Symposium on New Frontiers in Dynamic Spectrum Access Networks (DySPAN)*, Singapore, Apr. 6–9, 2010, pp. 1–12.

A. Sahai, R. Tandra, S. M. Mishra, and N. Hoven, "Fundamental design tradeoffs in cognitive radio systems," *Proc. 1st Int. Workshop Technol. Policy for Accessing Spectrum (TAPAS 2006)*, Boston, MA: Aug. 2–5, 2006.

A. Saleh and R. A. Valenzuela, "A statistical model for indoor multipath propagation," *IEEE J. Sel. Areas Commun.*, **5**, no. 2, 128–137, Feb. 1987.

M. Sanna and M. Murroni, "Optimization of non-convex multiband cooperative sensing with genetic algorithms," *IEEE J. Sel. Topics Signal Process.*, **5**, no. 1, 87–96, Feb. 2011.

S. Sardellitti, M. Giona, and S. Barbarossa, "Fast distributed average consensus algorithms based on advection-diffusion processes," *IEEE Trans. Signal Process.*, **58**, no. 2, 826–842, Feb. 2010.

H. Sato, "Two user communication channels," *IEEE Trans. Inf. Theory*, **23**, no. 3, 295–304, May 1977.

H. Sato, "The capacity of the Gaussian interference channel under strong interference," *IEEE Trans. Inf. Theory*, **27**, no. 6, 786–788, Nov. 1981.

T. Schmidl and D. Cox, "Robust frequency and timing synchronization for OFDM," *IEEE Trans. Commun.*, **45**, no. 12, 1613–1621, Dec. 1997.

M. Schwartz, W. R. Bennett, and S. Stein, *Communication Systems and Techniques*. McGraw-Hill, 1966.

G. Scutari and D. P. Palomar, "MIMO cognitive radio: a game theoretical approach," *IEEE Trans. Signal Process.*, **58**, no. 2, 761–780, Feb. 2010.

S. Y. Seidel, T. S. Rappaport, S. Jain, M. L. Lord, and R. Singh, "Path loss, scattering and multipath delay statistics in four European cities for digital cellular and microcellular radiotelephone," *IEEE Trans. Veh. Technol.*, **40**, no. 4, 721–730, Nov. 1991.

S. Sesia, I. Toufik, and M. Baker, Eds., *LTE – The UMTS Long Term Evolution*. J. Wiley & Sons, 2009.

M. Shafi, M. Zhang, A. L. Moustakas, P. J. Smith, A. F. Molisch, F. Tufvesson, and S. H. Simon, "Polarized MIMO channels in 3-D: models, measurements and mutual information," *IEEE J. Sel. Areas Commun.*, **24**, no. 3, 514–527, Mar. 2006.

X. Shang, G. Kramer, and B. Chen, "A new outer bound and the noisy-interference sumrate capacity for Gaussian interference channels," *IEEE Trans. Inf. Theory*, **55**, no. 2, 689–699, Feb. 2009.

C. E. Shannon, "A mathematical theory of communication," *Bell Sys. Tech. Journal*, pp. 379–423, 623–656, 1948.

C. E. Shannon, "Communications in the presence of noise," *Proc. IRE*, **37**, pp. 10–21, 1949.

C. E. Shannon, "Two-way communication channels," *Proc. Berkeley Symposium on Math, Statistics, and Probability*, **1**, 1961, 611–644.

C. E. Shannon and W. Weaver, *The Mathematical Theory of Communication*, University of Illinois Press, 1949.

Shared Spectrum Company, "Comprehensive spectrum occupancy measurements over six different locations," Aug. 2005. www.sharedspectrum.com/?section=nsf_summary.

Shared Spectrum Company, "General survey of radio frequency bands – 30 MHz to 3 GHz," Tech. Rep., Sep. 2010.

F. Shayegh and M. R. Soleymani, "Rateless codes for cognitive radio in a virtual unlicensed spectrum," *Proc. IEEE Sarnoff Symposium*, Princeton, NJ, May 2–4, 2011.

Y. Shei and Y. T. Su, "A sequential test based cooperative spectrum sensing scheme for cognitive radios," *Proc. 19th Int. Symp. Personal, Indoor, and Mobile Radio Communications (PIMRC 2008)*, Cannes, France, Sep. 15–18, 2008.

S. Shetty, M. Song, C. Xin, and E. K. Park, "A learning-based multiuser opportunistic spectrum access approach in unslotted primary networks," *Proc. IEEE Conf. Computer Communications (INFOCOM 2009)*, Rio de Janeiro, Brazil, Apr. 19–25, 2009.

A. N. Shiryaev, "On optimum methods in quickest detection problems," *Theory Probab. Appl.*, **8**, no. 1, 22–46, Jan. 1963.

A. N. Shiryaev, "Quickest detection problems: fifty years later," *Sequential Anal.*, **29**, no. 4, pp. 345–385, Oct. 2010.

Y. Shoham and K. Leyton-Brown, *Multiagent Systems: Algorithmic, Game-Theoretic, and Logical Foundations*, Cambridge University Press, 2009.

D. Siegmund, *Sequential Analysis*, Springer-Verlag, 1985.

M. K. Simon and M.-S. Alouini, *Digital Communication over Fading Channels* (2nd edn.). J. Wiley & Sons, 2005.

S. P. Singh, T. Jaakkola, M. L. Littman, and C. Szepesvári, "Convergence results for single-step on-policy reinforcement-learning algorithms," *Mach. Learn.*, **39**, 287–308, 2000.

B. Sklar, "Rayleigh fading channels in mobile digital communication systems. I. Characterization," *IEEE Commun. Mag.*, **35**, no. 9, 136–146, Sep. 1997.

A. Somekh-Baruch, S. Shamai (Shitz), and S. Verdú, "Cognitive interference channels with state information," *Proc. IEEE International Symposium on Information Theory*, Toronto, Canada, July 6–11, 2008.

I. M. Sonin, "A generalized Gittins index for a Markov chain and its recursive calculation," *Technical Report*, Department of Mathematics, University of North Carolina at Charlotte, Charlotte, NC, 2005.

E. S. Sousa, V. M. Jovanovic, and C. Daigneault, "Delay spread measurements for the digital cellular channel in Toronto," *IEEE Trans. Veh. Technol.*, **43**, no. 4, 837–847, Nov. 1994.

"Spatial channel model for multiple input multiple output (MIMO) simulations (Rel. 7)," 3GPP TR 25.996 V7.0.0, 2007.

Q. H. Spencer, B. D. Jeffs, M. A. Jensen, and A. L. Swindlehurst, "Modeling the statistical time and angle of arrival characteristics of an indoor multipath channel," *IEEE J. Sel. Areas Commun.*, **18**, 347–360, Mar. 2000.

S. Sridharan, A. Jafarian, S. Vishwanath, S. A. Jafar, and S. Shamai (Shitz), "A layered lattice coding scheme for a class of three user Gaussian interference channels," *Proc. Allerton Conference on Communication, Control and Computing*, Monticello, IL, Sep. 24–26, 2008, pp. 531–538.

M. Steinbauer and A. F. Molisch (chapter eds.), "Directional channel models," in L. Correia (ed.), *Flexible Personalized Wireless Communications*, J. Wiley & Sons, 2001.

M. Steinbauer, A. F. Molisch, and E. Bonek, "The double-directional radio channel," *IEEE Antennas Propag. Mag.*, **43**, no. 4, 51–63, Aug. 2001.

G. Stüber, *Principles of Mobile Communications*. Kluwer, 1996, 2nd edition, 2001.

H. A. Suraweera, P. J. Smith, and M. Shafi, "Capacity limits and performance analysis of cognitive radio with imperfect channel knowledge," *IEEE Trans. Veh. Technol.*, **59**, no. 4, 1811-1822, May 2010.

R. S. Sutton and A. G. Barto, *Reinforcement Learning: An Introduction*, MIT Press, 1998.

C. Szepesvári, *Algorithms for Reinforcement Learning*, Synthesis Lectures on Artificial Intelligence and Machine Learning, Morgan & Claypool, 2010.

S. S. Szyszkowicz, H. Yanikomeroglu, and J. S.Thompson, "On the feasibility of wireless shadowing correlation models," *IEEE Trans. Veh. Technol.*, **59**, no. 9, pp. 4222–4236, Nov. 2010.

A. Taherpour, Y. Norouzi, M. Nasiri-Kenari, A. Jamshidi, and Z. Zeinalpour-Yazdi, "Asymptotically optimum detection of primary user in cognitive radio networks," *IET Commun.*, **1**, no. 6, 1138–1145, Dec. 2007.

L. Talbi and G. Y. Delisle, "Experimental characterization of EHF multipath indoor radio channels," *IEEE J. Sel. Areas Commun.*, **14**, no. 3, 431–440, Apr. 1996.

B. Talha and M. Patzold, "Channel models for mobile-to-mobile cooperative communication systems," *IEEE Veh. Technol. Mag.*, **6**, no. 2, 33–43, 2011.

R. Tandra, *Fundamental Limits on Detection in Low SNR*, MS Thesis, University of California Berkeley, Spring 2005.

R. Tandra and A. Sahai, "SNR walls for feature detectors," *Proc. 2nd IEEE Int. Symp. New Frontiers in Dynamic Spectrum Access Networks (DySpan 2007)*, Dublin, Ireland, pp. 559–570, Apr. 2007.

R. Tandra and A. Sahai, "SNR walls for signal detection," *IEEE Journal Sel. Topics Signal Process.*, **2**, no. 1, 4–17, Feb. 2008.

R. Tandra and A. Sahai, "Overcoming SNR walls through macroscale features," *Forty-Sixth Annual Allerton Conference*, Allerton House, UIUC, IL, pp. 583–590, Sep. 23–26, 2008.

S. Tantaratana and H. V. Poor, "Asymptotic efficiencies of truncated sequential tests," *IEEE Trans. Inf. Theory*, **28**, no. 6, 911–923, Nov. 1982.

H. F. A. Tarboush, S. Khan, R. Nilavalan, H. S. Al-Raweshidy, and D. Budimir, "Reconfigurable wideband patch antenna for cognitive radio," *Antennas Propagation Conference, 2009*, Nov. 2009, pp. 141–144.

G. Taricco, "Optimization of linear cooperative spectrum sensing for cognitive radio networks," *IEEE J. Sel. Topics Signal Process.*, **5**, no. 1, 77–86, Nov. 2009.

A. G. Tartakovsky and G. V. Moustakides, "State-of-the-art in Bayesian changepoint detection," *Sequential Anal.*, **29**, no. 2, 125–145, Apr. 2009.

A. G. Tartakovsky, M. Pollak, and A. S. Polunchenko, "Third-order asymptotic optimality of the generalized Shiryaev-Roberts changepoint detection procedures," *Teor. Veroyatnost. i Primenon.*, **56**, no. 3, 534–565, 2011, and *Theory Probab. Appl.*, to be published.

A. G. Tartakovsky and A. S. Polunchenko, "Quickest changepoint detection in distributed multisensor systems under unknown parameters," *Proc. 11th Int. Conf. Information Fusion*, Cologne, Germany, June 30–July 3, 2008.

A. G. Tartakovsky and V. V. Veeravalli, "Change-point detection in multichannel and distributed systems with applications," in N. Mukhopadhyay, S. Datta, and S. Chattopadhyay (eds.), *Applications of Sequential Methodologies*, pp. 331–363, Marcel Dekker, Inc., 2004.

A. G. Tartakovsky and V. V. Veeravalli, "Asymptotically optimal quickest change detection in distributed sensor systems," *Sequential Anal.*, **27**, no. 4, 441–475, 2008.

C. Tekin and M. Liu, "Online learning in opportunistic spectrum access: a restless bandit approach," *Proc. 30th IEEE Int. Conf. Computer Communications*, Shanghai, China, Apr. 10–15, 2011.

E. Telatar, "Capacity of multi-antenna Gaussian channels," *European Trans. Telecommun.*, **10**, no. 6, 585–596, Nov. 1999.

J. Tellado, ed., *Multicarrier Modulation with Low PAR: Applications to DSL and Wireless.* Kluwer Academic Publishers, 2000.

N. M. Temme, "Asymptotic and numerical aspects of the noncentral chi-square distribution," *Computers Math. Applic.*, **25**, no. 5, 55–63, 1993.

"The Universal Software Radio Peripheral," www.ettus.com/.

S. C. A. Thomopoulos, R. Viswanathan, and D. K. Bougoulias, "Optimal distributed decision fusion," *IEEE Trans. Aerosp. Electron. Syst.*, **25**, no. 5, 761–765, Sep. 1989.

Z. Tian and G. Giannakis, "Compressed sensing for wideband cognitive radios," *IEEE International Conference on Acoustics, Speech and Signal Processing, ICASSP.*, **4**, Apr. 2007, IV-1357–IV-1360.

M. Toeltsch, J. Laurila, K. Kalliola, A. F. Molisch, P. Vainikainen, and E. Bonek, "Statistical characterization of urban spatial radio channels," *IEEE J. Sel. Areas Commun.*, **20**, no. 3, 539–549, Apr. 2002.

M. Tokic, "Adaptive ϵ-greedy exploration in reinforcement learning based on value differences," in R. Dillmann, J. Beyerer, U. D. Hanebeck, and T. Schultz (eds.), *KI 2010: Advances in Artificial Intelligence*, Lecture Notes in Computer Science, vol. 6359, pp. 203–210, Springer, 2010.

P. Tortelier and M. M. Azeem, "Improving the erasure recovery performance of short codes for opportunistic spectrum access," *14th International Symposium on Wireless Personal Multimedia Communications (WPMC'11)*, Brest, France, Oct. 3–7, 2011.

S. Toumpis and A. J. Goldsmith, "Capacity regions for wireless ad hoc networks," *IEEE Trans. Wireless Commun.*, **2**, no. 4, 736–748, July 2003.

D. N. C. Tse and P. Viswanath, *Fundamentals of Wireless Communication*, Cambridge University Press, 2005.

J. N. Tsitsiklis, "Asynchronous stochastic approximation and Q-learning," *Mach. Learn.*, **16**, no. 3, 185–202, Sep. 1994.

Ultra-wideband wireless communications – theory and applications, *IEEE J. Sel. Areas Commun.*, **24**, no. 4, Apr. 2006.

J. Unnikrishnan and V. V. Veeravalli, "Algorithms for dynamic spectrum access with learning for cognitive radio," *IEEE Trans. Signal Process.*, **58**, no. 2, 750–760, Feb. 2010.

J. Unnikrishnan, V. V. Veeravalli, and S. Meyn, "Minimax robust quickest change detection," *IEEE Trans. Inf. Theory*, **57**, no. 3, 1604–1614, Mar. 2011.

"Urban transmission loss models for mobile radio in the 900 and 1800 MHz bands," European Co-operative in the Field of Science and Technical Research EURO-COST 231, Rev. 2, The Hague, Sep. 1991.

H. Urkowitz, "Energy detection of unknown deterministic signals," *Proc. IEEE*, **55**, no. 4, 523–531, Apr. 1967.

J. van de Beek, M. Sandell, and P. Borjesson, "ML estimation of time and frequency offset in OFDM systems," *IEEE Trans. Signal Process.*, **45**, no. 7, 1800–1805, July 1997.

P. P. Varaiya, J. C. Walrand, and C. Buyukkoc, "Extensions of the multiarmed bandit problem: the discounted case," *IEEE Trans. Autom. Control*, **AC–30**, no. 5, 426–439, May 1985.

P. K. Varshney, *Distributed Detection and Data Fusion*, Springer-Verlag, 1996.

V. V. Veeravalli, "Decentralized quickest change detection," *IEEE Trans. Inf. Theory*, **47**, no. 4, 1657–1665, May 2001.

V. V. Veeravalli, T. Basar, and H. V. Poor, "Decentralized sequential detection with a fusion center performing the sequential test," *IEEE Trans. Inf. Theory*, **39**, no. 2, 433–442, Mar. 1993.

V. V. Veeravalli and C. W. Baum, "Asymptotic efficiency of a sequential multihypothesis test," *IEEE Trans. Inf. Theory*, **41**, no. 6, 1994–1997, Nov. 1995.

S. Verdú, "On channel capacity per unit cost," *IEEE Trans. Inf. Theory*, **36**, no. 5, 1019–1030, Sep. 1990.

S. Verdú and T. S. Han, "A general formula for channel capacity," *IEEE Trans. Inf. Theory*, **40**, no. 4, pp. 1147–1157, July 1994.

J. Vermorel and M. Mohri, "Multi-armed bandit algorithm and empirical evaluation," in J. Ceama, R. Camacho, P. Brazdil, A. Jorge, and L. Torgo (eds.), *Machine Learning: ECML 2005*, Lecture Notes in Computer Science, **3720/2005**, 437–448, 2005.

W. Vickrey, "Counterspeculation, auctions, and competitive sealed tenders," *J. Financ.*, **16**, no. 1, 8–37, Mar. 1961.

S. Vishwanath, N. Jindal, and A. Goldsmith, "Duality, achievable rates, and sum-rate capacity of Gaussian MIMO broadcast channels," *IEEE Trans. Inf. Theory*, **49**, no. 10, 2658–2668, Oct. 2003.

R. Viswanathan and P. K. Varshney, "Distributed detection with multiple sensors: Part I – Fundamentals," *Proc. IEEE*, **85**, no. 1, 54–63, Jan. 1997.

M. Vu, N. Devroye, M. Sharif, and V. Tarokh, "Scaling laws of cognitive networks," *Proc. of the IEEE International Conference on Cognitive Radio Oriented Wireless Networks and Communications (CrownCom)*, Orlando, FL, July 31–Aug 3, 2007, pp. 2–8.

A. Wald, "Sequential tests of statistical hypotheses," *Ann. Math. Stat.*, **16**, no. 2, 117–186, June 1945.

A. Wald and J. Wolfowitz, "Optimum character of the sequential probability ratio test," *Ann. Math. Stat.*, **19**, no. 3, 326–339, 1948.

B. Wang and K. J. R. Liu, "Advances in cognitive radio networks: a survey," *IEEE J. Sel. Topics Signal Process.*, **5**, no. 1, 5–23, Feb. 2011.

B. Wang, Y. Wu, and K. J. R. Liu, "Game theory for cognitive radio networks: an overview," *Comput. Netw.*, **54**, no. 14, 2537–2561, Oct. 2010.

C.-C. Wang, S. R. Kulkarni, and H. V. Poor, "Bandit problems with side observations," *IEEE Trans. Autom. Control*, **50**, no. 3, 338–355, Mar. 2005.

C.-C. Wang, S. R. Kulkarni, and H. V. Poor, "Arbitrary side observations in bandit problems," *Adv. Appl. Math. — Special Issue Dedicated to Dr. David P. Robbins*, **34**, no. 4, 903–938, May 2005.

C.-J. Wang and W. T. Tsai, "A slot antenna module for switchable radiation patterns," *IEEE Antennas and Wireless Propag. Lett.*, **4**, 202–204, 2005.

H. Wang, J. Lee, S. Kim, and D. Hong, "Capacity enhancement of secondary links through spatial diversity in spectrum sharing," *IEEE Trans. Wireleless Commun.*, **9**, no. 2, 494–499, Feb. 2010.

H. Wang, G. Noh, D. Kim, S. Kim, and D. Hong, "Advanced sensing techniques of energy detection in cognitive radios," *J. Commun. Netw.*, **12**, no. 1, 19–29, Feb. 2010.

J. Wang, G. Scutari, and D. P. Palomar, "Robust MIMO cognitive radio via game theory," *IEEE Trans. Signal Process.*, **59**, no. 3, 1183–1201, Mar. 2011.

K. Wang and L. Chen, "On optimality of myopic policy for restless multi-armed bandit problem: an axiomatic approach," *IEEE Trans. Signal Process.*, **60**, no. 1, 300–309, Jan. 2012.

C. J. C. H. Watkins, *Learning from Delayed Rewards*, Ph.D. Dissertation, University of Cambridge, Cambridge, UK, 1989.

C. J. C. H. Watkins and P. Dayan, "Q-learning," *Mach. Learn.*, **8**, 279–292, 1992.

R. R. Weber and G. Weiss, "On an index policy for restless bandits," *J. Appl. Prob.*, **27**, no. 3, 637–648, Sep. 1990.

R. R. Weber and G. Weiss, "Addendum to 'On an index policy for restless bandits'," *Adv. Appl. Prob.*, **23**, no. 2, 429–430, June 1991.

H. Weingarten, Y. Steinberg, and S. Shamai, "The capacity region of the Gaussian multiple-input multiple-output broadcast channel," *IEEE Trans. Inf. Theory*, **52**, no. 9, 3936–3964, Sep. 2006.

T. Weiss and F. Jondral, "Spectrum pooling: an innovative strategy for the enhancement of spectrum efficiency," *IEEE Commun. Magazine*, **42**, no. 3, S8–14, Mar. 2004.

J. Weitzen and T. J. Lowe, "Measurement of angular and distance correlation properties of log-normal shadowing at 1900 MHz and its application to design of PCS systems," *IEEE Trans. Veh. Technol.*, **51**, no. 2, 265–273, Mar. 2002.

P. Whittle, "Arm-acquiring bandits," *Ann. Probab.*, **9**, no. 2, 284–292, Apr. 1981.

P. Whittle, "Restless bandits: activity allocation in a changing world," *J. Appl. Prob.*, **25**, 287–298, 1988.

J. Wiart, P. Pajusco, A. Levy, and J. C. Bic, "Analysis of microcellular wide band measurements in Paris," *Proc. IEEE International Symposium on Personal, Indoor and Mobile Radio Communications*, pp. 144–147, 1995.

B. Wild and K. Ramchandran, "Detecting primary receivers for cognitive radio applications," *Proc. 1st IEEE Int. Symp. New Frontiers in Dynamic Spectrum Access Networks (DySPAN 2005)*, Baltimore, MD, USA, Nov. 8–11, 2005, pp. 124–130.

M. Z. Win and R. A. Scholtz, "Ultra-wide bandwidth time-hopping spread-spectrum impulse radio for wireless multiple-access communications," *IEEE Trans. Commun.*, **48**, no. 4, 679–689, Apr. 2000.

L. A. Wolsey, *Integer Programming*. Wiley, 1998.

C. Wu, K. Chowdhury, M. Di Felice, and W. Meleis, "Spectrum management of cognitive radio using multi-agent reinforcement learning," *Proc. 9th Int. Conf. Autonomous Agents and Multiagent Systems (AAMAS 2010)*, Toronto, Canada, May 10–14, 2010.

G. Wu, S. Talwar, K. Johnsson, N. Himayat, and K. Johnson, "M2M: from mobile to embedded internet," *IEEE Commun. Magazine*, **49**, no. 4, 36–43, Apr. 2011.

W. Wu, S. Vishwanath, and A. Arapostathis, "Capacity of a class of cognitive radio channels: interference channels with degraded message sets," *IEEE Trans. Inf. Theory*, **53**, no. 11, 4391–4399, Nov. 2007.

Z. Wu and B. Natarajan, "Interference tolerant agile cognitive radio: maximize channel capacity of cognitive radio," *Proc. IEEE Consumer Communications and Networking Conference (CCNC)*, Las Vegas, NV, Jan. 11–13, 2007, pp. 1027–1031.

S. Wyne, T. Santos, A. P. Singh, F. Tufvesson, and A. F. Molisch, "Characterisation of a time-variant wireless propagation channel for outdoor short-range sensor networks," *IET Communications*, **4**, 253–264, 2010.

Y. Xie and A. J. Goldsmith, "Diversity-multiplexing-delay tradeoffs in MIMO multi-hop networks with ARQ," *Proc. IEEE International Symposium on Information Theory*, Austin, TX, June 13–18, 2010, pp. 2208–2212.

Y. Xie, D. Gündüz, and A. J. Goldsmith, "Multihop MIMO relay networks with ARQ," *Proc. IEEE Global Telecommunications Conference (GLOBECOM)*, Honolulu, HI, Nov. 30 - Dec. 4, 2009. pp. 1–6.

Y. Xin, H. Zhang, and S. Rangarajan, "SSCT: a simple sequential spectrum sensing scheme for cognitive radio," *Proc. IEEE Global Communications Conf. (GLOBECOM 2009)*, Honolulu, HI, Nov. 30–Dec. 4, 2009.

F. Xue and P. R. Kumar, "Scaling laws for ad-hoc wireless networks: an information theoretic approach," *NOW J. Found. Trends Netw.*, **1**, no. 2, 145–270, 2006.

K. Yamazaki, *Essays on Sequential Analysis: Multi-Armed Bandit with Availability Constraints and Sequential Change Detection and Identification*, Ph.D. Dissertation, Princeton University, Princeton, NJ, Apr. 2009.

E. Yang and D. Gu, "Multiagent reinforcement learning for multi-robot systems: a survey," *Technical Report CSM-404*, Department of Computer Science, University of Essex, UK, 2004.

J. Yang, R. Brodersen, and D. Tse, "Addressing the dynamic range problem in cognitive radios," *IEEE International Conference on Communications, ICC.*, June 2007, pp. 5183–5188.

K. Yao and E. M. Biglieri, "Multidimensional error bounds for digital communication systems," *IEEE Trans. Inf. Theory*, **IT-26**, no. 4, 454–464, July 1980.

K. Yao and R. M. Tobin, "Moment space upper and lower error bounds for digital systems with intersymbol interference," *IEEE Trans. Inf. Theory*, **IT-22**, no. 1, 65–74, Jan. 1976.

K.-L. A. Yau, P. Komisarczuk, and P. D. Teal, "A context-aware and intelligent dynamic channel selection scheme for cognitive radio networks," *Proc. Conf. Cognitive Radio Oriented Wireless Networks and Communications (CROWN-COM 2009)*, Hanover, Germany, June 2009.

K.-L. A. Yau, P. Komisarczuk, and P. D. Teal, "Context awareness and intelligence in distributed cognitive radio networks: a reinforcement learning approach," *Proc. 11th Australian Communications Theory Workshop (AusCTW 2010)*, Canberra, Australia, Feb. 2010.

K.-L. A. Yau, P. Komisarczuk, and P. D. Teal, "Achieving efficient and optimal joint action in distributed cognitive radio networks using payoff propagation," in *Proc. IEEE Int. Conf. Communications (ICC 2010)*, Cape Town, South Africa, May 23–27, 2010.

C. Yin, L. Gao, and S. Cui, "Scaling laws for overlaid wireless networks: a cognitive radio network versus a primary network," *IEEE/ACM Trans. Netw.*, **18**, no. 4, 1317–1329, Aug. 2010.

W. Yuan, H. Leung, S. Chen, and W. Cheng, "A distributed sensor selection mechanism for cooperative spectrum sensing," *IEEE Trans. Signal Process.*, **59**, no. 12, 6033–6044, Dec. 2011.

Y. Yuan, P. Bahl, R. Chandra, *et al.*, "KNOWS: Cognitive radio networks over white spaces," *IEEE International Symposium on New Frontiers in Dynamic Spectrum Access Networks, 2007. DySPAN 2007*, Apr. 2007, pp. 416–427.

T. Yücek and H. Arslan, "A survey of spectrum sensing algorithms for cognitive radio applications," *IEEE Comm. Surveys & Tutorials*, **11**, no. 1, 116–130, first quarter 2009.

L. Zacharias and R. Sundaresan, "Decentralized sequential change detection using physical layer fusion," *IEEE Trans. Wireless Commun.*, **7**, no. 12, 4999–5008, Dec. 2008.

Y. Zeng, C. L. Koh, and Y. C. Liang, "Maximum eigenvalue detection: theory and applications," *Proc. IEEE International Conference on Communications (ICC)*, Beijing, China, pp. 4160–4164, May 2008.

Y. Zeng and Y. C. Liang, "Covariance based signal detections for cognitive radios," *Proc. 2nd IEEE Int. Symp. New Frontiers in Dynamic Spectrum Access Networks (DySpan 2007)*, Dublin, Ireland, pp. 202–207, Apr. 2007.

Y. Zeng and Y. C. Liang, "Maximum–minimum eigenvalue detection for cognitive radio," *Proc. 18th IEEE Int. Symp. Personal, Indoor, and Mobile Radio Communications (PIMRC 2007)*, Athens, Greece, Sep. 2007.

Y. Zeng and Y. C. Liang, "Eigenvalue-based spectrum sensing algorithms for cognitive radio," *IEEE Trans. Commun.*, **57**, no. 6, 1784–1793, June 2009.

Y. Zeng, Y. C. Liang, E. C. Y. Peh, and A. T. Hoang, "Cooperative covariance and eigenvalue based detections for robust sensing," *Proc. IEEE Global Communications Conf. (GLOBECOM 2009)*, Honolulu, HI, Nov. 30–Dec. 4, 2009, pp. 1–6.

D. Zhang and N. Mandayam, "Bandwidth exchange for fair secondary coexistence in TV white space," *Proceedings of International ICST Conference on Game Theory for Networks (GameNets)*, Apr. 2011.

D. Zhang, R. Shinkuma, and N. Mandayam, "Bandwidth exchange: an energy conserving incentive mechanism for cooperation," *IEEE Trans. Wireless Commun.*, **9**, no. 6, 2055–2065, June 2010.

R. Zhang and Y.-C. Liang, "Exploiting multi-antennas for opportunistic spectrum sharing in cognitive radio networks," *IEEE J. Sel. Topics Signal Process.*, **2**, no. 1, 88–102, Feb. 2008.

Y. Zhang, J. Zhang, D. Dan, et al., "A novel spatial autocorrelation model of shadow fading in urban macro environments," *Proc. IEEE GLOBECOM 2008*, 2008.

Y. P. Zhang and Y. Hwang, "Characterization of UHF radio propagation channels in tunnel environments for microcellular and personal communications," *IEEE Trans. Veh. Technol.*, **47**, no. 1, 283–296, Feb. 1998.

Q. Zhao, L. Tong, and A. Swami, "Decentralized cognitive MAC for opportunistic spectrum access in ad hoc networks: A POMDP framework," *IEEE J. Sel. Areas Commun.*, **25**, no. 3, 224–232, Apr. 2007.

Q. Zhao, B. Krishnamachari, and K. Liu, "On myopic sensing for multi-channel opportunistic access: structure, optimality, and performance," *IEEE Trans. Wireless Commun.*, **7**, no. 12, 5431–5440, Dec. 2008.

Q. Zhao, L. Tong, A. Swami, and Y. Chen, "Decentralized cognitive MAC for opportunistic spectrum access in ad hoc networks: a POMDP framework," *IEEE J. Sel. Areas Commun.*, **25**, no. 3, 589–600, Apr. 2007.

Q. Zhao and J. Ye, "Quickest detection in multiple on-off processes," *IEEE Trans. Signal Process.*, **58**, no. 12, 5994–6006, Dec. 2010.

X. Zhao, J. Kivinen, P. Vainikainen, and K. Skog, "Propagation characteristics for wideband outdoor mobile communications at 5.3 GHz," *IEEE J. Sel. Areas Commun.*, **20**, no. 3, 507–514, Apr. 2002.

K. Zheng, H. Li, S. M. Djouadi, and J. Wang, "Spectrum sensing in low SNR regime via stochastic resonance," *Proc. 44th Annu. Conf. Information Sciences and Systems (CISS 2010)*, Princeton, NJ, Mar. 17–19, 2010.

L. Zheng and D. N. C. Tse, "Diversity and multiplexing: a fundamental trade-off in multiple antenna channels," *IEEE Trans. Inf. Theory*, **49**, no. 5, 1073–1096, May 2003.

F. Zhu and S.-W. Seo, "Enhanced robust cooperative spectrum sensing in cognitive radio," *J. Commun. Netw.*, **11**, no. 2, 122–133, Apr. 2009.

Q. Zou, S. Zheng, and A. H. Sayed, "Cooperative sensing via sequential detection," *IEEE Trans. Signal Process.*, **58**, no. 12, 6266–6283, Dec. 2010.

Index

access policy, *see* sensing and access policy
ADC, 21
alphabets
   input,output, 56
antennas, 20
   frequency reconfigurable, 20
   narrowband, 20
   radiation pattern reconfigurable, 20
   wideband, 20
autocorrelation function, 122, 125, 126, 131
   cyclic, 162
autocorrelation-based detection, *see* detection, distributed detection
automatic-repeat-request, 54
awareness
   location, 242
   situational, 184–186

bandit problems, 220–227, *see also* reinforcement learning
   arm-acquiring multiarmed bandit problems, 222–223
   contextual bandit problems, 222–223
   goal (objective function), 221
   multiarmed bandit problems, 220–227, 246
      Gittins index, 221, 223
   multiarmed bandit problems with availability constraints, 222–223
   multiarmed bandit problems with dependent arms, 222–223
   multiarmed bandit problems with multiple plays, 222–223
   multiarmed bandit problems with side observations, 222–223
   multiarmed bandit problems with switching penalties, 222–223
   one-armed bandit, 220
   regret, 221, 225–227, 235
   restless multiarmed bandit problems, 222–227
      indexable, 223, 224
      Whittle index, 223–225
   upper confidence bound (UCB), 226
bandwidth exchange, 26

Bayesian detection, *see* detector, distributed detection
Bellman equation, 215
Bellman optimality equation, 215, 228
binning, 86, 88
bits
   per channel use or dimension, 56

capacity, 55
   AWGN channel, 56
   ergodic, 49, 58
   multiuser channels, 59
   outage, 49
   overlay channel, 89
   parallel set of channels, 57
   per unit energy, 54
   pre-log, 53
   random-switch channel, 81
   region, 48, 60, 62
      cut-set bound, 61, 62, 64
      sum-rate, 61, 66
      sum-rate point, 52
   scaling laws, 51, 62
   Shannon, 48
   versus outage probability, 50
channel
   broadcast, 48
   cognitive radio, 52
   downlink, 48
   interference, 51
   memoryless, 55
   MIMO, 57
   model
      inaccurate, 174
      spatial, 144
   multiple access, 48
   multiuser, 48
   point-to-point, 47
   random-switch, 80
   relay, 51
   state, 49
   time-varying, 49
   uplink, 48

with random interference state, 87
with random state, 87
codebook, 43
   Gaussian, 67
codes
   structured, 68
codeword, 60
cognitive radio bands, 4
   cellular, 5
   fixed wireless access, 5
   UHF, 4
coherence
   bandwidth, 132
common information, 91
constraint
   average received power, 71
   interference, 71
   received peak power, 71
   spectral mask, 71
convergence, 7
cooperation, 88
cooperative detection, *see* distributed detection
CORNET, 32
correlation
   distance, 111, 112, 122, 136
   function, 112
   time, 122
cost
   per-symbol, 70
cost function, 70
cross-layer design, 30
cross-pol discrimination, 139
curse of dimensionality, 215
cyclostationarity, 162
   wide-sense, 162
cyclostationarity-based detection, *see* detection, distributed detection

DARPA XG program, 31
data sequence, 59
DDIR, 138
decoding error, 60
decoding function, 60
degrees of freedom, 53
denial-of-service attacks, 18
detection, *see also* distributed detection, quickest detection, sensing, sequential detection
   autocorrelation-based, 163, 169
   coherent, 160, 178
      with pilot tone, 161
   cyclic prefix, 170
   cyclostationarity-based, 161, 162
   energy, 216, 218, 220
   sliding window, 170
detector
   cyclostationarity-based, 173

nonrobust, 171
optimization, 155
   Bayesian criterion, 155
   generalized likelihood-ratio test, 157
   likelihood-ratio test, 156
   minimax criterion, 155, 156
   Neyman–Pearson criterion, 155, 156, 167
robust, 171
DHCP, 16
dirty paper coding, 44
dispersion
   angle, 105, 134
   angular, 136
   delay, 105, 127, 128
      models, 141
   frequency, 127
      models, 142
distributed detection, *see also* detection, fusion rules, quickest detection, sensing, sequential detection
   autocorrelation-based, 196
   bit-error probability (BEP) wall, 201–202
   centralized, 189
   cyclostationarity-based, 196–199
   decentralized, 189–192
   energy, 188, 194–196, 205, 206, 217
   fusion center, 189–190
   optimization
      Bayesian criterion, 192
      Neyman–Pearson criterion, 191, 193, 201
   quantization, 200–202
   reporting-channel errors, 200–202
   spatial diversity order, 188
   topology
      ad hoc configuration, 190
      parallel, 190–192
      serial, 190
      tree, 190
distribution
   chi-square, 158
   noncentral chi-square, 158, 176
diversity, 53
   gain, 53
diversity versus multiplexing tradeoff, 53
DOA, 135, 139
DOD, 135, 139
Domain name service, 14
Doppler
   frequency, 123–125
   spectrum, 104, 122, 123, 125–127, 130
   spread, 104
DPC, 87, 88, 90, 92
DSA, 2, 9
dynamic programming, 210, 213–216, 219, 246, *see also* reinforcement learning
   value function, 214

Index

optimal, 215

encoder
    cognitive, 85, 86, 88
encoding
    Markov block, 88
encoding function, 60
energy
    energy–rate tradeoff, 54
    minimum per-bit, 54
energy detection, *see* detection, distributed detection
energy scaling law, 54
erasures, 79
exploitation vs. exploration dilemma, 227

fading, 152
    multipath, 102, 119
    Nakagami, 121
    Rayleigh, 119–121
    shadow, 110, 111, 113, 114
    small-scale, 103, 105, 118, 119, 122, 123
    temporal, 122, 123
false alarm, 79, 185, 191, *see also* probability
feature detector, 157
field strength, 118
frequency
    correlation function, 131
    cyclic, 162, 163
frequency reuse, 51
FTTH, 11
function
    autocorrelation, 162, 196
        cyclic, 162
    Euler's Gamma, 158
    Gaussian tail, 160, 161, 196
    generalized Marcum Q, 159, 176
    incomplete Gamma, 158
    Kronecker delta, 221
    modified Bessel, 159
fusion rules, *see also* distributed detection
    belief propagation (BP), 193
    Boolean rules, 192
    Dempster–Shafer theory, 193
    generalized likelihood ratio test (GLRT), 193, 197
    K-out-of-N rule, 192, 193, 201, 202, 206
        AND rule, 192, 193
        MAJORITY rule, 192, 193
        OR rule, 192, 193, 201
    likelihood ratio test (LRT), 191–192
    selection combining, 194, 197
    weighted linear combining, 193–196
        equal gain combining (EGC), 188, 194
        maximal ratio combining (MRC), 194

gain
    antenna, 114
    reduction factor, 114
game theory, 237–242
    auction games, 241
        second-price sealed-bid auction, 241
    cooperative games, 237, 239–241
        coalitional games, 239–241
        core, 240
        grand coalition, 239
    noncooperative games, 237–239
        Nash equilibrium, 238, 239
    normal form game, 237
    stochastic games, 230, 241–242
    strategic game, 237
    strategy, 237
Gelfand–Pinsker
    encoding, 86–88, 91
    problem, 87
geolocation, 242
global control plane, 14

impulse response
    double-directional, 135, 136
interference
    PSD, 43
    strong, 65, 89
    temperature, 43
    threshold, 43
    very strong, 66
    weak, 66, 89, 90
interference alignment, 68, 92
interference cancellation, 88
interference channel
    AWGN, 65
    K-user, 63
    two-user, 52
interference channels, 68
interference constraint
    average power, 75
interference temperature, 151, 152
interweave, 45, 68
    paradigm, 45
interweave channel
    capacity versus outage, 83
interweave network
    capacity region, 77
    scaling laws, 83
interweaving, 105, 150, 184

KNOWS, 33
KUAR, 32

last mile problem, 11
location awareness, *see* awareness

M2M communications, 10
MAC protocol, 46, 68, 77, 78

Index

Markov decision process (MDP), 214, 227
   partially observable Markov decision process (POMDP), 225
medium access
   time-sharing, 77
medium access control, 42
MEMS technology, 21
message, 43, 60
MIMO, 134, 138
missed detection, 79, 185, 191, *see also* probability
moment space, 174
moment-bound theory, 174
multicasting, 50
multihop routing, 51, 63
multipath
   components, 119, 120, 122, 127, 129–131, 133–136, 138
multiplexing, 53
   gain, 53
mutual information, 55

Nakagami
   pdf, 121
Nakagami m-factor, 121
Neyman–Pearson detection, *see* detector, distributed detection
noise
   nonwhite, 164
null space, 45

OFDM, 22, 104, 166
   noncontiguous, 23
   PAPR, 24
   spectrum sensing with, 166
opportunistic communication, 45
overlay, 43
   paradigm, 43
overlay channel
   two-user, 86
overlaying, 104, 105, 150, 184

parallel channels, 59
   interference, 68
path loss, 102, 105, 107, 112, 113, 118, 120, 121, 123
   exponent, 109
   free-space, 105, 107, 114
   median, 109, 111, 112
   models, 108, 113, 115, 141
performance
   metrics, 48
   region, 52
phase uncertainty, 92
polarization, 105, 138
power
   delay profile, 104, 130

gain, 106
   median, 109
   spectrum
      angular, 136
precoding against interference, 86, 87, 92
primary exclusive region, 84
probability
   of correct detection, 154
   of false alarm, 154, 159, 160, 192, 195
   of missed detection, 154, 159, 160, 192, 195
probability of error, 60
process
   cyclostationary, 166
   wide-sense cyclostationary, 162

quickest detection, 207–212
   Bayesian, 207–209
   cooperative, 210–212
   cumulative sum (CUSUM) test, 207
   Shiryaev–Roberts (SR) stopping rule, 209
   Shiryaev–Roberts–Pollak (SRP) stopping rule, 210
   SR-r stopping rule, 209

rate
   achievable, 60
rate-splitting, 67, 88, 91
Rayleigh
   pdf, 119
receiver operating characteristic (ROC), 170, 188, 195
reinforcement learning, 227–237, *see also* bandit problems, dynamic programming
   Q-learning, 229–231, 234–236
   action selection
      ϵ-greedy algorithm, 229–231
      softmax, 229, 234
   action-value function, 228, 229, 237
      optimal, 228
   goal (objective function), 221, 228, 230–231
   off-policy algorithm, 229
   on-policy algorithm, 229
   Sarsa, 229, 230, 235
   value function, 214
      optimal, 215
Rice
   K-factor, 104, 121, 122, 141
      temporal, 122
   factor, 120
   pdf, 120, 121

sensing, *see also* detection, sensing and access policy
   autocorrelation-based, 153
   coherent, 153, 172
   cooperative, 80, 151, 153

malicious users, 207
cyclostationarity-based, 153
energy, 153, 157, 171, 176
energy-efficient, 188, 198–200
  censoring, 198–200
maximizing throughput, 211, 214–220
passive primary receivers, 243
periodicity, 151
time, 154, 160, 170
sensing and access policy, 185–186, 193, 212–213
  multiuser
    auction games, 241
    cooperative game theory, 240–241
    multiagent reinforcement learning-based, 234–237
    myopic (greedy), 225
    noncooperative game theory, 238–239
    restless multiarmed bandit-based, 225–227
    single-agent reinforcement learning-based, 231–232
  single-user
    dynamic programming-based, 210
    myopic (greedy), 224–225
    partially observable Markov decision process (POMDP), 225
    restless multiarmed bandit-based, 223–226
    single-agent reinforcement learning-based, 232–234
sequential detection, 203–207
  Bayesian, 205
  composite hypotheses, 204–205
  cooperative, 205–207
  decentralized sequential probability ratio test (D-SPRT), 205
  M-ary sequential probability ratio test (MSPRT), 206
  sequential probability ratio test (SPRT), 203–205
  truncated test, 205, 206
  weighted sequential probability ratio test (WSPRT), 207
shadowing, 152
side information, 41, 46
  network, 41, 46
situational awareness, *see* awareness
SNR
  low-SNR regime, 68
spatial diversity
  in spectrum sensing, 187–188
spectral density
  cyclic, 162
spectral mask, 43, 69
spectral opportunity, 184–187, 202, 213, 234, 235, 241
spectral pooling, 78
spectrum
  agility, 2, 4

allocation, 1, 5, 29, 31
broker, 16
coexistence, 3, 25, 28
governance, 1, 5, 6
hole, 4, 9, 19, 20
policy, 2, 5
pooling, 26
property rights, 6
sensing, 3, 18–20, 28, 33, 35, 150
servers, 15, 29
space, 150
utilization, 1, 5, 18, 31
spectrum hole, 45, 184
spread
  angular, 134, 136, 137
  delay, 131, 132, 137
spread spectrum, 43
structured codes, 92
superposition coding, 67, 88, 89, 91
supervised learning, 227

two-switch model, 80

ultra-wideband, 43
  systems, 129
  channel, 130, 133
underlay, 42
  paradigm, 42, 69
underlay channel capacity
  AWGN, 72
  ergodic, 74
  MIMO, 73
underlaying, 105, 133, 150, 151, 184
unsupervised learning, 227
usage scenario, 3
  interweave, 3, 4
  underlay, 4
user hierarchy, 3
  capable users, 3
  primary users, 3
  secondary users, 3
USRP, 31

vehicle-to-vehicle
  propagation, 139

WARP, 32
waterfilling, 57
white space, 3, 4, 45
wireless
  fixed, 113, 122, 123, 126, 127
  sensor networks, 140
wireless network, 47
  ad hoc, 48
  two-tier, 48

Printed in the United States
by Baker & Taylor Publisher Services